MULTIVARIATE DATA ANALYSIS

MULTIVARIATE

JOHN WILEY & SONS, INC.

DATA ANALYSIS

William W. Cooley
Paul R. Lohnes

New York . London . Sydney . Toronto

Library of Congress Catalogue Card Number: 70-127661

ISBN 0-471-17060-7

Printed in the United States of America

10 9 8 7 6 5 4 3

Preface

IIIIIIIIIIIIIIIIIIIIIIIIIIIIIIIIIIIIIIIIIIIIIIII

In 1962, the year of publication of our previous book on multivariate procedures, John Tukey published a long and marvelous discussion on "The Future of Data Analysis." His gift to us was our professional identity. We knew we were not statisticians. That was too painfully obvious. We were young educational researchers who had been thoroughly persuaded by Professor Phillip J. Rulon that multivariate quantitative procedures had to be the keys that would unlock the secrets of human social behaviors. Rulon transmitted primarily the vision of Truman L. Kelley, although other pioneers in behavioral research, including Cyril Burt, Louis L. Thurstone, and Raymond B. Cattell, had similar visions. The multivariate heritage we acquired was the work of great statisticians, such as Pearson, Fisher, Hotelling, Bartlett, and Wilks. Still, we were not statisticians. Tukey told us we were data analysts, and we have since learned to accept and even enjoy the identification.

Tukey argued that there have to be people in the various sciences who concentrate much of their attention on methods of analyzing data and of interpreting the results of statistical analysis. These have to be people who are more interested in the sciences than in mathematics, who are temperamentally able to "seek for scope and usefulness rather than security," and who are "willing to err moderately often in order that inadequate evidence shall more often *suggest* the right answer." They have to use scientific judgment more than they use mathematical judgment, but not the former to the exclusion of the latter. Especially as they break into new fields of sciencing, they must be more interested in "indication procedures" than in "conclusion procedures." Tukey made our interest in multivariate heuristics rather than multivariate inference sound respectable. Most of the attitudes he ascribed to data analysts we found in ourselves. So now we can say that data analysis is what this book is about.

v

Our literary debts are testified to in the bibliography. Our personal debts are numerous, including those to our teachers, Palmer O. Johnson, Phillip J. Rulon, Fletcher Watson, John B. Carroll, and David V. Tiedeman; to our colleagues, John C. Flanagan, Barry G. Wingersky, Thomas O. Marshall, Sigmund Tobias, S. David Farr; to our students, Eugene C. Lee, Richard L. Ferguson, Silas Halperin, Fred Hockersmith, Richard E. Sass; to our editors, Joseph F. Jordan and Ronald St. John; to Charleye Dyer, who labored hard with Snopake and typewriter; and to our wives, Cynthia and Kathleen, to whom we dedicate this book.

William W. Cooley
Paul R. Lohnes

Contents

IIIIIIIIIIIIIIIIIIIIIIIIIIIIIIIIIIIIIIIIIIIIIIIIIIII

MULTIVARIATE DATA ANALYSIS

PART I

Introduction

||||||||||||||||||||||||||||||||||||||||||||||||||||||||||||||||

CHAPTER ONE

Multivariate Analysis and Computer Programming

1.1 OVERVIEW OF MULTIVARIATE ANALYSIS

Multivariate analysis is the branch of statistics concerned with analyzing multiple measurements that have been made on one or several samples of individuals. The important distinction is that the multiple variates are considered in combination, as a system of measurement. The multivariate analyst's concern with the *jointness* of his p measures on N subjects has been expressed by Kendall (1957, p. 5):

> The variates are dependent among themselves so that we cannot split off one or more from the others and consider it by itself. The variates must be considered together.

Technically, multivariate analysts do research in which as much attention is paid to the $p(p-1)/2$ different covariances among the variates as is paid to the p means and the p variances, if not more. It is no accident that the three types of statistics just mentioned—means, variances, and covariances—are the natural parameters of multivariate normal distributions. The mathematical model on which multivariate statistical procedures are based is the multivariate normal distribution (hereafter m.n.d.). Most of our procedures

3

are concerned with estimating or making inferences about parameters of an m.n.d. Anderson has pointed out that the assumption of the m.n.d. for our multiple measures can be justified by the same central limit theorem argument that leads to the assumption of normality for a univariate measurement, which is that "the multivariate normal distribution often occurs because the multiple measurements are sums of small independent effects." (Anderson, 1958, p. 2.) The m.n.d. has additional attractions for us, in that it is mathematically tractable, especially by matrix algebra methods, and inference procedures based on it are usually powerful.

Francis Galton deserves recognition as the founder of multivariate statistics, as well as the founder of the great movement of trait and factor psychology for which this methodology has been crucial. As early as 1889 he was applying the mathematics of the bivariate normal distribution to studies of inheritance. He discovered the correlation coefficient and the regression lines as expressions of the linear relatedness of two variables. Karl Pearson refined and amplified correlation analysis in several ways during the first quarter of the twentieth century, and another Englishman, Ronald Fisher, developed the analysis of variance and discriminant analysis. In America, Harold Hotelling developed the principal components and canonical correlation models, and Samuel Wilks developed the lambda statistic for multivariate analysis of variance tests, for which an Englishman, M. S. Bartlett, then derived useful distribution properties. These are the names of some of the pioneers in a field which is worked by a large group of creative mathematical statisticians today. The literature of multivariate statistics has grown to huge proportions and it is an intrinsically difficult literature for the nonmathematical researcher. The reader should be aware that the present authors are researchers, not statisticians, and that they do not claim to be generally knowledgeable with respect to the literature of mathematical multivariate statistics. This book represents a collection of multivariate procedures with which the authors are intimate, in the ways a journeyman is intimate with the tools of his trade. The collection is *basic* in that only certain of the most broadly useful procedures for the behavioral sciences are included. Many special tools for fancy work have been excluded, in keeping with the notion of an introductory text.

This book tells a little about the *why* of the models it presents, but mostly it tells about the *how* of them. The emphases are on the research acts of selecting a method of analysis, computing the analysis, and interpreting the outcomes. Perhaps the best teaching about research procedures is in the examples which are presented and which should be studied to see what numbers are useful to report and what kinds of statements may be made in interpretation of reported numbers.

The researcher who is serious about learning multivariate statistical procedures needs a multivariate statistics library. Our procedures fall into four clusters:

(1) factor analysis,

(2) canonical, multiple, and partial correlation,

(3) multivariate analysis of variance,

(4) discriminant and classification functions, for each of which an applied statistics book is desirable. Horst's *Factor Analysis of Data Matrices* covers the first area handsomely, as does DuBois' *Multivariate Correlational Analysis* the second area. Bock is preparing a textbook on multivariate analysis of variance, and in the interim his two long papers in recent handbooks fill the gap (Bock, 1966; Bock and Haggard, 1966). Rulon and Tiedeman cover the fourth area with their *Multivariate Statistics for Personnel Classification.* Anderson gives the basic mathematical statistics coverage with his *An Introduction to Multivariate Statistical Analysis*, which may be supplemented with Rao's *Advanced Statistical Methods in Biometric Research.* Harman's *Modern Factor Analysis* provides the complete history of his subject. Harris' (editor), *Problems in Measuring Change* demonstrates some of the fancier current doings in the field. A good matrix algebra book with which to round out a basic bookshelf is Searle's *Matrix Algebra for the Biological Sciences.*

There are several principles by which multivariate procedures might be classified. One is provided by consideration of the structure of the process of science. Tatsuoka and Tiedeman have described sciencing as a hypothetico-deductive-observational procedure, as displayed in Figure 1.1, which is their graphic. In the creative invention stage at which theory is initiated, mostly by use of analogies, they see that statistics may have a heuristic role in the discovery and refinement of constructs. "The statistical techniques of factor analysis, discriminant analysis, and certain other species of multivariate analysis ... are heuristic tools available to the researcher for producing constructs which may prove to be theoretically fruitful" (1963, p. 144). They go on to point out that the usual role of statistics comes toward the end of the process in the testing of hypotheses deduced from theory against empirical results.

We depend very heavily on linear functions fitted to complexes of variables by multivariate procedures to specify the details of theoretical constructs and of relationships among constructs as the examples in the chapters ahead reveal. In a young science the heuristic uses of these procedures are far more important than the hypothesis testing uses (although the two uses are usually intertwined).

Another principle of classification of procedures is provided by considering whether the research involves deliberate manipulation of variables in experiments, or involves the observation of subjects in their natural environments

Initial observational data ("educated" familiarity with phenomena in the field of study)	Creative invention, perhaps assisted by certain statistical techniques; also assisted by analogies from principles in related sciences or already established principles in the field	$\{C_1, C_2, C_3, \ldots\}$ (Set of theoretical constructs embedded in network of hypotheses, $H_{1,2}$, $H_{2,3}$, etc.; stating the interrelationships among them)

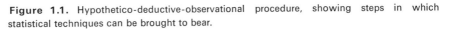

		Deduction, plus interpretation (coordinating defs relating C_i to O_i)
Empirical testing and decisions on hypotheses, guided by statistical techniques	(Set of empirically testable hypotheses, $H'_{1,2}$, $H'_{3,4,5}$, etc., stating interrelationships among various observational variables $O_1, O_2, O_3, \ldots$)	$O_1 \longleftrightarrow O_2$ $(H'_{1,2})$ $O_3 \longleftrightarrow O_4, O_5$ $(H'_{3,4,5})$

Figure 1.1. Hypothetico-deductive-observational procedure, showing steps in which statistical techniques can be brought to bear.

in survey researches. Recently Cattell (1966b) has argued strongly for the latter approach, calling it the "new psychology" (the same name T. L. Kelley gave it in 1928), and claiming that it is more potent than experimentation. Cattell also calls survey psychology "the Galton-Thurstone movement," and characterizes it as multivariate rather than bivariate and freely observing rather than manipulative. He claims the potency of the new approach results because it "took life's own manipulations, in clinical, social, and physiological data, and by more intricate, non-interfering, statistical finesse teased out the causal connections among data which could not be manipulated" (1966b, p. 8).

To demonstrate the vigor of the new psychology, Cattell has helped to launch the *Multivariate Behavioral Research* journal and has edited the *Handbook of Multivariate Experimental Psychology* (1966). He can employ the term "experimental" in the title of his handbook because he has redefined it in a way which embraces multivariate surveys as well as multivariate manipulative designs.

Although the procedures in this book are applicable to experiments and a few examples of experimental analyses are included, for the most part the examples are from surveys. The emphasis on heuristic rather than hypothesis testing uses of the procedures follows from this preoccupation with survey sciencing. In fact, the theory of statistical hypothesis testing for multivariate procedures is much more highly developed than the treatment in this text suggests.

One way of classifying multivariate techniques is to place each procedure in a quadrant of a lattice depending on how many sets of variables and how many populations are included in the design. The lattice and the placement of procedures within it are displayed in Figure 1.2. Kendall has pointed out that the models in the first quadrant are interdependence models, while those in the other three quadrants are dependence models. With the dependence models (in $Q2$, $Q3$, and $Q4$) it is possible to specify which variables are the predictors, which the criteria, and which the control variables (if any). Also, all the dependence types can be subsumed under general regression theory, since clever use of binary variates makes it possible to treat the nominal variables of population membership as regressed variables. Tatsuoka and Tiedeman show how this is done (1963, p. 148). Such a " general linear model " approach to multivariate analysis of variance designs provides the ultimate in flexibility, as can be seen by studying R. D. Bock's papers (1966; Bock and Haggard, 1966). However, the flexibility is purchased at the expense of considerable sophistication in the mathematics of the model. Our approach, which may be described as Fisherian, is not general and flexible but it is relatively easy to learn and the simple factorial designs it provides are broadly serviceable.

Tatsuoka and Tiedeman express a strong prejudice against the use of control variables, as in the multiple partial correlation and covariance design.

POPULATIONS

		One Population		*Two or More Populations*
V	One set	Principal components		Multivariate analysis of variance
a				Discriminant functions
r		Factor analysis		Classification functions
i			$Q1$	$Q2$
a			$Q3$	$Q4$
b	Two or three sets	Polynomial fit		Multivariate covariance
l		Multiple correlation		
e		Canonical correlation		
s		Multiple partial correlation		

Figure 1.2. Classification of multivariate procedures.

"In a certain sense, the use of control variables is an admission of defeat: it is tantamount to saying that our theory holds only on condition that certain relevant factors are held in abeyance" (1963, p. 167). Procedures for using control variables need to be available for those occasions when the establishment of a restricted theory is a reasonable objective.

This book has been organized so that the designs in $Q1$ and $Q3$ for studying sets of variables on a single sample from a single population form Part II which, with the fundamentals of Part I and the support of lectures and outside readings (as well as the exercises provided), may make a one semester course. Where Part II is concerned with multivariate correlational analyses, Part III discusses multivariate analyses of among-groups variance, and may outline a second semester course.

Chapter Two deals with the estimation of the parameters of a vector random variable which is assumed to have a multivariate normal distribution. Some algebra applied to vector variables is introduced because, while derivations are not to be presented in general, the authors believe that it is useful for researchers to be able to manipulate multivariate models somewhat by the method of matrix algebra in order to derive particular interpretive results for which the mathematicians may not see the need. For example, in Chapter Three this method is used to get the factor pattern for viewing a multiple regression function as a factor of the predictor battery. This is a view of the regression function which is attractive to the reasearcher but would not occur to the mathematician. The computer techniques for reducing observational vectors to basic vectors and matrices introduced in Chapter Two are then used in all the programs throughout the book. Part II begins with a consideration of multiple correlation because it is the best-known multivariate procedure and is one which many students will already have studied. Even so, our matrix algebra formulation of the model will probably be new to students.

Multiple correlation is used to examine the relation between a single criterion or dependent variable and two or more predictors or independent variables. Thus grade point average might be predicted from several aptitude scores. For the simplest case of two predictor scores, the problem can be visualized by using a geometric concept called *the test space model*. This conceptualization is very useful throughout multivariate statistics. If p measurements have been made on N-individuals, the test space model represents each individual as a point in an p-dimensional Cartesian space. The p variates become the p orthogonal axes spanning the space (each axis at right angles to all the others). Each point representing an individual has a location determined by taking the p scores for the individual as Cartesian coordinates.

Consider the case when two test scores are obtained for ten students; $p = 2$ and $N = 10$. This case results in ten points in a two-dimensional space, which is represented by the familiar bivariate score roster (Table 1.1) and the

TABLE 1.1   A 10 by 2 Data
Matrix or Score Roster

Subject	Test 1	Test 2
1	3	3
2	1	2
3	2	2
4	1	1
5	2	1
6	−1	−1
7	−2	−1
8	−2	−2
9	−1	−2
10	−3	−3

two-dimensional scatter plot (Figure 1.3). Students having similar test score combinations will occupy similar regions of the test space. In the bivariate examples each student can be depicted as a vector from the origin to his point. Student 2 is so represented in Figure 1.4. Symbolically, the row vector for Student 2 may be represented as [1 2]. Thus a score vector is a row or column of numbers, the numbers being the p scores or p coordinates for the p axes of the test space. The ten rows of the roster of Table 1.1 represent ten vectors. Together they form a *data matrix*, an N by p or 10 by 2 matrix in this case. The size, or order, of a matrix is notated (r, c) or r by c, signifying r rows and c columns. In multivariate analysis, the bivariate techniques are extended so that more than two variables can be considered, but of course it is difficult to visualize or to diagram spaces of more than two or three dimensions.

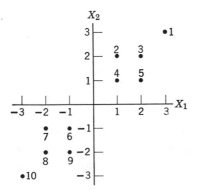

Figure 1.3. Two-dimensional test space.

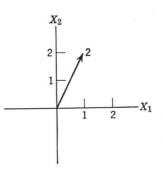

Figure 1.4. Vector for student 2.

Fortunately, matrix algebra makes it possible to consider p-dimensional systems. Most mathematics of multivariate analysis involve the manipulation of matrices. For example, the data matrix can be manipulated by a few simple matrix operations to produce a correlation matrix, and these operations are presented in Chapter Two.

Applying the test space model to the two-predictor multiple correlation case requires visualizing a three-dimensional test space in which a regression plane is to be located. The regression plane is placed so that the sum of the squared distances of all the score points from the plane, when these distances are measured parallel to the criterion axis, is smaller than for any other plane that could be located in the test space. The sum of the squared distances divided by the appropriate number of degrees of freedom is an estimate of the variance of the criterion scores about the regression plane, the square root of which provides the standard error of estimate. The estimate of an individual's criterion score is computed as a linear function of his two predictor scores. When $p - 1$ predictors are involved, the test space is p dimensional and the plane becomes a hyperplane. A hyperplane lying in the p-space is analogous to the plane lying in the three-dimensional space. It is a subspace of $p - 1$ dimensions.

The weights for the different predictor variables can be scaled to indicate the relative contributions of the variables to the prediction of the criterion, and a coefficient of multiple correlation is available as an indication of the strength of relationship between the criterion and the linear function of the predictors. The square of this coefficient represents the proportion of variance on the criterion that is predictable from, or explained by, the known variance on the linear combination of predictors. There is an analysis of variance test of the statistical significance of the multiple correlation coefficient.

When the investigator is interested in determining how one characteristic of a set of objects can be predicted from other characteristics, when all the measures are continuously distributed variables, multiple correlation provides an analytic tool. Whether or not it is the best tool depends on cases. Presumably, before the scientist applies the model to his data, he will review its several assumptions. In fact, this reservation regarding the suitability of a model's assumptions applies to all the techniques discussed in this book, and considerable emphasis will be placed on assumptions.

The term *model* has been used so liberally with reference to the multivariate statistical procedures discussed in this text that a word about our usage of it is in order. By a model we mean the basic matrix algebra specification of a procedure for analyzing data. In this usage a model is a matrix equation. For example, in Chapter Three the fundamental equation for multiple correlation is

$$\mathbf{R}_{pp}^{-1}\mathbf{R}_{pc} = \mathbf{b}$$

which specifies that the inverse of the intercorrelations of predictors post-multiplied by the column vector of predictor-criterion correlations yields the coefficients **b** for the standardized regression function. Learning to read and understand these fundamental equations is part of the task of studying multi-variate procedures. The job of the multivariate analyst is in part the selection of appropriate matrix equations for particular sets of data and of hypotheses. However, an algebraic model would be of little use if its fundamental equation were not set in a rich context of interpretable relations among its components. The analyst using multiple correlation, for example, needs to know the matrix algebra for expressing the multiple correlation coefficient and the regression function as a factor of the predictor battery.

There is a secondary referent of model which is the geometric representation of the functions fitted to the data. Some authors choose to emphasize these geometric interpretations, but we are convinced that the algebraic view of multivariate methods is easier to learn, easier to remember and manipulate, more concise and precise. We will usually sketch in the geometric interpretation, but we definitely emphasize the matrix algebra of our methods.

Procedures of multivariate analysis are often concerned with the problem of reducing the original test space to the minimum number of dimensions needed to describe the relevant information contained in the original observations. Models differ in the types of information they preserve. Usually a model preserves information in the predictor battery which is systematically related to some criterion variable or battery. One model which is not criterion oriented is principal components, described in Chapter Four. Psychometricians have been very active in recent years in the study of the logical implications of systematic intercorrelations within sets of measures. They have published many approaches to the reduction of dimensionality in correlated systems of measurements, and these approaches fall under the rubric of *factor analysis*. This text does not offer general coverage of factor analysis, but confines its scope to the particular factoring procedure of principal components which has the advantage of being easily integrated into multivariate statistics. Chapter Five does discuss some of the basic equations of factor analysis and the Quartimax and Varimax rotations of principal components to more interpretable positions. The geometric interpretation of factor analysis is presented in terms of *the sample space model*, which erects vectors representing the measures in an N-dimension sample space for which the Cartesian reference axes stand for the N subjects measured. The N scores of the sample subjects on a test locate a point in the N space which is the terminus of a vector for that test which begins at the origin of the Cartesian axes. In this model the cosine of the angle of separation of two test vectors is the correlation between the tests. The factor analysis problem is to select a smaller set of orthogonal reference axes to span the swarm of points, and to project the tests onto these new axes, which are the factors. The geometry sounds complicated because it is, but we shall see that the basic matrix algebra is quite manageable.

The canonical correlation model which is presented in Chapter Six is essentially a procedure for factoring two batteries simultaneously, in order to extract factors which are uncorrelated within their batteries but which provide maximum correlations of pairs of factors across batteries. That is, the first factor of each battery is located so that the canonical correlation of the first factors is maximized. The resulting coefficient is the largest product-moment correlation that can be developed between linear functions of the two batteries. It expresses the maximum redundancy of a pair of factors, one from each of the batteries. Then a second factor of each battery is located so that these second factors are orthogonal to the first and have maximum correlation between them. This process continues until all the orthogonal dimensions of common variance or redundancy between the two batteries have been located. Significance tests for hypotheses about the canonical correlations are provided. The typical application of this model is illustrated by a researcher who has a battery of ability measures and a battery of interest measures on a sample of subjects, and who wants to know the overlap in measurement variance between the two systems of measurement. Chapter Six provides an opportunity to compute on such data.

Sometimes the researcher has reason to remove from his predictor and criterion measurement batteries the predictable variance in them related to the variance in a third measurement battery, called the *control* battery. The method of multiple partial correlation described in Chapter Seven permits him to do this. After he has partialled out the variance explained by the control measures, he will go ahead and develop the predictor-criterion relationship in residuals by either multiple or canonical correlation.

The generalization of simple analysis of variance to the case of a multivariate criterion is treated in Chapter Eight. Generalizations of both the customary homogeneity of means tests and the not so customary homogeneity of variances test are given. The research issue behind these generalized tests is whether two or more sample groups should be thought of as arising from a single multivariate population or from two or more multivariate populations. The null hypothesis for each test asserts that the sample statistics arose from two or more samplings of a single population, or single swarm of points in the multivariate space. The generalized means test is called Wilks' lambda (Λ) test; it determines a probability level for the null hypothesis of equality of population centroids (means vectors) on the assumption of equality of dispersion (variance-covariance matrices). The assumption is analogous to that of homogeneity of variance in the univariate F-ratio test of equality of means.

A test of the null hypothesis of equality of population dispersions due to Bartlett is given.

Related to the multivariate analysis of variance model is the multiple group discriminant model presented in Chapter Nine. This model is useful to the scientist who is interested in examining or predicting the group membership of individuals on the basis of a set of continuously scaled attributes of those individuals. The distinction of this problem is that the criterion variable of membership is categorical and nominal rather than continuous in its scaling. The approach is to locate a line in the space of the p attributes of the persons for which the separation of the groups is optimized when the individual points of the different sample groups are projected onto it. Actually, the ratio of the among-groups to the within-groups sums of squared deviations from group means on this discriminant function is maximized. Since in general this first discriminant function may not exhaust the power of the test battery to separate the groups, additional functions, mutually orthogonal (or uncorrelated), may be fitted. The limit to the number of discriminant functions that may be required in an analysis is the smaller of p and $g - 1$, where g is the number of groups sampled.

A test of the statistical significance of the separation of the g groups in the p-space is standard; an approximate test of the significance of separation of the groups by individual discriminants is also available. It is possible to scale the weights for the variables in each discriminant function to show the relative contributions of the variables to the discriminant. Computing and plotting the centroids (vectors of group means) in the discriminant space often provides a helpful map of the location of the groups in the reduced space, particularly if only two or three discriminant functions are required to describe the observed group differences.

It is sometimes desirable to compute actual predictions for new samples of subjects in order to test previous findings in replication studies, or to provide a basis for practical decisions in applying the findings. Programming the computation of linear functions that yield estimated scores in factor, regression, and discriminant analyses is described in the text. Chapter Ten presents classification models which permit the estimation of the probability of an individual being a member of each of a number of groups on the basis of his test scores, factor scores, or discriminant scores. Probabilities of this type have been found especially useful in the study of occupational choice. These classification procedures make it possible to estimate at any point in the measurement space the relative densities of the different groups.

Attention is directed in Chapter Eleven to a procedure for introducing multiple covariance controls into multivariate analysis of variance which is seldom encountered in published reports of educational or psychological research, but which seems to have great promise for analysis of experiments in these fields. In curriculum research this design makes it possible to test the separation of a number of treatment groups on a set of p achievement variables after covariance adjustment for differences on a set of c aptitude variables. This procedure is much like the method of multiple partial correlation adjustment. An example described in Chapter Eleven concerns whether students taught by two different approaches differ significantly on five criterion measures after adjustment has been made for differences in aptitude and initial knowledge.

The multivariate analysis of variance model is capable of extension in all the directions in which the univariate analysis of variance has been extended. The complexities of such generalizations are not suitable matters for an introductory survey of multivariate procedures, however. In order to suggest how the model can be extended, but especially to provide one of the more useful extensions of the model, Chapter Twelve treats the factorial discriminant analysis for the case of a two-factor balanced design with interaction. This is a fixed effects model in which disciminant functions are fitted for the separation of the levels on the rows effect, for the separation of the levels on the column effect, and for the separation of the interaction effect, and significance tests for all effects are provided.

Project TALENT

Many of the examples of multivariate procedures provided in subsequent chapters have been drawn from the research at Project TALENT. Also, a set of Project TALENT data has been provided in the Appendix, and each chapter includes a computing assignment on that set of data, in order to provide students with exercises in the computing and interpreting of multivariate statistics. The authors have drawn on Project TALENT examples so heavily, not simply as a matter of convenience, but because the Project represents a major psychometric sampling survey the design and execution of which depends completely on multivariate methods and the data from which provide an outstanding opportunity to demonstrate the power of multivariate procedures, especially their heuristic power. Since the Project TALENT examples and exercises are among the most instructive aspects of this text, a word of introduction to the Project is in order.

In 1959, the U.S. Office of Education agreed to support a massive census of the abilities and other personality characteristics of American high school youth, and of the characteristics of their high schools. The project was conceived and initiated by John C. Flanagan, and has been a joint undertaking of the American Institutes for Research in the Behavioral Sciences and the University of Pittsburgh. The primary purpose of the study is to survey the talents of youth, in order to estimate "the size of the manpower pool qualified for training in science, engineering, and other professional fields."

The relationships among the abilities, interests, and other characteristics of American youth are under study, and the relations of these personality variables to relevant environmental variables, especially the characteristics of the schools and families of the youths, are being determined. The Project staff is conducting periodic follow-up studies on the sample, and will continue to for at least twenty years, in order that their ultimate life and career adjustment outcomes may become known, and the degrees of predictability of these adjustments from the measurement variables of the original survey may be computed. Follow-up information obtained one year and five years after high school graduation is already available. The scope of the undertaking is evidenced by the sample sizes involved: over 440,000 youths in over 1300 high schools filled out over two million answer sheets in two days of testing and inventorying. Meanwhile, school officials provided pages of information on each of the schools. A stratified random sample of the nation's high schools was obtained with great care to insure adequate representation of all types of schools, public and private, large and small, rural and urban, in all geographic areas. All students in grades nine through twelve in each of the selected schools were tested, yielding a probability sample of approximately five percent of the nation's high school youth in 1959–1960. From the test and inventory results on this five percent sample it has been possible to make estimates of measurement parameters for the entire population of high school youth. The data in the Appendix on which the chapter exercises are based is a two percent random sample of the TALENT file for twelfth grade males and females. A selection of 1960 ability and interest measures for these boys and girls is presented, as well as their 1960 socioeconomic status, and several college and career plan variables.

1.2 ELEMENTARY MATRIX ALGEBRA

Vectors and matrices are arrays or tables of numbers. The algebra for symbolic operations on them is different from the algebra for operations on *scalars*, or single numbers. For example, there is no operation of division in matrix algebra, although there is an operation called multiplying by an inverse which is similar in purpose to dividing in scalar algebra. There is always a way to write out in scalar algebra expressions, involving the elements of vectors and matrices, the exact equivalents of matrix algebra equations, so that matrix algebra notation may be construed as shorthand for the corresponding scalar longhand. Once you have learned to use this shorthand to symbolize manipulations of tables of numbers you will seldom want to substitute the scalar longhand. Matrix algebra is usually elegant in its simplicity while the corresponding scalar algebra is involved and messy. For example the scalar notation of the vector product $c = \mathbf{a}'\mathbf{b}$ is

$$c = \sum_{i=1}^{p} a_i b_i$$

A *vector* is a column of numbers

$$\mathbf{a} = \begin{bmatrix} a_1 \\ a_2 \\ \vdots \\ a_p \end{bmatrix}$$

The *transpose* of $\mathbf{a}$, denoted $\mathbf{a}'$, is the row arrangement of the elements of $\mathbf{a}$

$$\mathbf{a}' = [a_1 \quad a_2 \quad \cdots \quad a_p]$$

The sum of two vectors is the vector of sums of corresponding elements

$$\mathbf{a} + \mathbf{b} = \begin{bmatrix} a_1 + b_1 \\ a_2 + b_2 \\ \vdots \\ a_p + b_p \end{bmatrix}$$

The difference of two vectors is the vector of differences of corresponding elements. Although the product of $\mathbf{a}'\mathbf{b}$ is a scalar

$$\mathbf{a}'\mathbf{b} = a_1 b_1 + a_2 b_2 + \cdots + a_p b_p$$

the product $\mathbf{a}\mathbf{b}'$ is a square matrix

$$\mathbf{a}\mathbf{b}' = \begin{bmatrix} a_1 b_1 & a_1 b_2 & \cdots & a_1 b_p \\ a_2 b_1 & a_2 b_2 & \cdots & a_2 b_p \\ \vdots & \vdots & & \vdots \\ a_p b_1 & a_p b_2 & \cdots & a_p b_p \end{bmatrix}$$

The product of a scalar, k, times a vector, $\mathbf{a}$, is k times each element of $\mathbf{a}$:

$$k\mathbf{a} = \mathbf{a}k = \begin{bmatrix} ka_1 \\ ka_2 \\ \vdots \\ ka_p \end{bmatrix}$$

A *matrix* is a rectangular table of numbers, with p rows and n columns. It may be viewed as an assembly of n column vectors of length p. Thus

$$\mathbf{A} = \begin{bmatrix} a_{11} & a_{12} & \cdots & a_{1n} \\ a_{21} & a_{22} & \cdots & a_{2n} \\ \vdots & \vdots & & \vdots \\ a_{p1} & a_{p2} & \cdots & a_{pn} \end{bmatrix}$$

is a p by n matrix. The typical element of **A** is a_{ij}, signifying the element of the ith row and jth column. Matrices are added and subtracted by adding or subtracting corresponding elements of the matrices. Thus

$$\mathbf{A} + \mathbf{B} = \begin{bmatrix} a_{11} + b_{11} & a_{12} + b_{12} & \cdots & a_{1n} + b_{1n} \\ a_{21} + b_{21} & a_{22} + b_{22} & \cdots & a_{2n} + b_{2n} \\ \vdots & \vdots & & \vdots \\ a_{p1} + b_{p1} & a_{p2} + b_{p2} & \cdots & a_{pn} + b_{pn} \end{bmatrix}$$

Matrix multiplication involves the computation of the sum of the products of elements from a row of the first matrix (the premultiplier) and a column of the second matrix (the postmultiplier). This sum of products must be computed for every combination of rows and columns. For example, if **A** is a 2×3 matrix and **B** is a $3 + 2$ matrix, the product **AB** is

$$\mathbf{AB} = \begin{bmatrix} a_{11}b_{11} + a_{12}b_{21} + a_{13}b_{31} & a_{11}b_{12} + a_{12}b_{22} + a_{13}b_{32} \\ a_{21}b_{11} + a_{22}b_{21} + a_{23}b_{31} & a_{21}b_{12} + a_{22}b_{22} + a_{23}b_{32} \end{bmatrix}$$

Thus the product **AB** is a 2×2 matrix. On the other hand, the product **BA** is a 3×3 matrix

$$\mathbf{BA} = \begin{bmatrix} b_{11}a_{11} + b_{12}a_{21} & b_{11}a_{12} + b_{12}a_{22} & b_{11}a_{13} + b_{12}a_{23} \\ b_{21}a_{11} + b_{22}a_{21} & b_{21}a_{12} + b_{22}a_{22} & b_{21}a_{13} + b_{22}a_{23} \\ b_{31}a_{11} + b_{32}a_{21} & b_{31}a_{12} + b_{32}a_{22} & b_{31}a_{13} + b_{32}a_{23} \end{bmatrix}$$

In general, if **A** is a $k \times p$ and **B** is a $p \times n$ matrix, the product **AB** is a $k \times n$ matrix, and assuming $k \neq n$ it is impossible to form the product **BA**. We say that matrices *conform* for the operations of addition, subtraction, or multiplication when their respective *orders* (numbers of rows and columns) are such as to permit the operations. Matrices that do not conform for addition or subtraction cannot be added or subtracted. Matrices that do not conform for multiplication cannot be multiplied.

Numerical examples may clarify these operations. If

$$\mathbf{A} = \begin{bmatrix} 5 & 6 \\ 3 & 7 \end{bmatrix} \quad \text{and} \quad \mathbf{B} = \begin{bmatrix} 3 & 2 \\ 1 & 5 \end{bmatrix}$$

then

$$\mathbf{A} + \mathbf{B} = \begin{bmatrix} 8 & 18 \\ 4 & 12 \end{bmatrix} \quad \mathbf{A} - \mathbf{B} = \begin{bmatrix} 2 & 4 \\ 2 & 2 \end{bmatrix}$$

and

$$\mathbf{AB} = \begin{bmatrix} 21 & 40 \\ 16 & 41 \end{bmatrix} \quad \mathbf{BA} = \begin{bmatrix} 21 & 32 \\ 20 & 41 \end{bmatrix}$$

To multiply a matrix by some scalar, we multiply each element of the matrix by that scalar

$$2\mathbf{A} = \begin{bmatrix} 10 & 12 \\ 6 & 14 \end{bmatrix} \qquad \tfrac{1}{2}\mathbf{B} = \begin{bmatrix} 1.5 & 1.0 \\ .5 & 2.5 \end{bmatrix}$$

Premultiplying a matrix that has p rows and n columns by the transpose of an p-element vector yields an n-element vector transpose:

$$\mathbf{c}' = \mathbf{a}'\mathbf{B} = [a_1 \quad a_2] \begin{bmatrix} b_{11} & b_{12} & b_{13} \\ b_{21} & b_{22} & b_{23} \end{bmatrix} = [c_1 \quad c_2 \quad c_3]$$

Postmultiplying a matrix that has p rows and n columns by an n-element vector yields an n-element vector

$$\mathbf{c} = \mathbf{Ba} = \begin{bmatrix} b_{11} & b_{12} & b_{13} \\ b_{21} & b_{22} & b_{23} \end{bmatrix} \begin{bmatrix} a_1 \\ a_2 \\ a_3 \end{bmatrix} = \begin{bmatrix} c_1 \\ c_2 \end{bmatrix}$$

It is not possible to premultiply a matrix by a column vector, nor to postmultiply a matrix by a row vector. The matrix product $\mathbf{a}'\mathbf{Ba}$ yields a scalar, and is called a *quadratic form*. Note that $\mathbf{B}$ must be a square matrix if $\mathbf{a}'\mathbf{Ba}$ is to conform for multiplication. Here is a worked example of the quadratic form:

$$\mathbf{a}'\mathbf{Ba} = [2 \quad 3] \begin{bmatrix} 1 & 2 \\ 3 & 1 \end{bmatrix} \begin{bmatrix} 2 \\ 3 \end{bmatrix} = [11 \quad 7] \begin{bmatrix} 2 \\ 3 \end{bmatrix} = 43$$

The matrix algebra analog of division involves an operation called inverting a matrix. Only square matrices can be inverted, and the numerical analysis by which a square matrix is inverted is a tedious chore for which computers are better suited than human beings. Chapter Three presents a computer method for inverting matrices. For the moment we merely introduce the notion and notation of the *inverse*. $\mathbf{A}^{-1}$ (A-inverse) is a unique matrix that satisfies the relation

$$\mathbf{A}^{-1}\mathbf{A} = \mathbf{A}\mathbf{A}^{-1} = \mathbf{I}$$

where $\mathbf{I}$ is a matrix of the form:

$$\mathbf{I} = \begin{bmatrix} 1 & 0 & 0 & \cdots & 0 \\ 0 & 1 & 0 & \cdots & 0 \\ 0 & 0 & 1 & \cdots & 0 \\ \vdots & \vdots & \vdots & & \vdots \\ 0 & 0 & 0 & \cdots & 1 \end{bmatrix}$$

$\mathbf{I}$ is called an *identity* matrix, and is a special case of a diagonal matrix. Any matrix that has zeros in all the off-diagonal positions is a *diagonal* matrix.

Unfortunately, every square matrix does not have an inverse, although most do. Associated with any square matrix is a single number that represents a unique function of the numbers in the matrix. This scalar function of a square matrix is called the *determinant*. Chapter Three presents a computer method for evaluating determinants. The determinant of A is denoted $|A|$. If the determinant of a matrix is exactly zero the matrix is said to be *singular* and it has no inverse. We will be much occupied with the theory of singular matrices in the chapters ahead.

Of particular interest in statistics is the determinant of a square symmetric matrix D whose diagonal elements are sample variances and off-diagonal elements are sample covariances:

$$D = \begin{bmatrix} s_1{}^2 & s_1 s_2 r_{12} & \cdots & s_1 s_p r_{1p} \\ s_2 s_1 r_{21} & s_2{}^2 & \cdots & s_2 s_p r_{2p} \\ s_p s_1 r_{p1} & s_p s_2 r_{p2} & \cdots & s_p{}^2 \end{bmatrix}$$

The generalized variance in a set of p tests may be represented as $|D|$.

Besides a determinant and possibly an inverse, every square matrix has associated with it a *characteristic equation*. The characteristic equation of a matrix is formed by subtracting some one value λ from each of the diagonal elements of the matrix, where λ is selected so that the determinant of the resulting matrix is equal to zero. Thus the characteristic equation of a second order matrix A may be written:

$$|A - \lambda I| = \begin{vmatrix} a_{11} - \lambda & a_{12} \\ a_{21} & a_{22} - \lambda \end{vmatrix} = 0$$

For a matrix of order p there may be as many as p different values for λ (lambda) that will satisfy the characteristic equation. These different values are called the *eigenvalues* of the matrix.

Associated with each eigenvalue is a vector, v, called the *eigenvector*. The eigenvector is selected to satisfy the equation:

$$Av = \lambda v$$

If the complete set of eigenvalues of A is arranged in the diagonal positions of a diagonal matrix L, and the complete set of corresponding eigenvectors is placed as columns of a matrix V, the following relation holds:

$$AV = VL$$

This equation specifies the complete eigenstructure of A. We shall see that eigenstructure theory figures heavily in multivariate procedures, and that the numerical evaluation of L and V is a central computing problem. Chapter

Four presents two methods for computing eigenstructures of symmetric matrices, and Chapter Six presents a method for computing eigenstructures of nonsymmetric matrices.

1.3 COMPUTER USAGE AND FORTRAN

The spreading application of multivariate statistical procedures in the behavioral sciences is an aspect of the computer revolution. The numerical analyses required by these procedures simply are not feasible without the incredible speed and accuracy of the digital computer. The researcher who wants to learn to design and execute multivariate studies must engage in some sort of relationship with the computer. There are many ways this relationship can be structured, with various degrees of involvement. The independent reasearcher probably needs to become very intimate with a computer. The researcher who participates in team research may be able to lean heavily on one or more other team members for the computer expertise required by the enterprise, so that the division of labor permits him to keep his distance from the machine. Nevertheless, he will have to have insights into computer operations if he is to think clearly about multivariate research problems.

This text contains computer programs for all the procedures covered. Features of the programs are discussed, test problems are provided, and exercises to be computed on the Project TALENT data in the Appendix are included. Students and other readers who will compile these programs locally and run the test problems and assignments will gain considerable experience in computing, and will acquire a serviceable library of basic multivariate routines. Those who will take a local FORTRAN course or study a FORTAN text will find that a few hours devoted to such effort will equip them to read the programs with understanding. More important, they will become able to modify the programs to suit their needs, or to adapt them to local computing center conventions.

The computer programs have been kept as simple and transparent as possible, in keeping with their role in the pedagogy of this text. They are set up to read input data from punched cards and to write output for off-line printing. They have no tape handling ability. They are not intended to represent the elements of a program system such as any large research organization would want to develop. Project TALENT, for example, has a very sophisticated program system developed by Mr. Bary Wingersky, and it is remarkably different from what appears between these covers. Wingersky's system is magnetic tape oriented, and its elements are interlocked in chaining sequences. It is much more powerful and efficient than our programs, but it is also more complex and opaque.

The programs in this book are written in a coding language called FORTRAN. This language is acceptable to almost all computers. There are some modifications of the language for each machine, and indeed for each installation. It is necessary to check the listings of the programs locally and correct some details, particularly in the input and output conventions employed. The smaller computers accept only a subset of the FORTRAN language statements used in these programs. Thus the most severe modifications are required to adapt the programs for small computers. Besides language restrictions, there are memory restrictions in small machines which may make it necessary to segment some of the longer analyses and run them in parts, and/or to assemble the programs with much smaller dimension statements controlling the maximum sizes of matrices and vectors, and thus of the number of variables that can be analyzed. The computer used by the authors has 32,000 words in its core memory, whereas many of the machines in the IBM 1620 range are limited to 4000 memory words. Of course there are now machines in some university computer centers with much larger core memories than the programs were developed for, so that much larger problems could be dimensioned and run, and the spreading practice of attaching disk storage mechanisms of slow speed but large capacity to smaller computers makes it possible for some readers to adapt the programs to involve disk storage of intermediate results. Where smaller computers have tape drives, intermediate storage on tape is not too hard to learn to do, either.

The great advantage of FORTRAN is that it closely approximates the everyday usage of English-speaking mathematicians, scientists, engineers, and statisticians. Anyone who has explored the ways in which modern digital computers "think" realizes that their internal machine language usually consists exclusively of binary-coded numbers, which are very difficult for human beings to comprehend or generate. Also, instead of a single instruction sentence which the human being perceives as a simple command (such as, "Read the following list of things: A, B, C, D"), a computer must be presented with an extremely long sequence of binary-coded instructions to accomplish even simple operations. If we had to communicate with computers in their own natural language most of us would turn to problems capable of solution on desk calculators, and would shun multivariate procedures. Fortunately, programs called *compilers* have been devised which enable the computers to translate English-like language terms into machine instructions, and to assemble programs the machine can obey from statements the human user can easily write and understand. FORTRAN is such a compiler.

The following list pairs FORTRAN statements on the right with English explanations on the left to show the similarities. This sample of FORTRAN statements illustrates the major features of the language.

English Statement	FORTRAN Statement
Reserve memory cells for a matrix A of maximum 50 x 50 size, and for a vector x of maximum 50 elements.	DIMENSION A(50,50), X(50)
Read from file 7 according to format 3, values for variables PROBNO, N, and m.	READ (7,3) PROBNO, N, M
Specification as format 3 a card containing 3 5-column integer fields.	3 FORMAT (3I5)
Write on file 6 a label and the problem number, using format 4.	WRITE (6,4) PROBNO 4 FORMAT ('PROBLEM NO.'I3)
Place B in storage location A, i.e., set $A = B$.	A=B
Set $A = B + C(1/D)$.	A=B + C * (1.0 / D)
Set $A = B^n$.	A=B ** N
Set $A = \sqrt{B}$.	A=SQRTF (B).
Set A to the absolute value of A.	A=ABSF (A)
Set $x_i = a\sqrt{y_i}$, for $i = 1, 2, 3, \ldots, N$.	DO 2 I=1, N 2 X(I)=A * SQRTF (Y(I))
Set $x_i = \sqrt[3]{ni}$ for $i = 1, 2, 3, \ldots, n$. Note that EN = N converts an integer number to a floating point number for floating point arithmetic.	EN=N DO 5 I=1, N EI=I 5 X(I)=(EN * EI) ** (1.0/3.0)
Transfer control to statement 7.	GO TO 7
Test n; if it is negative go to 2, if it is zero go to 3, if it is positive go to 4.	IF (N) 2, 3, 4

English Statement	FORTRAN Statement
Call subroutine MATINV, which will operate on matrix X of order m to compute its inverse, which will replace X, and its determinant, which will be placed in location DET.	CALL MATINV (X, M, DET)

The most important feature of the FORTRAN language from the point of view of multivariate analysis is the easy handling of subscripted variables which facilitates matrix manipulations. The "DO loop" and nest of DO loops are the key to this facility. A DO loop instruction causes a subscript to take on each of its possible values successively. A DO nest causes two or more subscripts to vary systematically. For example, the following nest causes the writing in the cells reserved for matrix $B(n \times m)$ of the transpose of matrix $A(m \times n)$:

```
DO 10   J=1, M
DO 10   K=1, N
10 B(K,J)=A(J,K)
```

The outside (or first) DO sets J = 1. This is held constant while the inside DO loop causes statement 10 to be executed for values of K from 1 to N. Then the outside DO changes J from 1 to 2 and the inside DO is repeated, running K from 1 to N again. This sequence continues through J = M. As a further example, note how the following DO nest creates the lower triangle of a symmetric matrix A from the established upper triangle:

```
DO 12   J=1, M
DO 12   K=J, M
12 A(K,J)=A(J,K)
```

In the following triple DO nest a very important procedure, matrix multiplication, is illustrated. Some students find that study of this DO nest is the best way to understand what matrix multiplication really consists of.

```
DO 20   I=1, M
DO 20   J=1, N
C(I,J)=0.0
DO 20   K=1, L
20 C(I,J)=C(I,J)+A(I,K)*B(K,J)
```

Can you see that $C(m \times n)$ is formed as the product of $A(m \times l)$ times $B(l \times n)$? The furthest inside (or third) loop has to be run through for the computation of each cell entry in the product **C**, so that it is run a great many times if m and n are sizeable. As this example illustrates, "DO 20" actually means "execute each of the following instructions through the statement numbered 20."

Among the other significant advantages of FORTRAN are its great flexibility of input and output procedures, achieved through the FORMAT statement and the "variable format" feature (which allows input and output card or tape formats to be read as data, and thus varied for each use of a program), and its subroutine feature, which allows the writing and testing of analytical building blocks out of which a main program can be assembled. Our programs rely very heavily on two utility subroutines, one called HOW which computes the eigenstructure of a symmetric matrix, and one called MATINV which computes the inverse and determinant of a square matrix. The authors did not write these workhorse subroutines, but instead were able to borrow them out of the productivity of professional programmers. A subroutine the authors did write, the DIRNM routine which computes the eigenstructure of a nonsymmetric matrix product of symmetric matrices, i.e., $\mathbf{B}^{-1}\mathbf{A}$, has proven useful in programs written by others. The subroutine package actually turns out to be much more transportable than does the main program package, because subroutines do standard tasks for main programs, and they seldom include input and output statements, which is the most idiosyncratic class of FORTRAN statements.

Format for raw data can be rearranged by use of FORTRAN format statements. This is important because some programs specify a certain arrangement of your variables. For example, in MULTR the criterion variable should be the last row and column of the input R matrix. If it does not happen to be the last variable in your observation vector, the T (transfer) format can be used. That is, if the criterion is in columns 15–16, and the five predictors are as follows:

Columns	11–12	13–14	15–16	17	18	21
Variable	X_1	X_2	Y	X_3	X_4	X_5

the required format is: (10X,2F2.0,2X,3F1.0,T15,F2.0). Thus "T15" means transfer to column 15. Any number of transfers can be used in one format statement.

A complete FORTRAN program for computing an average score for each of N subjects, given M scores for each subject as input, is illustrated in the statements that follow. The computer is directed to read a setup card from which it learns the values of N and M, and then to read one or a set of score cards for each subject, getting the subject's identification number and his

M scores. The M scores are averaged and a print-line image is written on tape 6 (for off-line printing) giving the identification number and the average. Note that the dimensioning of X assumes that the maximum value of M in any run will be 1000. A card with C in the first column is a "comment card" which is ignored by the compiler. A 1 in column 6 indicates the continuation of the previous FORTRAN statement.

```
$   COMPILE FORTRAN, EXECUTE
C     COMPUTE AVERAGE OF M SCORES FOR EACH OF N
C     SUBJECTS AND PRINT.
C     INPUT
C     FIRST CARD CONTAINS N IN COLS 1-5 and M IN COLS
C     6-10.
C     FOLLOWING CARDS ARE SCORE CARDS, ONE SET PER
C     SUBJECT.
C     FIRST CARD PER SET CONTAINS ID IN COLS 1-5 AND
C     UP TO 25 SCORES
C        IN 3 COLUMN FIELDS. ID FIELD (COLS 1-5) MAY BE
C        BLANK IN
C        ADDITIONAL CARDS IN A SET.
C
      DIMENSION  X(1000)
      READ (7, 1)    N,  M
  1   FORMAT (2I5)
      WRITE(6, 2)     M,  N
  2   FORMAT( 'AVERAGES ON'I6, 'TESTS FOR EACH OF 'I6,
     1 SUBJECTS')
      EM=M
      DO  5  J=1, N
      READ (7, 3)  ID, (X(K), K=1, M)
  3   FORMAT (I5, 25F3.0 / (5X, 25F3.0))
      SUM=0.0
      DO  4  K=1, M
  4   SUM=SUM+X(K)
      AV=SUM / EM
  5   WRITE(6, 6)  J,    ID,   AV
  6   FORMAT (I6, 3X, 'SUBJECT'I6, 3X, 'AV='F9.2)
$ DATA
     5     5
   821  3  7  9  4  7
   812  1  4  3  3  2
   813  3  2  3  1  1
   824  7  9  9  9  9
   825  6  9  8  8  5
```

The output from the program run on the test problem given appears in Table 1.2, which is the actual computer printout. Your many questions

TABLE 1.2   Computer Printout

AVERAGES ON	5 TESTS FOR EACH OF	5 SUBJECTS		
1	SUBJECT	821	AV=	6.00
2	SUBJECT	812	AV=	2.60
3	SUBJECT	813	AV=	2.00
4	SUBJECT	824	AV=	8.60
5	SUBJECT	825	AV=	7.20

about the details in the program, especially those about format details, can best be answered by consulting a FORTRAN manual. You will have to write formats for your input data if you are going to use our programs, since they generally read input formats as part of their setups. Your computer center will have a programmer consulting service to help you with this. The same service can help you get the appropriate monitor control cards placed in your program-plus-data deck. The cards with a $-sign in column 1 are examples of monitor control cards. The *monitor* is an excutive program that actually operates the computer, and control cards are devices that enable the user to describe his job to the monitor. Specifications for control cards vary greatly from computing center to computing center, so you will have to have local guidance regarding them.

The intention of this brief orientation to computer usage and FORTRAN has been to whet your appetite and give you courage. We hope you are persuaded that you can penetrate the mysteries of the cult of computer users with no great effort and with real advantage. We believe that every researcher needs an initiation to these mysteries.

1.4 EXERCISES

1. Evaluate the quadratic form:

$$[2 \quad 3 \quad 1] \cdot \begin{bmatrix} 2 & 4 & 3 \\ 4 & 1 & 1 \\ 3 & 1 & 2 \end{bmatrix} \cdot \begin{bmatrix} 2 \\ 3 \\ 1 \end{bmatrix}$$

2. Test the assertion that R^{-1} is as given:

$$R = \begin{bmatrix} 1.0 & .5 \\ .5 & 1.0 \end{bmatrix} \qquad R^{-1} = \begin{bmatrix} 1.3333 & -.6667 \\ -.6667 & 1.3333 \end{bmatrix}$$

3. What matrix algebra is the following FORTRAN performing?

```
      DO   5     J=1,  M
      Y(J)=0.0
      DO   5     K=1,  M
    5 Y(J)=Y(J)+X(K)  *  D(K,J)
      DO   6     J=1,  M
    6 Z=Y(J)  *  X(J)
```

4. If **a** is a column vector with four elements all equal to 1.00, and **b**′ = [1 2 3 4], what are the differences among the products **ab**′, **a**′**b**, **b**′**a**, and **ba**′?

5. If **A** = **BC** and you know the values for matrices **A** and **B**, how would you solve for **C**?

1.5 SOME BASIC BACKGROUND REFERENCES

1. Anderson, T. W. *An Introduction to Multivariate Statistical Analysis*. New York: Wiley, 1958.

2. Ayres, F., Jr. *Theory and Problems of Matrices*. New York: Schaum, 1962. (Schaum's Outline Series, paper)

3. Bock, R. D. "Contributions of Multivariate Statistical Methods to Educational Research." In R. B. Cattell (editor), *Handbook of Multivariate Experimental Psychology*. Chicago: Rand McNally, 1966.

4. Bock, R. D., and Haggard, E. A. "The Use of Mutilvariate Analysis of Variance in Behavioral Research." In W. Whitla (editor), *Handbook of Measurement in Education, Psychology and Sociology*. Boston: Addison-Wesley, 1966.

5. Borko, H. (editor). *Computer Applications in the Behavioral Sciences*. Englewood Cliffs, N.J.: Prentice-Hall, 1962.

6. Cattell, R. B. (editor). *Handbook of Multivariate Experimental Psychology*. Chicago: Rand McNally, 1966.

7. Cattell, R. B. "Multivariate Behavioral Research and the Integrative Challenge." *Multivariate Behavioral Research*, 1966, **1**, 4–23.

8. DuBois, P. H. *Multivariate Correlational Analysis*. New York: Harper, 1957.

9. Flanagan, J. C., Dailey, J. T., Shaycoft, M. F., Gorham, W. A., Orr, D. B., and Goldberg, I. *Design for a Study of American Youth*. Boston: Houghton Mifflin, 1962.

10. Flanagan, J. C., Cooley, W. W., Lohnes, P. R., Schoenfeldt, L. F., Holdeman, R. W., Combs, J., and Becker, S. J. *Project TALENT: One-Year Followup Studies*. Pittsburgh: University of Pittsburgh, 1966.

11. *General Information Manual: FORTRAN*. White Plains, N.Y.: IBM Data Processing Division, 1961.

12. Harman, H. H. *Modern Factor Analysis*. Chicago: University of Chicago Press, 1960.

13. Harris, C. W. (editor). *Problems in Measuring Change*. Madison, Wisconsin: University of Wisconsin Press, 1963.

14. Horst, P. *Factor Analysis of Data Matrices*. New York: Holt, Rinehart and Winston, 1965.

15. Kendall, M. G. *A Course in Multivariate Analysis*. New York: Hafner, 1957.

16. McCracken, D. D. *Digital Computer Programming*. New York: Wiley, 1957.

17. McCracken, D. D. *A Guide to FORTRAN Programming*. New York: Wiley, 1961.

18. Rao, C. R. *Advanced Statistical Methods in Biometric Research*. New York: Wiley, 1952.

19. Rulon, P. J., Tiedman, D. V., Langmuir, C. R., and Tatsuoka, M. M. *Multivariate Statistics for Personnel Classification*. New York: Wiley, 1967.

20. Searle, S. R. *Matrix Algebra for the Biological Sciences*. New York: Wiley, 1966.

21. Tiedeman, D. V. and Tatsuoka, M. M. "Statistics as an Aspect of Scientific Method in Research on Teaching." In N. L. Gage (editor), *Handbook of Research on Teaching*. Chicago: Rand McNally, 1963.

CHAPTER TWO

||||||||||||||||||||||||||||||||||||||||||||||||||||||||||||||||||||||||||

The Basic Vectors and Matrices

2.1 VECTOR RANDOM VARIABLES

Statistics is many things to many people, depending on the fields of application and the types of problems usually encountered by the groups of people who deal in statistics. A relatively small group of these people deals with theoretical problems in statistics. They are the mathematical statisticians. A larger group is composed of workers who know the theory of statistics and labor to bring it to bear on the various scientific and engineering fields. These are the applied statisticians. Then there is a congeries of groups of researchers for whom various collections of statistical procedures are everyday tools of their trade. Sometimes it is difficult to see what common faith binds all these groups together, particularly if the viewer is out in the fringe among the researchers who use statistics. If there is an intellectual kernel to which all this crowd of statisticians and users of statistics give allegiance, it is the notion of a random variable. A statistical approach to the study of any set of events is premised on the assumption that there is a stochastic element involved in the determination of the events. This assumption of an element of randomness in the causation of phenomena is perhaps the only infallible proofmark of a statistical inclination. A statistical theory of data always incorporates at least one random variable.

A conventional definition of a *random variable* states that it is a rule for assigning numbers to events in such a way that for every assigned number χ there is a probability $P(\chi)$ which represents the likelihood of occurrence of a score equal to or less than χ in magnitude. The mystery is in how a variable can be random when the assignment of values actually follows a rule. The answer is that the individual event is not assigned by the rule. Only the cumulative probability function for the population of all events is known. A random variable follows a probability law, or we might say a statistical law.

We are all familiar with the role of the scalar random variable and its cumulative probability function in univariate statistics. In multivariate statistics we deal with a *vector random variable* χ which is a p-element column vector. The p elements of χ are the coordinates of a random point from the theoretical distribution of χ, which is an infinite swarm of points in p-dimensional space. We will assume that the distribution of χ is the multivariate normal distribution, which we discuss in the next section of this chapter. Corresponding to the vector random variable χ we have the vector *sample* variable **x**, the p elements of which are scores on p measures for a subject in a random sample from the population. It is important to recognize that χ is a single point in the theoretic distribution even though it is specified by p elements, so that $P(\chi)$ is one number. Likewise, **x** is a single event in a random sample, resulting from a single draw on the population, even though **x** is described by p scores.

Where mathematical statistics texts are concerned with theory about χ, we tend to ignore χ and to concentrate on **x** in this text. Why do we introduce χ at all in a book about data manipulations? Simply to acknowledge that the ubiquitous vector random variable always lurks in the back of our minds as we make interpretations of vector data variables. There is statistical theory governing our inferences from data, even if we fail to exposit that theory.

Given the vector random variable χ, its *centroid* or vector of means is μ and its *dispersion* or variance-covariance matrix is Δ. Correspondingly in the sample space the centroid or vector of means is **m** where

$$\mathbf{m} = \frac{1}{N} \sum_{i=1}^{N} \mathbf{x}_i$$

and the dispersion or variance-covariance matrix is **D** where

$$\mathbf{D} = \frac{1}{N} \sum_{i=1}^{N} (\mathbf{x}_i - \mathbf{m})(\mathbf{x}_i - \mathbf{m})'$$

Be sure to notice that since **x** is a column vector and **m** is also a column vector the definition of **D** involves a column vector postmultiplied by its transpose, which gives a square matrix. For each subject this "squares and crossproducts of deviation scores" matrix is symmetric around its major diagonal, and thus the **D** matrix is the "sums of square and crossproducts of deviation scores" matrix divided by the number of subjects. The diagonal elements of **D** are sample variances (which we denote d_{jj}) and the off-diagonal elements are sample covariances (which we denote d_{jk}).

An interesting and useful simplification of the dispersion matrix occurs when the elements of the vector variable are standardized so that all the means are zero and all the variances are unity. Given that $m_j = 0$ and $d_{jj} = 1$ for $j = 1, 2, \ldots, p$ then $d_{jk} = r_{jk}$. That is, the dispersion for a standardized vector variable is a correlation matrix. We will use the symbol **z** for the standardized vector variable throughout this text. For the sample vector variable **z**

$$\mathbf{D} = \mathbf{R}$$

Notice that substituting **z** for **x** and **0** for **m** in the definition equation for **D** leads to

$$\mathbf{R} = \frac{1}{N} \sum_{i=1}^{N} \mathbf{z}_i \mathbf{z}_i'$$

The typical element of **R** is the maximum likelihood estimator of the correlation coefficient

$$r_{jk} = \frac{1}{N} \sum_{i=1}^{N} z_{ji} z_{ki}$$

which expresses that the sample correlation coefficient is the average cross-product of standard scores. Note that when $j = k$ the value of r_{jk} is unity, so that the main diagonal of the **R** matrix contains ones. These ones represent the total observed variances of the elements of the standardized sample vector variable.

It is the observed sample correlation matrix **R** on which are performed regression, factor, and canonical analyses in the chapters of Part II. All methods such as ours that analyze **R** may be called *components* methods to distinguish them from a large and popular class of methods called *communalities* methods that remove estimated error variance from **R** before analyzing the intercorrelations it contains. Component methods yield derived measures that are computable as linear functions of observed test score vectors. Statisticians refer to a linear function y of a vector variable **z** as a component of **z**. Such a component is specified by the algebraic notation

$$y = \mathbf{v}'\mathbf{z}$$

where **v** is a vector of coefficients. The scalar algebra for the component score for the ith individual is

$$y_i = v_1 z_{1i} + v_2 z_{2i} + \cdots + v_p z_{pi}$$

When a set of linear functions or components of z, say n of them, are to be defined, the n column vectors of coefficients with p coefficients in each vector are arrayed in the $p \times n$ coefficients matrix $\mathbf{V}$. Then the set of n transformations of z is specified by the equation

$$\mathbf{y} = \mathbf{V}'\mathbf{z}$$

For the ith individual the derived or transformed score vector is

$$\mathbf{y}_i = \mathbf{V}'\mathbf{z}_i$$

where $\mathbf{y}_i$ is an n element vector.

Two important theorems concern the centroid and dispersion of derived or transformed vector variables. The first states that when $\mathbf{y} = \mathbf{V}'\mathbf{x} + \mathbf{b}$, where $\mathbf{b}$ is a vector of n constants, then

$$\mathbf{m}_y = \mathbf{V}'\mathbf{m}_x + \mathbf{b}$$

and the second states that

$$\mathbf{D}_y = \mathbf{V}'\mathbf{D}_x\mathbf{V}$$

Special cases of these general theorems state that when $\mathbf{f} = \mathbf{V}'\mathbf{z}$, then

$$\mathbf{m}_f = \mathbf{m}_z = 0$$

and

$$\mathbf{D}_f = \mathbf{V}'\mathbf{R}_z\mathbf{V}$$

These theorems about linear transformations of vector variables are immensely useful to us because the central, organizing principle of the multivariate procedures presented in this text is that the differential calculus of systems of linear equations is applied to define linear functions of the input vector variable that maximize some payoff criterion or minimize some error criterion. The synthesizing principle for this collection of multivariate procedures is that they are all examples of components analysis (with the exception of the classification procedures of Chapter Ten, which are quadratic functions of the vector variable). What is usually done is to seek a set of components of the vector variable that will do a specific job better than any other components can. The relative contributions of the different elements of the vector variable to the components become very important in the interpretation of data. It is in this sorting out of the contributions of elements of the variable to the "best" component that the heuristic powers of these methods show up.

A component of a vector variable has arbitrary mean and variance. When the component is defined on a standardized vector variable in such a way that it will have zero mean and unit variance we call the component a *factor* of the measurement variable. Throughout the text we will so scale our components that they will become factors of our input measures. The representation of functions (principal component, canonical, multiple partial, discriminant, and covariance) as factors of the measurement battery provides the cord which binds our bundle of methods into a coherent set and gives us the unity of a single strategy for attacking multivariate research problems.

The theorems stated above lead to a simple solution to the problem of how to scale any arbitrary component of a vector variable so that it will be a factor of that variable. Let $\mathbf{v}$ be the coefficients defining a component of $\mathbf{z}$ as

$$y_i = \mathbf{v}'\mathbf{z}_i$$

We compute the standard deviation of the component as

$$s_y = \sqrt{\mathbf{v}'\mathbf{D}_z\mathbf{v}}$$

and divide out this standard deviation from the coefficients

$$\mathbf{c} = \frac{1}{s_y}\mathbf{v}$$

The standardized component or factor with zero mean and unit variance is now

$$f_i = \mathbf{c}'\mathbf{z}_i$$

The factor is most easily understood in terms of its correlations with the elements of the vector variable on which it is defined. What we want is the column vector of variable-factor correlations $\mathbf{r}_{zf}$ where

$$\mathbf{r}_{zf} = \frac{1}{N}\sum_{i=1}^{N}\mathbf{z}_i f_i$$

$$= \frac{1}{N}\sum \mathbf{z}_i(\mathbf{c}'\mathbf{z}_i)'$$

$$= \frac{1}{N}\sum \mathbf{z}_i\mathbf{z}_i'\mathbf{c}$$

$$= \mathbf{R}_z\mathbf{c}$$

That is, the vector of variable-factor correlations is obtained by post multiplying the intercorrelation matrix for the vector variable by the vector of factor-score coefficients. These correlations are known as the structure of the factor.

As an example of the interpretation of a factor through its correlations with the elements of the vector variable, let us look at two factors located in a study of 217 educably retarded children in a county school conducted by S. Seidman. This researcher had two vector variables. His predictor variable contained four elements: Chronological Age (substantial range in this group of subjects), I.Q. (very limited range), Manifest Anxiety, and a Lie score from the anxiety inventory. His criterion variable had three elements, all based on an achievement test battery: Spelling, Arithmetic, and Reading. Seidman used canonical correlation to develop a factor of each vector variable such that the pair of factors have maximum correlation for any pair of factors possible. His pair of factors actually correlated .66. The correlations of his factors with their defining variables are given in Table 2.1.

TABLE 2.1 Seidman's Canonical Factors for 217 Educably Retarded Youths from a County School

Predictors $r_{z_p f_p}$		Criteria $r_{z_c f_c}$	
Age	.95	Spelling	.80
I.Q.	.11	Arithmetic	.97
Anxiety	−.07	Reading	.85
Lie	−.46		
	$r_{f_p f_c} = .66$		

Looking first at the factor of the criterion variable, f_c, we observe that it is almost perfectly correlated with the Arithmetic element, but also highly correlated with Spelling and Reading. It is interpretable as a General Academic Achievement factor. The factor of the predictor variable, f_p, is interesting in that it correlates almost perfectly with Age. The older the youth the greater the academic achievement. This finding is a tribute to the teachers of these retarded youths. The low correlation of I.Q. with the predictor factor undoubtedly stems from the strong restriction in range for this element. The Anxiety scale doesn't relate to this factor, and the modest negative relationship of the Lie score is hard to comment on since we don't know anything about this scale.

Is the canonical correlation of .66 between the two factors statistically significant? Since this was an available group of subjects, not a random sample from any population of educably retarded youths, the question is meaningless. However, if we had a random sample (some researchers do), the question would be valid. There is an analysis-of-variance type significance test for the canonical correlation which is computed by the CANON program. The justification for the test (assuming a random sample) involves an assumption about the theoretical distribution of the vector variables involved, namely that they are multivariate normal in distribution. We now take a look at this assumption which underlies all the significance tests in this book.

2.2 MULTIVARIATE NORMAL DISTRIBUTIONS

All of the multivariate models presented in this book are based on the assumption of a multivariate normal distribution for each population that has been sampled. Our inference tests for null hypotheses depend on this assumption, although we will not derive or defend the tests we present. This book is not concerned very much with distribution theory, nevertheless it is worth while to take a look at some of the properties of a multivariate normal distribution (m.n.d.) that our models for data analysis depend on.

The density function of a p-element normal distribution is

$$p(\chi) = Ke^{-.5(\chi-\mu)'\Delta^{-1}(\chi-\mu)}$$

where K is chosen so that the integral over the entire p-dimensional Cartesian space is unity. We write $N(\mu, \Delta)$ to specify that χ is normal with centroid μ and dispersion Δ. For $N(\mu, \Delta)$

$$K = (2\pi)^{-.5p}|\Delta|^{-.5}$$

The cumulative density function is $P(\chi)$ where

$$P(\chi) = Pr(x_1 \leq \chi_1, x_2 \leq \chi_2, \ldots, x_p \leq \chi_p)$$

where Pr stands for *probability*. A discussion of the evaluation of $P(\chi)$ in the bivariate case, along with useful nomographs, occurs in Abramowitz and Stegun, Chapter 26. It is noteworthy that this ambitious source on computation and tables of probability functions is unwilling to tackle the numerical evaluation of $P(\chi)$ for $p > 2$. A significant feature of the m.n.d. is that the surface in p-dimensional Cartesian space on which the density function is constant is an *ellipsoid* (a p-dimensional generalization of the ellipse) specified by the quadratic form

$$d^2 = (\chi - \mu)'\,\Delta^{-1}(\chi - \mu)$$

The center of each ellipsoid is at μ, the shape and orientation of each ellipsoid is determined by Δ, and the size of each ellipsoid is determined by its unique value for d^2. We have to conceptualize a nest of congruent ellipsoids around a common centroid. The nest contains an infinite number of layers, since there is an infinity of values for d^2.

For all points on the surface of a given ellipsoid, d^2 and $P(\chi)$ are constant. In Chapter Ten we present a rationale for considering d^2 a chi-square with

p degrees of freedom and a method for evaluation $P(\chi)$ on that basis. Note that d^2 can also be considered as a *distance function*, since it tells how far the surface of the ellipsoid is from the center of the swarm relative to distances from the centroid to surfaces of other ellipsoids in the nest.

Three classes of distributions that are derivative from a parent m.n.d. are of great interest to us, namely

    (1) marginal distributions,

    (2) conditional distributions

    (3) component distributions.

A *marginal* distribution is the univariate distribution for any single element of a vector variable. If the vector variable is multivariate normal in distribution, then every one of its marginal distributions is normal. However, even if all the marginals are normal, it is not necessarily true that the vector variable is multivariate normal. We are interested in normal marginals because after we have established a predictive validity for a vector variable we will want to make inferences about the predictive validities for the specified criterion of each of the elements of the vector variable separately. For example, we will follow up on a significant multivariate analysis of variance centroids separation test by inspecting univariate F-ratios for the means separation on each element of the vector variable. Figure 2.1 illustrates the two marginal distributions for a bivariate normal variable.

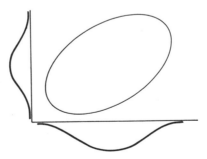

Figure 2.1. Marginal distributions.

A *conditional* distribution is the predicted distribution for a particular marginal element z_j given the known distribution of the remainder of the vector random variable z. We speak of $\hat{z}_j$ as the *regression* of z_j on z. The regressed variable is a special linear function of the remainder of the vector random variable

$$\hat{z}_j = \mathbf{b}'\mathbf{z}$$

with the regression coefficients in **b** determined to minimize the error variance in the prediction of z_j. Figure 2.2 illustrates a conditional distribution.

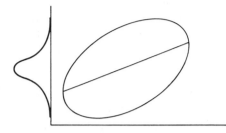

Figure 2.2. A conditional distribution.

An important theorem is that if a vector variable is multivariate normal in distribution, then every conditional distribution defined on it is normal.

A *component* distribution is the distribution of any arbitrary linear function of a vector variable. We have seen that factors are simply standardized components. If a vector variable is multivariate normal in distribution, then every component or factor defined on it is normal. Figure 2.3 illustrates this

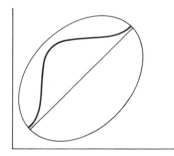

Figure 2.3. A principal component distribution.

for the principal component of a bivariate normal distribution. Since conditional distributions are special components of a vector variable, the normality of all conditional distributions is a special application of this more general theorem. Likewise, since each marginal of a vector variable is simply that special component defined by setting a unit weight for the assigned element and zero weights for all other elements, the normality of all marginals is a special case of this general theorem. The very general theorem that all linear components defined on a vector variable that has an m.n.d. are normal is especially important because it is reversible. That is, it is also true that any vector variable for which *every possible* linear component is normal is a vector variable that is multivariate normal in distribution. In his recent mathematical statistics text, Rao (1965, p. 437) uses this theorem to provide the basic characterization of the m.n.d. Anderson, whose Chapter Two is an excellent reference on the m.n.d., summarizes:

> One of the reasons the study of normal multivariate distributions is worthwhile is that marginal distributions and conditional distributions are also normal distributions. Moreover, linear combinations of normal variates are again normally distributed. (Anderson, 1958, p. 19)

How much fretting should researchers who are going to use our multivariate models do about the issue of whether their data have approximately an m.n.d. form? Very little, in our judgment. It is useful and in some cases necessary to inspect the marginal distributions. This is where a computerized cathode tube display or plotter diagram of a grouped-data frequency polygon for each marginal comes in handy. If one or more of the marginals is really out of whack, normalizing transformations may be sought, and perhaps found. But we have said that normal marginals do not themselves guarantee an m.n.d., and we do not know of any useful test for multivariate normality. It is true that we can test the particular linear functions of our vector variable that we create through data analyses for normality as univariate distributions and, of course, if one of them is out of whack the vector variable on which it is defined is then known to be other than multivariate normal. It is also true that the multivariate central limit theorem insures us that a linear component of a vector variable will have an enhanced approximation to normality over that of the vector variable itself. Basically, it would seem that those researchers who most need to fret over normality are the ones whose work is most dependent on statistical inference outcomes.

If it is possible to work with large enough and representative enough samples, sizable effects and contrasts that appear in data analyses will have compelling implications for behavioral theory and educational practice without the buttressing of propositions about the probabilities of null hypotheses. This strategy requires showing what the factors extracted from the predictor batteries can do to reduce the unexplained variances in socially significant criterion variables. Prediction on replication samples is protection against capitalization on chance. The hazards of overfitting in multivariate analysis are great. Although significance tests, when appropriate, can help to protect against reporting results that can never be replicated, we tend to treat our multivariate models as primarily heuristic rather than inferential procedures.

2.3 REDUCTION OF DATA TO SAMPLE ESTIMATES

One problem of estimation in multivariate analysis is that the maximum likelihood estimators of variances and covariances are not the same as the unbiased estimators. Most of us have learned to favor unbiased estimators in statistics because of the role they play in the analysis of variance, and we will not be surprised to find that multivariate analysis of variance also depends on unbiased estimators. On the other hand, maximum likelihood estimators are particularly useful in regression and factor analyses because maximum likelihood estimators of parameters of linear function distributions are very straightforward transformations of maximum likelihood estimators for the original vector variable. In his discussion of this issue, Anderson speaks of maximum likelihood estimators as usually having some optimum properties for problems involving the m.n.d. (Anderson, 1958, p. 44). In the first part of this text, which is concerned with regression and factor theories for data, we will always compute the maximum likelihood estimators, but will shift gears to unbiased estimators in the second half, which is concerned with multivariate analysis of variance models.

The maximum likelihood estimator of the centroid is an unbiased estimator, and is given by the column vector $\mathbf{m}$ where:

$$\mathbf{m} = \frac{1}{N} \sum_{i=1}^{N} \mathbf{x}_i$$

The maximum likelihood estimator of the dispersion, or variance-covariance matrix, is

$$\mathbf{D} = \frac{1}{N} \sum_{i=1}^{N} (\mathbf{x}_i - \mathbf{m})(\mathbf{x}_i - \mathbf{m})'$$

As you know, the elements of $\mathbf{D}$ are biased on the small side when N is small. The correction to compute the unbiased estimate $\mathbf{D}_u$ is

$$\mathbf{D}_u = \frac{N}{N-1} \mathbf{D}$$

A useful computation formula for D is

$$\mathbf{D} = \frac{1}{N} \left[\sum_{i=1}^{N} \mathbf{x}_i \mathbf{x}_i' \right] - \mathbf{m}\mathbf{m}'$$

The standard deviations are obtained by taking the square root of the diagonal elements of $\mathbf{D}$. The set of p standard deviations are frequently represented as $D_{\text{diag}}^{1/2}$. The elements of the maximum likelihood estimator of the correlation matrix are computable as:

$$r_{jk} = \frac{d_{jk}}{\sqrt{d_{jj} d_{kk}}}$$

It is noteworthy that **m** is distributed normally and independently of **D**. In fact, the sample centroid is distributed $N[\mu, (1/N)\Delta]$ when χ is distributed $N(\mu, \Delta)$. The unbiased estimate of Δ is required in the testing of hypotheses about μ. Hotelling (1931) introduced the statistic T^2 as the multivariate extension of Fisher's t.

$$T^2 = N(\mathbf{m} - \mu)'\mathbf{D}_u^{-1}(\mathbf{m} - \mu)$$

This is an example of a quadratic form, so that T^2 is a scalar. On the null hypothesis $H: \mu = \mu_0$ the following function of T^2 is F-distributed:

$$F_{N-p}^p = \frac{N-p}{p} \cdot \frac{T^2}{N-1}$$

The requirement for $\mathbf{D}_u^{-1}$, which is the inverse of the unbiased estimator of the dispersion, in the computation of T^2 illustrates the need for knowledge of matrix inverses and how to compute them; this will be discussed in the next chapter.

2.4 THE CORREL PROGRAM

The first task in almost any multivariate analysis is to accumulate the sums and sums of squares and cross products of scores for the sample group. This preliminary reduction of data is necessary to establish sample estimates of **m** and **D** or **R** regardless of the type of multivariate analysis to follow. From these accumulations of raw sums and raw sums of squares and cross products (hereafter s.s.c.p.) the means vector, the deviations s.s.c.p. matrix, the dispersion, and correlation matrix may be computed as required. Because these first steps are similiar in all multivariate analyses, a routine called CORREL has been written as a source for the reductions required in many of the main programs of this book.

The fundamental trick in data reduction is to read and dispose of one score vector at a time, so that it is unnecessary to have the entire score roster stored in memory. This trick makes N, the number of subjects, irrelevant for planning memory requirements in programming. All the storage requirement in core memory for scores is one p-element vector into which score vectors are read sequentially, each new one replacing the preceding one. Of course, the magnitude of N is very important in the determination of the time the correlation reductions will take for execution (for example, on the

IBM 7090 we find that it takes about one hour to do correlation reductions for 8000 subjects and 50 tests), but in theory there is no limitation on the number of subjects the CORREL routine can process. The way this works is that each score vector is read, its contributions to the sums vector and the s.s.c.p. matrix are registered, and the vector is replaced by the next score vector in the data deck. In FORTRAN, we initialize the accumulator vector and matrix (this is just like clearing a desk calculator before adding or cumulative multiplying):

```
C    M IS NUMBER OF TESTS
     DO 8 J=1, M
     Y(J)=0.0
     DO 8 K=1, M
  8  A(J,K)=0.0
```

Then we set up a loop in which each of the N score vectors is read into $X(J)$, its elements added into $Y(J)$, and the sums of squares and cross products of its elements added into $A(J,K)$:

```
     DO 9   L=1, N
     READ   (5,FMT)   (X(J),   J=1, M)
C  NOTE THAT FMT CONTAINS THE VARIABLE FORMAT FOR
C  A SCORE VECTOR.
     DO 9 J=1, M
     Y(J)=Y(J)+X(J)
     DO 9   K=1, M
  9  A(J,K)=A(J,K)+X(J) * X(K)
```

When the outside loop ending on statement 9 has been completed all N score vectors have been processed, and the reduction aspect of CORREL is completed. The routine then proceeds to compute and report the means and standard deviations, the dispersion, and the correlation matrix. These operations are straightforward and study of the computer program will reveal the specific formulas employed. Notice that when all estimates have been reported the program returns control to statement 1, in order to see whether another correlation reduction job is waiting to be done. If there is no data for another job the monitor will clock the user off the computer.

A feature of CORREL is the use of a simple utility subroutine called MPRINT to arrange the printing of the dispersion and correlation matrices. The attractive and convenient presentation of matrices, with row and column identifications and sectioning of the matrix if it is too large to fit on one page, is important in multivariate analyses. Sometimes we will want to photographically reproduce matrix printouts in reports. This MPRINT subroutine is used extensively in our programs.

A word of caution about punched output of matrices from our programs. Do not remove the blank card after any row containing exactly the number of elements specified by the format. This blank is expected by the corresponding read loops of other programs that take punched matrices as input.

2.5 NUMERICAL EXAMPLE

Numerical examples are presented in each chapter for two reasons. They help clarify the multivariate technique under consideration and they provide a means for checking your version of these programs at your computer center. It is important to develop the habit of skepticism regarding the veracity of computer output and the more onerous habit of checking on it. Often, checks can be made on almost trivial data sets by hand calculations. When the program ostensibly solves an equation system, a check can be programmed by having the computer plug the supposed solution back into the equation to see if the equation is satisfied. This is the way we will check our matrix inversion subroutine in the next chapter. In other cases where there is an awful lot of matrix algebra and no straightforward check, the best approach may be to seek out other programs that do the same job and see if there is agreement in results from different programs. From years of experience the authors can testify that no available version of any program should be trusted implicitly. There are too many ways for errors to creep into a program, even one that once, somewhere, was known to be free of errors. Every program in every library should be understood to bear the implicit warning, USE AT YOUR OWN RISK.

The example for this chapter is based on a one percent sample of the Project TALENT 12th grade females file. This TALENT file itself represents a 5 percent probability sample of the girls in the 12th grade in all the nation's secondary schools in 1960. Thus we are processing a random one percent draw from a major research data file (see Appendix B). Table 2.2 reports the intercorrelations among seven ability variables for the 271 girls of this sample. The table is typical of the correlation table format employed for journal articles or other research reports. A general principle to follow in arranging tables for publication is to include only those numerical results that the reader would probably need if he is to be able to test the authors' generalization against his own judgments of the numbers. Almost any quantitative research will lead to the computer generation of many numbers. It is the scientist's responsibility to boil the computer output down to the smallest assemblage of numbers to be reported that can do justice to the sciencing he is reporting, and to arrange those critical statistics in a readable format.

CORREL

```
C       CORRELATION DATA REDUCTION PROGRAM.  A COOLEY-LOHNES ROUTINE.
C
C       THIS PROGRAM COMPUTES CENTROID, STANDARD DEVIATIONS, DISPERSION,
C       AND CORRELATION MAXIMUM LIKELIHOOD ESTIMATORS.
C
C
C       INPUT
C
C       1) FIRST TEN CARDS OF DATA DECK DESCRIBE THE PROBLEM IN A TEXT THAT
C          WILL BE REPRODUCED ON THE OUTPUT. DO NOT USE COL 1 OF THESE CARDS.
C       2) CONTROL CARD (CARD 11)     COLS 1-2   M, NUMBER OF VARIABLES
C                                     COLS 3-7   N, NUMBER OF SUBJECTS.
C       3) FORMAT CARD (CARD 12),   FOR READING SCORE VECTORS.
C       4) N SETS OF SCORE CARDS.
C
C       PUNCHED OUTPUT
C
C       1) MEANS, OR CENTROID VECTOR
C       2) STANDARD DEVIATIONS VECTOR
C       3) R, THE CORRELATION MATRIX, UPPER-TRIANGULAR.
C          NOTE THAT ALL PUNCHED OUTPUT IS TO BE READ INTO OTHER PROGRAMS
C              BY FORMAT (10X, 7F10.3 / (10X, 7F10.3)).
C
C       SUBROUTINE MPRINT IS REQUIRED.
C
        DIMENSION   TIT(16),   X(100),  Y(100),  A(100,100)
C
1       WRITE(6,2)
2       FORMAT (1H1)
        DO 5   J = 1, 10
        READ(5,4)  (TIT(K),  K = 1, 16)
4       FORMAT (16A5)
5       WRITE(6,4)  (TIT(K),  K = 1,  16)
        READ(5,6)  M, N
6       FORMAT (I2, I5)
        WRITE(6,7) M, N
7       FORMAT(44H0CORRELATION REDUCTIONS VIA COOLEY-LOHNES,  I3,9H TESTS,
       C  I6,10H SUBJECTS.)
C
        READ(5,4)    (TIT(K),   K = 1, 16)
        EN = N
C
        DO 8   J = 1, M
        Y(J) = 0.0
        DO 8   K = 1, M
8       A(J,K) = 0.0
        DO 9   L = 1, N
        READ(5,TIT)    (X(J),   J = 1, M)
        DO 9   J = 1, M
        Y(J) = Y(J) + X(J)
        DO 9   K = 1, M
9       A(J,K) = A(J,K) + X(J) * X(K)
C    Y(J) NOW CONTAINS RAW SUMS. A(J,K) CONTAINS RAW S.S.C.P.
        DO 10   J = 1, M
        DO 10   K = 1, M
10      A(J,K) = (A(J,K) - Y(J) * Y(K) / EN) / EN
        DO 11   J = 1, M
        Y(J) = Y(J) / EN
11      X(J) = SQRT  (A(J,J))
        WRITE(6,12)
12      FORMAT (28H0TEST      MEAN      S. D.)
        DO 13   J = 1, M
13      WRITE(6,14)   J, Y(J),  X(J)
14      FORMAT (I4, 2F11.2)
        WRITE(6,15)
```

```
  15    FORMAT (18H0DISPERSION MATRIX)
        CALL MPRINT (A, M)
        DO 16   J = 1, M
        DO 16   K = 1, M
  16    A(J,K) = A(J,K) / (X(J) * X(K))
C
        WRITE(6,17)
  17    FORMAT(19H0CORRELATION MATRIX)
        CALL MPRINT (A, M)
C
C   FINALLY, PUNCH OUT CENTROID, S. D. VECTOR, AND R MATRIX.
        WRITE(7,19)      M,  (Y(J),   J = 1, M)
        WRITE(7,19)      M,  (X(J),   J = 1, M)
  19    FORMAT (7H VECTOR, I3, 5E14.7 / (10X, 5E14.7))
        DO 18   J = 1, M
  18    WRITE(7,3)      J,  (A(J,K),   K = J, M)
  3     FORMAT (4H ROW, I3, 3X, 7F10.7 / (10X, 7F10.7))
C
        GO TO 1
        END
>
```

Subroutines such as MPRINT must have matrices and vectors dimensioned to the same size as their counterparts in the main program. For example, the DIMENSION statement in MPRINT is correct for CORREL but must be reduced from 100 to 50 for use with the PARTL program on page 217.

MPRINT

```
        SUBROUTINE MPRINT(R, M)
C       R(I,J) = MATRIX TO BE PRINTED
C       M = ORDER
        DIMENSION  R(100,100),  J(100)
        L1 = 9
        N = 10
        J1 = 0
        J2=0
        JSEC = 0
        DO 8   I= 1,M
  8     J(I) = I
  9     J1 = J2 +1
        J2 = J1 +L1
        IF (J2 - M) 13,13,12
  12    J2=M
  13    JSEC = JSEC + 1
        IF (JSEC -1)  18, 18 , 19
  18    WRITE (6,17)   JSEC
  17    FORMAT (10H0 SECTION I3/)
        GO TO 21
  19    WRITE (6,20) JSEC
  20    FORMAT (10H1 SECTION I3/)
  21    WRITE (6,27) (J(I),I=J1,J2)
  27    FORMAT(6H0  ROW 3X10I12)
        DO 29 I=1,M
  29    WRITE (6,30) I,  (R(I,K),K=J1,J2)
  30    FORMAT(1X,I5,4X,10(F11.3,1X))
        IF (J2-M) 9,32,32
  32    RETURN
        END
>
```

TABLE 2.2 Intercorrelations Among Seven Ability Measures for 271 Project TALENT 12th Grade Female File

Test	1	2	3	4	5	6	7
1 Information, Part I	1.00	.84	.57	.71	.60	.44	.72
2 Information, Part II		1.00	.55	.71	.53	.43	.58
3 English Mechanics			1.00	.53	.34	.38	.60
4 Reading Comprehension				1.00	.58	.55	.58
5 Mechanical Reasoning					1.00	.51	.54
6 Abstract Reasoning						1.00	.44
7 Mathematics							1.00

Of course, he has the additional responsibility of maintaining archives from which all the numerical detail can be recovered upon demand. For example, Project TALENT already has a room full of bound volumes of computer outputs.

The interesting thing about the correlations in Table 2.2 is that they are all positive and all fairly large. The lowest correlation is .34 between English and Mechanical Reasoning, suggesting that these are the most disparate abilities in the battery for grade 12 girls. The highest correlation, that of .84 between the two parts of the Information test suggests a high degree of parallelism between the parts. That Reading Comprehension correlates the same with both parts of the Information test, whereas Mathematics correlates substantially higher with Part I than with Part II, is understandable when one is told that Part I contains a Mathematics Information subscale and Part II does not. These are some of the features of the correlation matrix, but we hope that as you examine the 21 different correlations it reports, you feel the need for further analysis of this system of correlations. Actually, these seven variables are a subset of a system of over 100 tests for which Project TALENT has scores on about 400,000 adolescents. A matrix of intercorrelations for a battery of 60 tests contains 1770 different correlations. Imagine staring at that matrix and trying to decide what to say about it.

2.6 EXERCISES

1. Modify CORREL so that it will report:
    (1) the deviation sums of squares and crossproducts matrix,
    (2) the unbiased estimates of the centroid, the dispersion, and the correlation matrix.

2.   Run both our CORREL and your modified version of it on the sample of
     TALENT 12th grade males given in Appendix B, processing all the measure-
     ment variables (variables 8 through 19). Compare the biased and unbiased
     estimates. Interpret the most interesting correlations. Save these printouts
     for future reference.

3.   If $\mathbf{D}$ is a variance-covariance matrix based upon a sample of the vector
     variable $\mathbf{x}$, and $\mathbf{c}$ is a vector of coefficients for $\mathbf{x}$ such that $\mathbf{c}'\mathbf{x} = y$, what can
     be said about $\mathbf{c}'\mathbf{D}\mathbf{c}$ in relation to the derived variable y ?

PART II

Studies of a Single Population

||||||||||||||||||||||||||||||||||||||||||||||||||||||||||||||||||||||

CHAPTER THREE

Multiple Correlation and the Method of Least Squares

3.1 MATHEMATICS OF MULTIPLE CORRELATION

Multiple correlation provides an analysis of the relations among two or more predictor measures and a single criterion measure. One result of the analysis is an equation for predicting the criterion score of a subject from his known set of predictor scores. The complete test space in the analysis is p-dimensional, with $p - 1$ predictors plus the criterion variable. In our treatment we will partition a p-element vector variable so that the first $p - 1$ elements are the predictor elements and the pth element is the criterion element. As with all the prediction schemes developed in this text, the researcher requires a complete set of p scores for N subjects in what we may call the norming sample or estimating group. On the basis of what is learned about the interrelationships among the p elements of the vector variable as they are estimated from this norming sample it becomes possible to take the $p - 1$ predictor scores for new subjects, in a replication sample, and compute for each new subject a predicted score on the criterion element. This predicted score is also called a regressed score, and the formula that yields it is called a regression formula. The regressed score is a linear component of the predictor scores.

As we have previously observed, most of the prediction schemes in multivariate statistics involve linear combinations, components, or functions of the predictor variables. The simple slope–intercept equation for a straight line

$$Y = bX + a$$

is the simplest linear function. It may be diagrammed as in Figure 3.1. Notice that the equation of a line involves coefficients specifying the location (i.e., intercept, a) and direction (i.e., slope, b) of the line, but no powers of X other than the first power. Of the infinite number of functional relations that might be established between X and Y, the subset of possible linear relations represents a very simple and tractable collection, and statisticians usually elect to restrict their explorations to it. The simplicity resulting from the imposition of the linear restraint on statistical models can be enhanced by translating the raw scores into deviation scores or standard scores, so that the fitted line always passes through the origin of the axis system. Then the equation has only to specify the direction of the line (its slope), since the location is fixed (the intercept is constantly zero). The regression equation for predicting z_{2i} from z_{1i}, where these are standard scores of subject i on tests 1 and 2, may be written

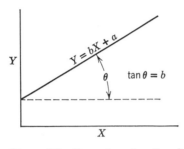

Figure 3.1. Y as a linear function of X.

$$\hat{z}_{2i} = r_{12} z_{1i}$$

The correlation between the tests is the slope coefficient; no intercept coefficient is required, and $\hat{z}_{2i}$ is the prediction for individual i on z_2.

When there are two or more predictors, additional terms are required for the regression equation to describe the orientation of a plane in the test space, which is now of higher dimensionality. In the case of a p-dimensional space, the standard score regression equation has the form

$$\hat{z}_{pi} = b_1 z_{1i} + b_2 z_{2i} + \cdots + b_{p-1} z_{p-1, i}$$

Geometrically, the general equation locates a hyperplane of $p - 1$ dimensions in the p-dimensional space from which projections are made to the criterion axis. The regression line in this case of multiple correlation becomes a line normal to, or perpendicular to, the regression hyperplane, which passes through the origin. The angle of separation of this regression line and the pth or criterion axis has as its cosine the multiple correlation coefficient. The regression weights, b_j, of the equation are a function of the correlations of the predictor elements with the criterion element and the intercorrelations among the predictor elements. These coefficients are called standard partial regression coefficients. For a given set of predictors, comparison of the absolute values of the regression weights is an aid in assessing the relative contributions of the corresponding elements to the prediction of the criterion element, given that particular set of coefficients.

It is possible to compute a coefficient of multiple correlation, $R_{p \cdot 1, 2, \ldots, p-1}$, or simply R, that is similar to a simple product moment correlation coefficient except that it is always a positive number with the range $0 \le R \le 1$. This coefficient is the actual product moment correlation between the criterion z_p and the linear regression component $\hat{z}_p$.

The coefficient R^2 provides an estimate of the proportion of the total variance in the criterion that can be predicted from the known variances of the predictors, and is a measure of the overall effectiveness of the multiple regression. The significance of R may be tested by an analysis of variance test of the null hypothesis that the population multiple correlation equals zero, as follows:

$$F_{N-p}^{p-1} = \frac{R^2(N - p)}{(1 - R^2)(p - 1)}$$

The standard error of estimate for $\hat{z}_{pi}$ is given by

$$S_{e\hat{z}} = \sqrt{1 - R^2}$$

It is often desirable to compute the regressed deviation score $\hat{x}_{pi}$ for each subject rather than the regressed standard score. The formula for regressed deviation scores is

$$\hat{x}_{pi} = \left(\frac{s_p}{s_1}\right) b_1 x_{1i} + \left(\frac{s_p}{s_2}\right) b_2 x_{2i} + \cdots + \left(\frac{s_p}{s_{p-1}}\right) b_{p-1} x_{p-1, i}$$

The formula for regressed raw scores is

$$\hat{X}_{pi} = \left(\frac{s_p}{s_1}\right) b_1 X_{1i} + \cdots + \left(\frac{s_p}{s_{p-1}}\right) b_{p-1} X_{p-1, i} + a$$

where a is an intercept constant computed from the means as follows:

$$a = m_p - \left[\left(\frac{s_p}{s_1}\right) b_1 m_1 + \left(\frac{s_p}{s_2}\right) b_2 m_2 + \cdots + \left(\frac{s_p}{s_{p-1}}\right) b_{p-1} m_{p-1}\right]$$

The deviation score and raw score regressions have the same standard error of estimate:

$$S_{e\hat{x}} = S_{x_p}\sqrt{1 - R^2}$$

Differential calculus is used in the derivation of the multiple regression model to obtain a solution for the weights in the linear function that minimizes the average squared error of prediction. That is, if an error e_i is defined as

$$e_i = z_{pi} - \hat{z}_{pi}$$

which is the discrepancy between the actual and predicted score for the ith individual, the purpose of multiple regression is to minimize the function of e

$$f(e) = \frac{\sum_{i=1}^{N} e_i^2}{N} = \frac{\sum_{i=1}^{N} (z_{pi} - \hat{z}_{pi})^2}{N}$$

If the linear combination of the predictors that defines $\hat{z}_{pi}$ is substituted in the error function, the function to be minimized becomes

$$f(e) = \left(\frac{1}{N}\right) \sum_{i=1}^{N} [z_{pi} - (b_1 z_{1i} + b_2 z_{2i} + \cdots + b_{p-1} z_{p-1,i})]^2$$

When the partial derivative of the function with respect to each unknown b_j is set to zero a system of $p - 1$ normal equations in $p - 1$ unknowns is reached. These normal equations have the form

$$
\begin{aligned}
b_1 \quad &+ \quad r_{12} b_2 \quad + \quad r_{13} b_3 \quad + \cdots + r_{1,p-1} b_{p-1} = \quad r_{1p} \\
r_{21} b_1 \quad &+ \quad 1.0 b_2 \quad + \quad r_{23} b_3 \quad + \cdots + r_{2,p-1} b_{p-1} = \quad r_{2p} \\
r_{31} b_1 \quad &+ \quad r_{32} b_2 \quad + \quad 1.0 b_3 \quad + \cdots + r_{3,p-1} b_{p-1} = \quad r_{3p} \\
&\quad\vdots \qquad\qquad \vdots \qquad\qquad \vdots \qquad\qquad\qquad \vdots \qquad\qquad \vdots \\
r_{p-1,1} b_1 &+ r_{p-1,2} b_2 + r_{p-1,3} b_3 + \cdots + \quad 1.0 b_{p-1} \quad = r_{p-1,p}
\end{aligned}
$$

Notice that besides the unknown b weights the normal equations involve all the intercorrelations among the p variables, which are known for the norming sample. On the left we see the intercorrelations among the $p - 1$ predictors and on the right the correlations of the predictors with the criterion.

There are several useful computer methods for solving such a system of simultaneous equations, but a very helpful first step is to translate the problem into matrix algebra. The collection of all intercorrelations among the p variables forms a p-square symmetric matrix as it is reported by CORREL, and may be called **R**. Matrix **R** is partitioned into the required segments as follows:

$$
\mathbf{R} =
\left[
\begin{array}{ccccc:c}
1.0 & r_{12} & r_{13} & \cdots & r_{1,p-1} & r_{1p} \\
r_{21} & 1.0 & r_{23} & \cdots & r_{2,p-1} & r_{2p} \\
r_{31} & r_{32} & 1.0 & \cdots & r_{3,p-1} & r_{3p} \\
\vdots & \vdots & \vdots & & \vdots & \vdots \\
r_{p-1,1} & r_{p-1,2} & r_{p-1,3} & \cdots & 1.0 & r_{p-1,p} \\
\hdashline
r_{p1} & r_{p2} & r_{p3} & \cdots & r_{p,p-1} & 1.0
\end{array}
\right]
=
\left[
\begin{array}{c:c}
\mathbf{R}_{11} & \mathbf{R}_{12} \\
\hdashline
\mathbf{R}_{21} & 1.0
\end{array}
\right]
$$

$\mathbf{R}_{11}$ is the matrix of intercorrelations among the predictors, and $\mathbf{R}_{12} = \mathbf{R}_{21}'$ is the column vector of the correlations of the predictors with the criterion variable. The order of $\mathbf{R}_{11}$ is $p - 1$. The required vector of $\mathbf{b}$ weights is computed from the relationship:

$$\mathbf{R}_{11}\mathbf{b} = \mathbf{R}_{12}$$

which expresses the system of normal equations in matrix form, and makes it clear that the operation required is to invert $\mathbf{R}_{11}$ and multiply through by this inverse:

$$\mathbf{R}_{11}{}^{-1}\mathbf{R}_{11}\mathbf{b} = \mathbf{R}_{11}{}^{-1}\mathbf{R}_{12}$$

Since

$$\mathbf{R}_{11}{}^{-1}\mathbf{R}_{11} = \mathbf{I}$$

we have

$$\mathbf{b} = \mathbf{R}_{11}{}^{-1}\mathbf{R}_{12}$$

The squared multiple correlation coefficient is computed as the vector product of $\mathbf{b}$ and the predictor-criterion correlations:

$$R^2 = \mathbf{b}'\mathbf{R}_{12} = \sum_{j=1}^{p-1} b_j r_{jp}$$

The problem of interpreting the regression function is a thorny one. It is interesting to speculate about the relative contributions of the predictor elements to the prediction of the criterion element, but we have to temper our interpretations with the realization that the obtained prediction results form a *system of predictors* in which the elements interact in complex fashion. Consider the variance of the regressed standard scores, which is the "explained variance" for the criterion, in the two-predictor system:

$$R^2 = s_{\hat{z}_3}{}^2 = \left(\frac{1}{N}\right) \sum_{i=1}^{N} \hat{z}_{3i}{}^2$$

$$= \left(\frac{1}{N}\right)\left[\sum_{i=1}^{N} (b_1 z_{1i} + b_2 z_{2i})^2 \right]$$

$$= \left(\frac{1}{N}\right)\left(b_1{}^2 \sum_{i=1}^{N} z_{1i}{}^2 + b_2{}^2 \sum_{i=1}^{N} z_{2i}{}^2 + 2b_1 b_2 \sum_{i=1}^{N} z_{1i} z_{2i} \right)$$

$$= b_1{}^2 s_{z_1}{}^2 + b_2{}^2 s_{z_2}{}^2 + 2b_1 b_2 r_{12}$$

$$= b_1{}^2 + b_2{}^2 + 2b_1 b_2 r_{12}$$

Discussing this equation in the course of his outstanding presentation on correlation, McNemar says:

> We thus see that the relative importance of the variables X_1 and X_2 in "explaining" or "causing" variation in X_3 can be judged by the magnitude of the (b) coefficients. The third term in the formula represents a joint contribution which, it will be seen, is a function of the amount of correlation between the two predicting variables. (1962, p. 176)

Many workers rely on the vector of squared regression weights to indicate the relative importance of the predictors in the prediction system, which is sensible as long as one bears in mind that there is a series of terms of the form $2b_j b_k r_{jk}$ which are not being considered but which are important in determining the magnitude of R^2. In the case that all the regression weights agree in sign with their corresponding predictor-criterion correlations, the vector of crossproducts of the form $b_j r_{jp}$ may provide assistance in interpreting the importances of the predictors. This vector is the basis for the computed magnitude of R^2, since

$$R^2 = \sum_{j=1}^{p-1} b_j r_{jp}$$

This vector is also interesting because a negligible element in it for which the regression weight is sizeable leads to the identification of a suppressor variable in the system. This is a predictor which has almost no correlation with the criterion itself, but which gets a sizeable b weight because it is correlated with the predictors in a useful fashion. McNemar has a good discussion of this strange phenomenon (1962, p. 185–187).

One of the most useful ways to look at the regression function is in terms of its correlations with the predictor elements on which it is defined. Since the variance of the regression function is R^2, we can make a standardized variable, or factor, out of $\hat{z}_p$ by dividing it by its standard deviation, R. That is, since $\hat{z}_p$ is $N(0, R^2)$ distributed, it follows that $(1/R)\hat{z}_p$ is $N(0, 1)$ distributed. The desired correlations with the $p - 1$ predictor elements in the vector $\mathbf{z}$ are

$$\mathbf{r}_{z\hat{z}_p} = \frac{1}{N} \sum_{i=1}^{N} \mathbf{z}\left(\frac{1}{R}\hat{z}_p\right)$$

where

$$\frac{1}{R}\hat{z}_p = \frac{1}{R}\mathbf{b'z}$$

Substituting, we have

$$r_{z\hat{z}_p} = \frac{1}{N} \sum_{i=1}^{N} z \left(\frac{1}{R} b'z \right)$$

$$= \frac{1}{N} \sum_{i=1}^{N} zz'b \frac{1}{R}$$

$$= R_{zz} b \frac{1}{R}$$

where R_{zz} is the square matrix of order $p - 1$ containing the intercorrelations among the predictors, denoted R_{11} in the partition of the p order correlation matrix. So we may write

$$r_{z\hat{z}_p} = R_{11} b \frac{1}{R}$$

But we note that

$$b = R_{11}^{-1} R_{12}$$

Substituting yields

$$r_{z\hat{z}_p} = R_{11} R_{11}^{-1} R_{12} \frac{1}{R}$$

$$= I R_{12} \frac{1}{R}$$

$$= R_{12} \frac{1}{R}$$

Thus, the vector of correlations of the predictors with the regression function, which may be called *regression factor structure coefficients*, is given by dividing the vector of predictor-criterion correlations by the multiple correlation coefficient. Note that this interpretive device places very little emphasis on the magnitudes of the b weights, since they only influence the scaling value, $(1/R)$, which is used to modify the predictor-criterion correlations upward. Does it make sense to you that the predictors should be more strongly correlated with the regression factor defined on them than they are with the criterion variable?

Our tendency to deemphasize the b weights stems from experience with the phenomenon of extreme fluctuation of regression weights from sample to sample when the sample size is small. Even when the sample size is moderate there is substantial fluctuation. One of the computing assignments for this chapter gives you an opportunity to experience this yourself. One prominent statistician has warned that "when there are collinearities in the independent variables . . . no reliance whatever can be put on individual coefficients in regression equations embodying all the variables" (Kendall, 1957, p. 74). Kendall speaks of the case where some of the correlations among the predictors approach unity, but the more general problem is that whatever the degrees of correlation among the predictors the regression weights are disturbed by all the errors in estimating these $(p - 1)(p - 2)/2$ correlation parameters. Another statistics text summarizes the problem by saying that in practice we may have to generalize our estimates of regression coefficients to the population, "but the researcher must be aware of the possibility of securing results that are altogether untrustworthy and invalid" (Johnson and Jackson, 1959, p. 384).

Perhaps the most dramatic demonstration of the fluctuation phenomenon occurs in a paper read at the American Psychological Association convention, titled "Two kinds of regression weights that are better than betas in crossed samples" (Marks, 1966). (What we have termed b weights are often called beta weights in the literature.) Marks reported on 1656 cross-validation studies involving three to ten predictors and sample sizes of 20, 80, and 200 subjects. In each study the correlation coefficient between the predictor function and the actual criterion variable for a replication sample was computed for three prediction functions: (1) the beta (or b) weights fitted to the norming sample, (2) unit weights, and (3) the norming sample predictor-criterion correlations (our $\mathbf{R}_{12}$ vector) employed as function weights. His finding was that the prediction function based on norming sample predictor-criterion correlations yielded a higher correlation with the criterion for the replication sample than did the b weights 75 percent of the time. Even the unit weights outperformed the b weights 65 percent of the time. Marks' conclusion is that "conventional least-squares regression should be dropped from the applied statistician's repertoire in favor of RV weights for $N < 200$." (RV refers to predictor-criterion correlations.) His rule of thumb is a little too pat, especially for people who are working with more than ten predictors, but the exposé of our best-known and most used multivariate procedure is provocative. We simply must plan for large samples (and good ones) if we are going to use any of the function-fitting methods and expect the functions to be taken seriously as generalizations to populations. We return to this problem of fluctuation in Chapter Four with a more positive suggestion. Meanwhile, the student may want to read the recent monographs by Burket (1964) and Herzberg (1967).

Incidentally, the reader will have heard about stepwise regression methods. These are methods which add or subtract one predictor at a time to the regression equation, seeking the "best" set of predictors. Variables are added or dropped according to the statistical significance of their contributions to the reduction of uncertainty about the criterion. All this fitting is done on one sample. An excellent presentation of a stepwise procedure is given by Efroymson (1960). We believe that stepwise regression is seldom appropriate in behavioral research because of the enormous hazards of capitalization on chance. At least the user of a stepwise procedure should demonstrate on a replication sample what the actual shrinkage of his multiple correlation is. McNemar gives a compelling example of capitalization on chance in the selection of "best" predictors (1962, p. 185), and also gives a formula for estimating the shrinkage of multiple R (p. 184). Since the formula predicts shrinkage only for ideal sampling conditions, which almost never prevail in behavioral research, its use is no substitute for a replication sample demonstration in small sample studies.

We will not repeat this dictum *ad nauseam*, but we believe that the researcher who wants his linear components taken seriously must either (1) base them on very large and very representative samples, or (2) demonstrate their validity on replication samples.

3.2 RANK, INVERSE, AND DETERMINANT

Every square matrix $\mathbf{A}$ can be characterized by a unique scalar number which is its determinant, $|\mathbf{A}|$. If $\mathbf{A}$ is of order p, $|\mathbf{A}|$ is the sum of alternately-signed products of elements a_{jk} from permutations of the integers $j = 1, 2, \ldots, p$ and $k = 1, 2, \ldots, p$. There are $p!$ (p-factorial) such products entering the summation, with p factors in each term. For $p = 2$

$$|\mathbf{A}| = \begin{vmatrix} a_{11} & a_{12} \\ a_{21} & a_{22} \end{vmatrix} = a_{11}a_{22} - a_{12}a_{21}$$

For $p = 3$

$$|\mathbf{A}| = a_{11}a_{22}a_{33} - a_{11}a_{23}a_{32} + a_{12}a_{23}a_{31} - a_{12}a_{21}a_{33}$$
$$+ a_{13}a_{21}a_{32} - a_{13}a_{22}a_{31}$$

To write out the determinant in this way, the first subscript (j) is put in numerical order in each term, with subscript k then assuming all possible sequences from term to term. The sign for each term is based upon the number of inversions of subscript k, $(+)$ if even and $(-)$ if an odd number of inversions are involved. An inversion is when a larger subscript precedes a smaller one.

For large p it is very tedious to write out these $p!$ products but the computer can always evaluate the determinant for us. We are very interested in determinants because the determinant of a dispersion matrix is a multivariate generalization of the concept of variance that figures in all multivariate normal distribution theory. We are also interested in determinants because the determinant is intimately related to both the rank and the inverse of the square matrix. Most of our theory of multivariate data involves assumptions about rank, and most of our procedures require computation of inverses. Since all our applications will be to square symmetric matrices of the dispersion and correlations types, we will specialize our discussion somewhat.

If the determinant of a matrix is zero it is said to be *singular*. The rank of a matrix is the order of the largest non-singular square matrix that can be obtained by deleting rows and columns. If a square matrix A of order p has a nonzero determinant, it is non-singular and is of full rank. If $|A| = 0.0$, it is singular and the rank r of A is less than p. If an r by r nonsingular matrix can be constructed from A by deleting $(p - r)$ rows and columns, and if all such matrices from A of order higher than r are singular, then the rank of A is r.

The rank of a matrix is dependent upon the number of independent rows (or columns) it contains. A row is independent if its elements cannot be computed as a linear combination of the elements of other rows. The matrix

$$A = \begin{bmatrix} 2 & 3 & 2 \\ 3 & 4 & 3 \\ 2 & 3 & 2 \end{bmatrix}$$

is singular because row three's elements are exactly 1 times the corresponding elements of row one. Note that

$$|A| = 16 - 18 + 18 - 18 + 18 - 16 = 0$$

The matrix

$$A = \begin{bmatrix} 1 & 1 & 1 \\ 1 & 0 & 2 \\ 1 & 2 & 0 \end{bmatrix}$$

is rank 2 because row three's elements are exactly 2 times the corresponding elements of row one minus 1 times the corresponding elements of row two, that is,

$$a_{3j} = 2a_{1j} - 1a_{2j}$$

Note that

$$|A| = 0 - 4 + 2 - 0 + 2 - 0 = 0$$

A zero determinant always indicates a singular matrix with at least one linear dependency among the rows. A singular matrix is said to be of *reduced rank*.

Chapter Four considers how to evaluate the exact rank of a reduced rank matrix. We will not stop to demonstrate the fact, but an important fact for multivariate researchers is that a linear dependency among the rows of a $p \times N$ score matrix (observation matrix or data matrix) always leads to reduced rank dispersion and correlation matrices for the data. They will also be of reduced rank if $p > N$.

Useful results are:

(a) $|\mathbf{AB}| = |\mathbf{A}| \cdot |\mathbf{B}|$
(b) $|\mathbf{A}^k| = |\mathbf{A}|^k$
(c) $|k A| = k^p |A|$, where $\mathbf{A}$ is order p.

Singular matrices do not have inverses. Thus one cannot use a set of variables that contains a linear dependency as a predictor set in multiple correlation analysis. Quite a few psychological inventories (e.g., the Allport-Vernon-Lindzey *Study of Values*) are scored in such a way that the sum of all the trait or scale scores assigned to a subject must be equal to a given constant. Such trait profiles contain a built-in dependency, and one of the scales should be dropped if the profile is to be used as a predictor set in multiple or canonical regression.

The inverse is a complicated function of the matrix elements, and we will settle for knowledge that the defining equation for $\mathbf{A}^{-1}$ is

$$\mathbf{A}^{-1}\mathbf{A} = \mathbf{A}\mathbf{A}^{-1} = \mathbf{I}$$

Useful results are:

(a) $|\mathbf{A}^{-1}| = 1/|\mathbf{A}|$
(b) $(\mathbf{A}^{-1})^{-1} = \mathbf{A}$
(c) $(\mathbf{A}')^{-1} = (\mathbf{A}^{-1})'$
(d) If $\mathbf{A}' = \mathbf{A}$, then $(\mathbf{A}^{-1})' = \mathbf{A}^{-1}$
(e) If $\mathbf{A}' = \mathbf{A}^{-1}$, so that $\mathbf{A}'\mathbf{A} = \mathbf{A}\mathbf{A}' = \mathbf{I}$, $\mathbf{A}$ is said to be *orthonormal*
(f) $(\mathbf{AB})^{-1} = \mathbf{B}^{-1}\mathbf{A}^{-1}$

Can you verify that if

$$\mathbf{A} = \begin{bmatrix} 1 & 3 & 3 \\ 1 & 3 & 4 \\ 1 & 4 & 3 \end{bmatrix}$$

then

$$\mathbf{A}^{-1} = \begin{bmatrix} 7 & -3 & -3 \\ -1 & 0 & 1 \\ -1 & 1 & 0 \end{bmatrix}?$$

3.3 THE MATINV SUBROUTINE

The matrix inversion method we employ is essentially an invention of Karl F. Gauss (1777–1855), as modified by Jordan. This Gauss–Jordan method has been described by Burton S. Garbow of the Applied Mathematics Division, Argonne National Laboratory, who originally coded it in FORTRAN, as follows:

> Jordan's method is used to reduce a matrix A to the identity matrix I through a succession of elementary transformations; $l_m \, l_{m-1} \ldots l_1 \, A = I$. If these transformations are simultaneously applied to I ... the result is A^{-1}.

An excellent discussion of this method may be found in Alex Orden (1960). For a discussion of elementary operators see S. R. Searle (1966, p. 121) or F. Ayres, Jr. (1962, p. 39). Figure 3.2 reveals that the two algebraic operations

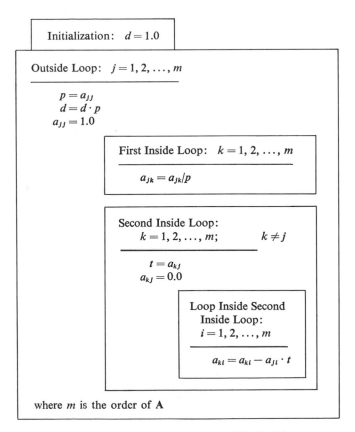

Figure 3.2. Diagram of simplified MATINV algorithm.

performed in the Gauss–Jordan method are dividing by a pivot (diagonal) element the elements of a pivotal row, and reducing every other row by products of its elements times the modified elements of the pivot row. The determinant is produced as the continuous product of the pivot elements. The whole job is done in one quick pass through the matrix. We give a teaching version of the algorithm to show how simple it is, and then a production version that is longer because it adds the sophisticated device of row and column interchanges to search for the largest elements of the matrix and bring them to the diagonal to serve as pivot divisors. These interchanges serve to guarantee that you will not lose precision by dividing very large numbers by very small numbers. Actually, most of our uses of matrix inversion involve the correlation matrix where the largest elements are already on the diagonal, so the teaching version is usually entirely practicable. In statistics, the interchange feature is only needed when deviation sums of squares and crossproducts or dispersion matrices for vector variables where there are large discrepancies in scale factors among the elements of the vector are being inverted. Even these cases can be avoided by preadjusting the scale factors. The researcher who works with a slow computer of limited memory capacity will prefer the teaching version, even if he occasionally has to rescale some of his variables.

To see how the simplified routine works, let's simulate the computer in tackling the following matrix:

$$A = \begin{bmatrix} 1.00 & .50 \\ .50 & 1.00 \end{bmatrix}$$

1. $d = 1.00$
2. $j = 1$ (Outside Loop)
    $p = a_{11} = 1.00$
    $a_{11} = 1.00$
    $d = d \cdot p = 1.00$

    2a. $k = 1$ (First Inside Loop)
    $a_{11} = a_{11}/p = 1.00/1.00 = 1.00$
    $k = 2$
    $a_{12} = a_{12}/p = .50/1.00 = .50$

    2b. $k = 2$ (Second Inside Loop)
    $t = a_{21} = .50$
    $a_{21} = .00$
    $i = 1$ (Loop Inside 2b)
    $a_{21} = a_{21} - a_{11} \cdot t = .00 - 1.00(.50) = -.50$
    $i = 2$
    $a_{22} = a_{22} - a_{12} \cdot t = 1.00 - .50(.50) = .75$

3. $j = 2$ (Outside Loop)

$p = a_{22} = .75$

$a_{22} = 1.00$

$d = d \cdot p = .75$

    3a. $k = 1$ (First Inside Loop)

        $a_{21} = a_{21}/p = -.50/.75 = -.667$

      $k = 2$

        $a_{22} = a_{22}/p = 1.00/.75 = 1.333$

    3b. $k = 1$ (Second Inside Loop)

        $t = a_{12} = .50$

        $a_{12} = .00$

        $i = 1$ (Loop Inside 3b)

           $a_{11} = a_{11} - a_{21} \cdot t = 1.00 - (-.667).50 = 1.333$

        $i = 2$

           $a_{12} = a_{12} - a_{22} \cdot t = .00 - 1.333(.50) = -.677$

Thus the subroutine returns the following values to the main program:

$$\mathbf{A}^{-1} = \begin{bmatrix} 1.333 & -.667 \\ -.667 & 1.333 \end{bmatrix}, \qquad |\mathbf{A}| = .75$$

The determinant obviously checks. Does the inverse? Let's try it.

$$\begin{bmatrix} 1.333 & -.667 \\ -.667 & 1.333 \end{bmatrix} \begin{bmatrix} 1.00 & .50 \\ .50 & 1.00 \end{bmatrix} = \begin{bmatrix} 1.000 & .000 \\ .000 & 1.000 \end{bmatrix}$$

and

$$\begin{bmatrix} 1.00 & .50 \\ .50 & 1.00 \end{bmatrix} \begin{bmatrix} 1.333 & -.667 \\ -.667 & 1.333 \end{bmatrix} = \begin{bmatrix} 1.000 & .000 \\ .000 & 1.000 \end{bmatrix}$$

Checking the subroutines on larger problems could be quite tedious, so we have prepared a program to test MATINV. Here are the listings of the two versions of MATINV, the TEST program, the data for our check problem, and the computer output for tests of both versions of the subroutine. (as Tables 3.1 and 3.2). This checking procedure is an example of plugging the supposed solution to a matrix equation into the equation to see if the solution satisfies the equation.

```
CMATINV
       SUBROUTINE MATINV (A, M, DET)
C
C      INVERSE AND DETERMINANT OF A BY THE GAUSS-JORDAN METHOD.
C      M IS THE ORDER OF THE SQUARE MATRIX, A.
C      A-INVERSE REPLACES A.
C      DETERMINANT OF A IS PLACED IN DET.
C      P. R. LOHNES, PROJECT TALENT, 1966.
C
       DIMENSION  A(100,100),  IPVT(100),   PVT(100),    IND(100,2)
C
       DET = 1.0
       DO 1   J = 1, M
  1    IPVT(J) = 0
       DO 10   I = 1, M
C      SEARCH FOR THE PIVOT ELEMENT.
       AMAX = 0.0
       DO 5   J = 1, M
       IF(IPVT(J) - 1)  2, 5, 2
  2    DO 5   K = 1, M
       IF (IPVT(K) - 1)  3, 5, 20
  3    IF (ABSF(AMAX) - ABSF(A(J,K)))   4,  5,  5
  4    IROW = J
       ICOL = K
       AMAX = A(J,K)
  5    CONTINUE
       IPVT(ICOL) = IPVT(ICOL) + 1
C      INTERCHANGE THE ROWS TO PUT THE PIVOT ELEMENT ON THE DIAGONAL.
       IF(IROW - ICOL)  6, 8, 6
  6    DET = -DET
       DO 7   L = 1, M
       SWAP = A(IROW,L)
       A(IROW,L) = A(ICOL,L)
  7    A(ICOL,L) = SWAP
  8    IND(I,1) = IROW
       IND(I,2) = ICOL
       PVT(I) = A(ICOL,ICOL)
       DET = DET * PVT(I)
C      DIVIDE THE PIVOT ROW BY THE PIVOT ELEMENT.
       A(ICOL,ICOL) = 1.0
       DO 9   L = 1, M
  9    A(ICOL,L) = A(ICOL,L) / PVT(I)
C      REDUCE NON-PIVOT ROWS.
       DO 10   L1 = 1, M
       IF (L1 - ICOL)  11, 10, 11
 11    SWAP = A(L1,ICOL)
       A(L1,ICOL) = 0.0
       DO 12   L = 1, M
 12    A(L1,L) = A(L1,L) - A(ICOL,L) * SWAP
 10    CONTINUE
C      INTERCHANGE THE COLUMNS.
       DO 20   I = 1, M
       L = M + 1 - I
       IF (IND(L,1) - IND(L,2))  13, 20, 13
 13    IROW = IND(L,1)
       ICOL = IND(L,2)
       DO 20   K = 1, M
       SWAP = A(K,IROW)
       A(K,IROW) = A(K,ICOL)
       A(K,ICOL) = SWAP
 20    CONTINUE
       RETURN
       END
 >
```

MATINV(S)

```
      SUBROUTINE MATINV (A,  M,  DET)
C
C  GAUSS REDUCTION INVERSION WITHOUT ROW AND COLUMN INTERCHANGES.
C  A TEACHING DEVICE.     P. R. LOHNES,  1966.
C     M IS THE ORDER OF THE SQUARE MATRIX, A.
C     A-INVERSE IS RETURNED IN A.
C     DETERMINANT IS RETURNED IN DET.
C
      DIMENSION     A(100,100)
      DET = 1.0
      DO 1   J = 1, M
      PVT = A(J,J)
      DET = DET * PVT
      A(J,J) = 1.0
      DO 2   K = 1, M
C  DIVIDE THE PIVOT ROW BY THE PIVOT ELEMENT.
   2  A(J,K) = A(J,K) / PVT
      DO 1   K = 1, M
C  REDUCE THE NON-PIVOT ROWS.
      IF (K - J)    3,  1,  3
   3  T = A(K,J)
      A(K,J) = 0.0
      DO 4   L = 1, M
   4  A(K,L) = A(K,L) - A(J,L) * T
   1  CONTINUE
      RETURN
      END
>
```

TABLE 3.1   Printout from Test of Simplified MATINV

INPUT MATRIX, R					
ROW	1	1.000	.753	.686	.778
ROW	2	.753	1.000	.736	.734
ROW	3	.686	.736	1.000	.633
ROW	4	.778	.734	.633	1.000
DETERMINANT=.620033E-01					
INVERSE OF R					
ROW	1	3.268	-.912	-.645	-1.464
ROW	2	-.912	3.211	-1.160	-.914
ROW	3	-.645	-1.160	2.403	-.169
ROW	4	-1.464	-.914	-.169	2.916
R-INVERSE * R					
ROW	1	1.000	-.000	-.000	-.000
ROW	2	-.000	1.000	.000	-.000
ROW	3	.000	-.000	1.000	-.000
ROW	4	.000	.000	.000	1.000

TEST MATINV

```
C TEST
C      TEST MATINV.     P.R. LOHNES
C
       DIMENSION R(50,50), B(50), S(50,50), T(50,50), FMT(16)
C
 1     READ(5,2)M
 2     FORMAT (I2)
       READ(5,3) (FMT(J), J=1,16)
 3     FORMAT (16A5)
       DO 4 J=1,M
 4     READ(7,FMT) (R(J,K), K=J,M)
       DO 5 J=1,M
       DO 5 K=J,M
       S(J,K)=R(J,K)
       S(K,J)=R(J,K)
 5     R(K,J)=R(J,K)
       WRITE(6,6)
 6     FORMAT (16H1INPUT MATRIX, R)
       DO 7 J=1,M
 7     WRITE (6,10) (J,(R(J,K), K=1,M))
 10    FORMAT (7H0 ROW I3, 7F10.3/(10X, 7F10.3))
C
       CALL MATINV (R,M, DET)
       WRITE(6,11) DET
 11    FORMAT (15H0DETERMINANT = E15.6)
       WRITE(6,12)
 12    FORMAT (13H0INVERSE OF R)
       DO 13 J=1,M
 13    WRITE(6,10) (J,(R(J,K), K=1,M))
       DO 14 J=1,M
       DO 14 K=1,M
       T(J,K)=0.0
       DO 14 L=1,M
 14    T(J,K) = T(J,K) +R(J,L) * S(L,K)
       WRITE(6,15)
 15    FORMAT (14H0R-INVERSE * R)
       DO 16 J=1,M
 16    WRITE(6,10) (J,(T(J,K),K=1,M))
       GO TO 1
       END
TEST DATA
(14X, 4F14.12)
ROW   1     1.0000000       .7531377       .6863721      .7777215
ROW   2     1.0000000       .7361974       .7340708
ROW   3     1.0000000       .6331719
ROW   4     1.0000001
>
```

TABLE 3.2   Printout from Test of Full MATINV

```
INPUT MATRIX, R
   ROW    1       1.000      .753      .686      .778
   ROW    2        .753     1.000      .736      .734
   ROW    3        .686      .736     1.000      .633
   ROW    4        .778      .734      .633     1.000
DETERMINANT=.620033E-01
INVERSE OF R
   ROW    1       3.268     -.912     -.645    -1.464
   ROW    2       -.912     3.211    -1.160     -.914
   ROW    3       -.645    -1.160     2.403     -.169
   ROW    4      -1.464     -.914     -.169     2.916
R-INVERSE * R
   R.W    1       1.000     -.000      .000     -.000
   ROW    2       -.000     1.000     -.000     -.000
   ROW    3       -.000     -.000     1.000     -.000
   ROW    4        .000      .000      .000     1.000
```

3.4 THE MULTR PROGRAM

Rather than provide a flow chart or lengthy verbal description of each of the programs, ample comment cards are provided in the programs themselves. You are urged to study the program listings carefully. Looking at the MULTR listing that follows, you will observe that it tells you exactly how to set up your data deck, incorporating a punched upper-triangular correlation matrix such as CORREL punches. You must also provide the punched means and standard deviations if you require the raw score or deviation score regression equations. Notice particularly that the MULTR program requires the criterion variable to be the last row and column of the correlation matrix. If you want to use some other element of the score vector as the criterion, this should be anticipated in running CORREL by use of the T format, as described in Chapter One.

Incidentally, all the programs in this book are written to read punched cards as input. Actually, more and more multivariate data files are being kept on magnetic tapes. You should know that a BCD (binary-coded decimal) tape contains records which are card images and are read under FORTRAN control just as cards are read. A little study of a FORTRAN manual will enable you to modify our READ statements to provide for reading BCD data tapes.

The project TALENT data of Appendix B can serve as a numerical example of applying the MULTR program. Using mechanical reasoning and mathematics ability as predictors of physical science interest for grade 12 males, the first step is to form the correlation matrix for these three variables using the program CORREL. Using the format (43X, F3.0, 3X, F3.0, 3X, F3.0) and the 234 data cards of Appendix B males, the results of Table 3.3 are obtained.

TABLE 3.3 Correlation Reductions
for 3 Tests, 234 Subjects

Dispersion Matrix			
Section 1			
Row	1	2	3
1	12.78	19.57	11.33
2	19.57	113.88	55.60
3	11.33	55.60	84.09

Test	Mean	Standard Deviation
1	13.57	3.57
2	26.25	10.67
3	21.30	9.17

Correlation Matrix			
Section 1			
Row	1	2	3
1	1.00	.51	.35
2	.51	1.00	.57
3	.35	.57	1.00

Given the correlation matrix $\mathbf{R}$, the multiple correlation analysis begins with

$$\mathbf{R}_{11} = \begin{bmatrix} 1.00 & .51 \\ .51 & 1.00 \end{bmatrix}$$

$$|\mathbf{R}_{11}| = .74$$

$$\mathbf{R}_{11}^{-1} = \begin{bmatrix} \dfrac{1.00}{.74} & -\dfrac{.51}{.74} \\ -\dfrac{.51}{.74} & \dfrac{1.00}{.74} \end{bmatrix} = \begin{bmatrix} 1.35 & -.69 \\ -.69 & 1.35 \end{bmatrix}$$

$$\mathbf{b} = \mathbf{R}_{11}^{-1}\mathbf{R}_{12} = \begin{bmatrix} 1.35 & -.69 \\ -.69 & 1.35 \end{bmatrix} \cdot \begin{bmatrix} .35 \\ .57 \end{bmatrix} = \begin{bmatrix} .07 \\ .53 \end{bmatrix}$$

```
C      MULTIPLE CORRELATION PROGRAM.  A COOLEY-LOHNES ROUTINE.
C
C      THIS PROGRAM READS IN THE UPPER-TRIANGULAR CORRELATION MATRIX,
C      AND COMPUTES MULT R AND ITS F, BETAS, SQUARED BETAS, REGRESSION
C      FACTOR CORRELATIONS WITH THE PREDICTORS (STRUCTURE COEFFICIENTS),
C      AND, IF USER SUBMITS CENTROID AND STANDARD DEVIATION VECTORS,
C      B AND C VALUES FOR RAW REGRESSION EQUATION.
C
C
C      NOTE THAT THE PROGRAM EXPECTS THE LAST, OR M-TH, VARIABLE TO BE
C      THE CRITERION VARIABLE.
C
C
C      INPUT
C
C      1) FIRST TEN CARDS OF DATA DECK DESCRIBE THE PROBLEM IN A TEXT
C      THAT WILL BE REPRODUCED ON THE OUTPUT. DO NOT USE COL 1 OF THESE
C      CARDS.
C      2) CONTROL CARD (CARD 11)     COLS 1-2   M, THE NUMBER OF VARIABLES
C                                    COLS 3-7   N, THE NUMBER OF SUBJECTS
C                                    COL  8     IT = 0 TO OMIT B WEIGHTS,
C                                    IT = 1 TO READ IN CENTROID AND S.D.S
C                                    AND COMPUTE B AND C VALUES.
C      3) FORMAT CARD (CARD 12),  FOR R MATRIX, UPPER-TRIANGULAR.
C       NOTE THAT IF R IS PUNCHED FROM CORREL, INPUT FORMAT IS
C          (10X, 7F10.7 / (10X, 7F10.7)).
C      4) UPPER-TRIANGULAR R MATRIX,
C      5) CENTROID VECTOR, FORMAT (10X,5E14.7/(10X,5E14.7)), OPTIONAL.
C      6) STANDARD DEVIATION VECTOR, FORMAT (10X,5E14.7/(10X,5E14.7)),
C       OPTIONAL.
C
C
C      PUNCHED OUTPUT
C
C      1) BETA WEIGHTS VECTOR, TO BE READ (10X,5E14.7/(10X,5E14.7)).
C
C
C      SUBROUTINES MATINV AND MPRINT ARE REQUIRED.
C
C
       DIMENSION   TIT(16),   X(100), Y(100), A(100,100), B(100), C(100),
      C D(100), E(100)
C
 1     WRITE (6,2)
 2     FORMAT (1H1)
C
       DO 5    J = 1, 10
       READ   (5,4) (TIT(K),   K = 1, 16)
 4     FORMAT (16A5)
 5     WRITE (6,4) (TIT(K),   K = 1, 16)
       READ   (5,6) M,   N,   IT
 6     FORMAT (I2, I5, I1)
       READ   (5,4) (TIT(K),   K = 1, 16)
       EN = N
       WRITE (6,7) M,   N
 7     FORMAT(44H0MULTIPLE CORRELATION   VIA COOLEY-LOHNES,    I3,9H TESTS
      C, I6, 10H SUBJECTS.)
C
 19    DO 21    J = 1, M
 21    READ   (5,TIT) (A(J,K),   K = J, M)
       DO 22    J = 1, M
       DO 22    K = J, M
 22    A(K,J) = A(J,K)
       WRITE (6,17)
 17    FORMAT (19H0CORRELATION MATRIX)
       CALL MPRINT (A, M)
```

```
C
 20    MP = M - 1
       CALL MATINV (A, MP, DET)
       WRITE (6,23) DET
 23    FORMAT (15HODETERMINANT = E16.5)
       DO 24    J = 1, MP
       B(J) = 0.0
       DO 24    K = 1, MP
 24    B(J) = B(J) + A(J,K) * A(K,M)
C      B NOW CONTAINS THE BETA WEIGHTS.
       RSQ = 0.0
       DO 25    J = 1, MP
       C(J) = B(J) * B(J)
       D(J) = B(J) * A(J,M)
 25    RSQ = RSQ + B(J) * A(J,M)
       WRITE (6,26) RSQ
 26    FORMAT (21HOMULTIPLE R SQUARE = F8.3)
       RMULT = SQRT  (RSQ)
       WRITE (6,27) RMULT
 27    FORMAT (14HOMULTIPLE R = F8.3)
       XNDF1 = M - 1
       XNDF2 = N - M
       F = (RSQ * XNDF2) / ((1.0 - RSQ) * XNDF1)
       WRITE (6,28) F
 28    FORMAT(35HOF FOR ANALYSIS OF VARIANCE ON R = F10.3)
       WRITE (6,29) XNDF1,  XNDF2
 29    FORMAT (11HON.D.F.1 = F3.0, 5X, 10HN.D.F.2 = F10.0)
       DO 30    J = 1, MP
 30    E(J) = A(J,M) / RMULT
       WRITE (6,31)
 31    FORMAT(65HOPREDICTOR  BETA   BETA SQ     R(CRITERION)   BETA*R    ST
      CRUCTURE R       )
 32    FORMAT(1H0I6,5X,F5.3,  4X,F5.3,  5X,F10.3,  4X,F6.3,  4X,F8.3)
       DO 33    J = 1, MP
 33    WRITE (6,32) J,  B(J),  C(J),  A(J,M),   D(J),  E(J)
C
C   NEXT, PUNCH OUT BETAS.
       WRITE (7,16) (B(J),   J = 1, MP)
 16    FORMAT (10H BETAS    , 5E14.7 / (10X, 5E14.7))
C
       IF (IT)   1, 1,  34
 34    CI = 0.0
C
       READ  (5,18) (Y(J),   J = 1, M)
       READ  (5,18) (X(J),   J = 1, M)
 18    FORMAT (10X, 5E14.7 /( 10X, 5E14.7))
       WRITE (6,12)
 12    FORMAT (34H0TEST       MEAN          S. D.     )
       DO 13    J = 1, M
 13    WRITE (6,14) J,  Y(J),  X(J)
 14    FORMAT (I4, 2F11.2)
C
       DO 35    J = 1, MP
       B(J) = B(J) * (X(M) / X(J))
 35    CI = CI + B(J) * Y(J)
       CI = Y(M) - CI
       WRITE (6,36)
 36    FORMAT (10H0B WEIGHTS)
       WRITE (6,37) (B(J),   J = 1, MP)
 37    FORMAT (10F8.3)
       WRITE (6,38) CI
 38    FORMAT (22H0INTERCEPT CONSTANT = F9.3)
       GO TO 1
       END
 >
```

Thus yielding the standardized partial regression coefficients, **b**. If the vector **b** is applied to each standardized score vector in the norming sample, a prediction $\hat{z}_3$ is obtained for each observation of the vector $[z_{1i} z_{2i}]$. This vector product

$$[z_{1i} z_{2i}] \cdot \begin{bmatrix} b_1 \\ b_2 \end{bmatrix} = \hat{z}_{3i}$$

correlates .57 with the observed z_{3i} for this norming sample. Of course, the multiple correlation coefficient is not computed that way, but is formed from the vector product $R^2 = \mathbf{R}_{21}\mathbf{b}$, which is

$$[.35 \quad .57] \cdot \begin{bmatrix} .07 \\ .53 \end{bmatrix} = .33 \quad \text{and} \quad \sqrt{.33} = .57$$

The variance ratio test of the hypothesis of zero relationship between science interest and these two ability measures is

$$F = \frac{.33(234 - 3)}{.77(3 - 1)} = 56$$

with 2 and 231 degrees of freedom. The $R^2 = .33$ indicates that about one-third of the variance in science interest can be predicted from these two ability measures.

The structure of this linear function of the predictor set is determined by multiplying the vector of predictor-criterion correlations by $1/R$, thus

$$\frac{1}{.57} \begin{bmatrix} .35 \\ .57 \end{bmatrix} = \begin{bmatrix} .60 \\ .99 \end{bmatrix}$$

The resulting structure indicates that $\hat{z}_3$ correlates .60 with z_1 and .99 with z_2. Mechanical reasoning seems to contain very little information regarding science interest that is not in the mathematics ability measure.

Although the raw score prediction equation for this example would not be of any interest in practice, for purposes of numerical illustration it is: $(.189)X_1 + (.456)X_2 + 6.775 = \hat{X}_3$.

The printout of the multiple correlation results is summarized in Table 3.4. This example allows you to check your version of our multiple correlation program and to relate the equations of Section 3.1 to an actual data analysis example.

TABLE 3.4   Multiple Correlation for 3 Tests on 234 Subjects

Correlation Matrix
  Section 1

Row	1	2	3
1	1.00	.51	.35
2	.51	1.00	.57
3	.35	.57	1.00

Determinant = .73679E00
Multiple R Square = .327
Multiple R = .572
F for Analysis of Variance on R = 56.062
N.D.F.1 = 2.   N.D.F.2 = 231.

Predictor	B	B Sq	R (Criterion)	$B*R$	Structure R
1	.074	.005	.346	.025	.605
2	.530	.281	.568	.301	.994

Raw Regression Weights
  .189   .456
Intercept Constant = 6.775

3.5 EXAMPLES OF MULTIPLE CORRELATION

In this first example the analysis is begun with a previously computed fourth order **R** matrix. The three predictors are Project TALENT abilities tests:

(1) R420, Table Reading
(2) R430, Clerical Checking
(3) R440, Object Inspection

All three are highly speeded psychomotor performance tests, and they are the three tests most highly correlated with a factor of Perceptual Speed and Accuracy in the TALENT battery of 60 abilities tests (c.f. Lohnes, 1966, Ch. 4). The factor itself is a linear function of all 60 tests, and the purpose of this analysis is to see how well the factor can be estimated from these three tests alone. In Table 3.5 the obtained R^2 suggests that 80 percent of the variance in the factor can be accounted for by the variance in this three-predictor regression function. The Clerical Checking test is the best of the three predictors.

Table 3.6 reports another in this series of studies of the predictability of factors of the 60 TALENT abilities tests from special subsets of the tests. Note that these studies are based on a sample size of 10,000, so we should not be surprised to see very large F ratios. The Verbal Knowledges factor which is

TABLE 3.5 Printout from Check on MULTR

SECOND CHECK PROBLEM FOR MULTR. R MATRIX INPUT.

MAIN PREDICTORS OF THE PERCEPTUAL SPEED AND ACCURACY
   FACTOR. P. R. LOHNES, PROJECT TALENT, 10–66.

THREE PREDICTORS ARE, R420, TBL, R430, CLR, R440,
   OBJ. CRITERION IS PSA (PERCEPTUAL SPEED AND
   ACCURACY)FACTOR FROM MAP ABILITIES DOMAIN.

DETERMINANT=.56517E–00
MULTIPLE SQUARE=.803
MULTIPLE R=.896
F FOR ANALYSIS OF VARIANCE ON R=13606.145
N.D.F.1=3   N.D.F.2=9996

PRE-DICTOR	BETA	BETA SQ	R (CRITERION)	BETA*R	STRUC-TURE R
1	.342	.117	.705	.241	.787
2	.462	.213	.760	.351	.848
3	.314	.099	.672	.211	.750

the criterion here is a general intelligence factor, with which every one of the 60 abilities tests is positively correlated (Lohnes, 1966, Ch. 3). The nine predictors are all specialized information tests:

(1) R102, Vocabulary
(2) R103, Literature
(3) R104, Music
(4) R105, Social Studies
(5) R108, Biology
(6) R109, Scientific Attitude
(7) R110, Aeronautics
(8) R111, Electricity
(9) R115, Sports

Observe in the " R(CRITERION)" column of the printout that all the predictors have healthy correlations with the factor, the highest being .70 for the Social Studies test which is the best predictor, yet the multiple correlation developed is only .78. The regression function accounts for only 61 percent of the variance in the Verbal Knowledges factor. Although we don't report the **R** matrix (you can find it as Table 3.10 in Lohnes, 1966), you can guess that the predictors must be highly intercorrelated themselves. What do you make of the two negative beta weights? There are no negative correlations in the **R** matrix.

Table 3.7 reports the multiple correlation analysis for the English Language factor of the TALENT battery regressed on five predictors:

(1) R231, Spelling
(2) R232, Capitalization
(3) R233, Punctuation
(4) R234, Usage
(5) R235, Expression

The best predictor is Capitalization, but the variance in the factor explained by the regression is only 55 percent.

TABLE 3.6   Multiple Correlation Printout

MAIN PREDICTORS OF VERBAL KNOWLEDGES. P. R. LOHNES, TALENT, 10–66.

VARIABLES ARE 1) R102, VOC, 2) R103, LIT,
 3) R104, MUS, 4) R105, SST, 5) R108, B10,
 6) R109, SCA, 7) R110, AER, 8) R 111, ELE,
 9) R115, SPO, 10) VKN MAP FACTOR (CRITERION)

DETERMINANT=.46420E–02
MULTIPLE R SQUARE=.613
MULTIPLE R=.783
F FOR ANALYSIS OF VARIANCE ON R=1760.328
N.D.F.1=9   N.D.F.2=9990

PRE-DICTOR	BETA	BETA SQ	R (CRITERION)	BETA*R	STRUC-TURE R
1	.091	.008	.659	.060	.842
2	.233	.054	.690	.160	.881
3	.244	.059	.651	.159	.831
4	.269	.072	.696	.187	.889
5	.000	.000	.508	.000	.649
6	-.018	.000	.466	-.008	.595
7	.193	.037	.503	.097	.642
8	-.124	.015	.364	-.045	.465
9	.007	.000	.479	.003	.612

TABLE 3.7 Multiple Correlation Printout

MAIN PREDICTORS OF THE ENGLISH FACTOR. P. R. LOHNES,
TALENT, 10—22.

VARIABLES ARE 1) R231, SPL, 2) R232, CAP,
3) R233, PNC. 4) R234, USG, 5) R235, EXP,
6) ENG MAP FACTOR (CRITERION)

DETERMINANT=.12313E—00
MULTIPLE R SQUARE=.550
MULTIPLE R=.741
F FOR ANALYSIS OF VARIANCE ON R=2440.740
N.D.F.1=5   N.D.F.2=9994

PRE—DICTOR	BETA	BETA SQ	R (CRITERION)	BETA*R	STRUC—TURE R
1	.215	.046	.584	.125	.788
2	.308	.095	.619	.191	.835
3	.108	.012	.604	.065	.815
4	.162	.026	.585	.095	.789
5	.138	.019	.531	.073	.716

Table 3.8 gives the results for regressing the Mathematics factor on five predictor tests:

(1) R106, Mathematics Information
(2) R107, Physics Information
(3) R311, Arithmetic Reasoning
(4) R312, 9th Grade Mathematics
(5) R333, Advanced Mathematics

The multiple regression accounts for 60 percent of the factor variance, and Advanced Mathematics is the best predictor. Note that Advanced Mathematics is correlated .92 with the regression function. The multiple regression of the Visual Reasoning factor on four predictor tests is given in Table 3.9. The predictors are:

(1) R270, Mechanical Reasoning
(2) R281, Visualization in Two Dimensions
(3) R282, Visualization in Three Dimensions
(4) R290, Abstract Reasoning

TABLE 3.8   Multiple Correlation Printout

MAIN PREDICTORS OF MATHEMATICS FACTOR. P. R. LOHNES,
    TALENT, 10—66.

VARIABLES ARE 1) R106, MAT, 2) R107, PHY,
    3) R311, ARR, 4) R312, MA9, 5) R333, ADV,
    6) MAT MAP FACTOR (CRITERION)

DETERMINANT=.57390E—01
MULTIPLE R SQUARE=.598
MULTIPLE R=.774
F FOR ANALYSIS OF VARIANCE ON R=2978.667
N.D.F.1=5   N.D.F.2=9994

PRE-DICTOR	BETA	BETA SQ	R (CRITERION)	BETA*R	STRUC-TURE R
1	.216	.047	.623	.135	.805
2	-.005	.000	.422	-.002	.546
3	-.212	.045	.344	-.073	.445
4	.307	.094	.611	.188	.790
5	.494	.244	.710	.351	.918

TABLE 3.9   Multiple Correlation Printout

MAIN PREDICTORS OF THE VISUAL REASONING FACTOR.
    P. R. LOHNES, TALENT, 10—66.

VARIABLES ARE 1) R270, MCR, 2) R281, VS2,
    3) R282, VS3, 4) R290, ABS, 5) VIS MAP FACTOR
    (CRITERION)

DETERMINANT=.28834E—00
MULTIPLE R SQUARE=.644
MULTIPLE R=.802
F FOR ANALYSIS OF VARIANCE ON R=4510.813
N.D.F.1=4   N.D.F.2=9995

PRE-DICTOR	BETA	BETA SQ	R (CRITERION)	BETA*R	STRUC-TURE R
1	.119	.014	.591	.071	.737
2	.304	.093	.627	.191	.782
3	.414	.171	.712	.295	.888
4	.153	.023	.572	.087	.713

R is .80, and the regression accounts for 64 percent of the variance in the factor. Visualization in Three Dimensions is the best predictor. Taken as a set of studies, these outcomes indicate that the factor regressions rank as follows:

Factor	R	Number of Predictors	Best Predictor
Perceptual Speed and Accuracy	.90	3	CLR
Visual Reasoning	.80	4	VS3
Verbal Knowledges	.78	9	SST
Mathematics	.77	5	ADV
English Language	.74	5	CAP

This table was very helpful in deciding whether it would be feasible to compute estimated factor scores for research subjects from subsets of the tests in the full battery of 60 tests on which the factors were defined. In general, these multiple correlations are not high enough to warrant such a strategy.

3.6 POLYNOMIAL CURVE FITTING

The purpose of this section is to demonstrate the ubiquity of the method of least squares as applied in multiple regression. The multiple correlation model is a normal theory model. It is totally appropriate when applied to samples from populations with an m.n.d., because all regression of any element on any of the other elements of a vector random variable from an m.n.d. is exhausted by the linear regression equation. This condition of linearity of all regressions is almost the best definition of an m.n.d. Nevertheless, the linear least squares regression can be very useful in many situations where an m.n.d. is not assumed. The entire domain of analysis of variance models can be developed as regression models, for example, as is done in the " general linear model" approach (c.f. Bock, 1966). That approach is beyond the scope of this text, so we have chosen a demonstration example in the area of curve fitting.

Suppose a researcher has strong reason for believing the relationship between two continuously measured variables is nonlinear, or as we say, is curvilinear. Perhaps inspection of a scatter plot has suggested this hypothesis. Or, he may have a theoretical position that requires a curvilinear relationship between X and Y. Eysenck (1965) has used the example of a researcher who is so impressed with the seemingly contradictory claims of the two sayings,

(1) "Absence makes the heart grow fonder,"

(2) "Out of sight, out of mind,"

that he decides both must be true in turn. Note that if (1) is always true, the graph of X (absence) against Y (fondness) would be of the form

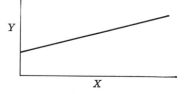

If (2) were always true, the form of the graph would be

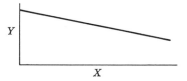

However, if (1) holds true for the early period of an absence and (2) holds true for the later period, a nice curvilinear graph of this form would pertain:

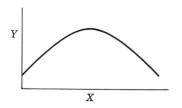

Although psychology does provide examples of curvilinear relations in some of its corners, the authors clutch at Eysenck's entertaining straw because the individual differences variables they tend to study simply do not provide real examples of nonlinear regressions.

Given a case of suspected curvilinearity, the researcher may want to do three things about it:

(1) Make statistical tests of the inferences of linearity and curvilinearity.

(2) Given statistical evidence for curvilinearity, fit constants by least squares method to a polynomial equation of appropriate degree, making $Y = f(X)$ a curvilinear function.

(3) Given a fitted polynomial, solve for values of $f(X)$ to replace X in the data set, since these will have a linear relationship with Y.

There are, of course, many equation forms a curvilinear relationship may take besides that of a general polynomial of degree k. Fortunately, a general polynomial of high enough degree will usually give a decent fit to any data. The form of the general polynomial is

$$\hat{Y}_i = b_1 X_i + b_2 X_i^2 + \cdots + b_k X_i^k + c$$

The polynomial is fitted by forming the correlation matrix for the $k + 1$ variables $(X, X^2, \ldots, X^k, Y)$ and entering the multiple regression procedure with it. Means and standard deviations for the $k + 1$ variables are required to translate the beta weights into b weights and to compute the intercept constant, c.

Given that the analysis of variance for multiple R for the polynomial of degree k suggests that P (population multiple correlation coefficient) is non-zero, the trick is to test by analysis of variance the significance of the increase of R obtained by fitting another term (i.e., the significance of the reduction in error sum of squares for fitting $b_{k+1} X_i^{k+1}$). The POLY program fits the following three polynomials:

(1) linear model: $\hat{Y}_i = b_1 X_i + c$
(2) hyperbolic model: $\hat{Y}_i = b_1 X_i + b_2 X_i^2 + c$
(3) sinusoidal model: $\hat{Y}_i = b_1 X_i + b_2 X_i^2 + b_3 X_i^3 + c$

Note that the regression weights and intercept constant are computed anew for each model, and thus are different values for each model, so that b_1 in (1) is different from b_1 in (2), which in turn is different from b_1 in (3), etc. The program tests the significance of R for each fit, and the significance of the change in R from (1) to (2), and from (2) to (3). It also provides for computing values of $f(X)$ for each of the fits. The reader recalls that a quadratic or 2nd degree polynomial describes a hyperbolic curve such as

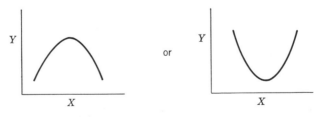

while a cubic or 3rd degree polynomial describes a curve with a cycle built into it, such as

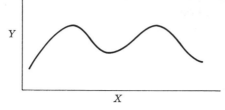

The differential calculus tells us that for model (1)

$$\frac{dY}{dX} = b_1$$

so that change in Y depends only on change in X and is constant everywhere (i.e., instantaneous velocity is constant throughout the domain of X). For (2),

$$\frac{dY}{dX} = 2b_2 X + b_1$$

so that change in Y for a given increment in X depends on the initial value of X. Also for (2),

$$d^2 Y/dX^2 = 2b_2$$

indicating constant acceleration for the quadratic model. For (3),

$$dY/dX = 3b_3 X^2 + 2b_2 X + b_1$$

and

$$d^2 Y/dX^2 = 6b_3 X + 2b_2$$

so that in the cubic model even acceleration varies as X varies. We do not undertake in POLY to fit higher order polynomials because it seems to us that even a finding of significance for a cubic term would so discourage the behavioral scientist that he would want to change his method of scaling one of his constructs, or change his research design, or his line of work.

The analysis of variance table reported by POLY is described symbolically in Table 3.10. The justification for this table may be studied in M. G. Natrella (1963), Chapter Six: "Polynomial and Multivariable Relationships Analysis by the Method of Least Squares." In that publication, by the way, the U.S. Government Printing Office provides a lot of statistics book for a small cost.

TABLE 3.10 Analysis of Variance for Polynomial Curve Fitting by Least Squares Method

Source	n.d.f.	Sum of Squares	F Ratio
Model (1), $\hat{Y}_i = b_1 X_i + c$, yielding $R_{(1)}$			$\dfrac{R_{(1)}^2}{(1.0 - R_{(1)}^2)/(N-2)}$
Reduction due to model	1	$R_{(1)}^2$	
Residual from linear fit	$N-2$	$1.0 - R_{(1)}^2$	
Model (2), $\hat{Y}_i = b_1 X_i + b_2 X_i^2 + c$, yielding $R_{(2)}$			$\dfrac{R_{(2)}^2/2}{(1.0 - R_{(2)}^2)/(N-3)}$
Reduction due to model	2	$R_{(2)}^2$	
Residual from quadratic fit	$N-3$	$1.0 - R_{(2)}^2$	
Reduction due to quadratic term alone	1	$R_{(2)}^2 - R_{(1)}^2$	$\dfrac{R_{(2)}^2 - R_{(1)}^2}{(1.0 - R_{(2)}^2)/(N-3)}$
Model (3), $\hat{Y}_i = b_1 X_i + b_2 X_i^2 + b_3 X_i^3 + c$, yielding $R_{(3)}$			$\dfrac{R_{(3)}^2/3}{(1.0 - R_{(3)}^2)/(N-4)}$
Reduction due to model	3	$R_{(3)}^2$	
Residual from cubic fit	$N-4$	$1.0 - R_{(3)}^2$	
Reduction due to cubic term alone	1	$R_{(3)}^2 - R_{(2)}^2$	$\dfrac{R_{(3)}^2 - R_{(2)}^2}{(1.0 - R_{(3)}^2)/(N-4)}$

3.7 THE POLY PROGRAM

POLY

```
C      LEAST SQUARES FIT OF FIRST, SECOND, AND THIRD DEGREE POLYNOMIALS.
C      A COOLEY-LOHNES ROUTINE.
C
C      SUBROUTINE MATINV IS REQUIRED.
C
C      INPUT
C
C      1) TEN TITLE CARDS, DESCRIBING THE JOB. DO NOT USE COL 1.
C      2) CONTROL CARD.   COL 1-5  N, THE NUMBER OF SUBJECTS)
C      3) VARIABLE NAMES, TWO 6-COLUMN FIELDS (COLUMNS 1-12).    NOTE
C         PROGRAM ASSUMES THAT THE SECOND VARIABLE IS THE CRITERION AND
C         FIRST VARIABLE IS THE PREDICTOR FROM WHICH THE POYNOMIALS ARE
C         TO BE FORMED.
C      4) FORMAT CARD
C      5) N SETS OF SCORE CARDS
C      6) PLOT TABLE PARAMETERS.
C               THESE PARAMETERS ARE START, XINCR, N (2F5.0, I5).
C      START IS LOWEST VALUE OF X FOR WHICH
C      POLYNOMIALS ARE TO BE SOLVED FOR Y(SHOULD
C      BE AN INTEGER). XINCR IS INCREMENT
C      FOR X VALUES, AND N IS NUMBER OF
C      Y VALUES DESIRED FOR PLOTTING CURVE.
C
       DIMENSION VNAME(4), X(4), SX(4), SS(4,4), SSD(4,4), D(4,4), XM(4),
      C R(4,4), RTP(4,4), SD(4), R11(4,4), R12(4), B(4), A(4), B1(4),
      C BSQ(4), X2(4),  TITLE(16)
       DATA RLIT1/6HXSQUAR/,  RLIT2/6HX CUBE/
C
  101  WRITE (6,904)
  904  FORMAT (1H1)
       DO 922  K = 1, 10
       READ  (5,903) (TITLE(J),   J = 1, 16)
  903  FORMAT (16A5)
  922  WRITE (6,903) (TITLE(J),    J = 1, 16)
       READ  (5,902) N
  902  FORMAT (I5)
       M = 2
       WRITE (6,905) M, N
  905  FORMAT(25H0POLYNOMIAL FITTING  FOR I3,10H TESTS,   I6,10H SUBJECTS
      C.)
       EN = N
       MT = M
       MA = 4
  901  FORMAT (13A6)
       READ  (5,901) (VNAME(J),   J = 1, M)
       READ  (5,903) (TITLE(J),   J = 1, 16)
       DO 907  J = 1, MA
       SX(J) = 0.0
       DO 907  K = 1, MA
       RTP(J,K) = 0.0
  907  SS(J,K) = 0.0
       VNAME(4) = VNAME(2)
       VNAME(2) = RLIT1
       VNAME(3) = RLIT2
       DO 908  L = 1, N
       READ  (5,TITLE) (X(J),   J = 1, M)
       X(4) = X(2)
       X(2) = X(1) * X(1)
       X(3) = X(2) * X(1)
       DO 908  J = 1, MA
       SX(J) = SX(J) + X(J)
       DO 908  K = 1, MA
  908  SS(J,K) = SS(J,K) + X(J) * X(K)
       M = M + 2
       DO 913   J = 1, M
       DO 913   K = 1, M
```

```
913    SSD(J,K) = SS(J,K) - SX(J) * SX(K) / EN
       DO 914  J = 1, M
       XM(J) = SX(J) / EN
914    SD(J) = SQRT  (SSD(J,J) / (EN-1.0))
       DO 915  J = 1, M
       DO 915  K = 1, M
       D(J,K) = SSD(J,K) / (EN-1.0)
915    R(J,K) = D(J,K) / (SD(J) * SD(K))
       WRITE (6,916)
916    FORMAT(35HOTEST MEANS AND STANDARD DEVIATIONS)
       DO 923  J = 1, M
923    WRITE (6,917) J,  VNAME(J),   XM(J),  SD(J)
917    FORMAT(13,3X,A6,3X,F9.3,3X,F9.3,  3X, F9.3,  3X, F9.3)
       WRITE (6,920)
920    FORMAT (19HOCORRELATION MATRIX)
       DO 921  J = 1, M
921    WRITE (6,917)   J, VNAME(J),  (R(J,K),   K = 1, M)
       WRITE (6,90)
90     FORMAT(81HO=========================================================
C======================          )
C
C
       ME = 0
       ME1 = 0
       M = 3
       DO 1   I = 1, M
       DO 1   J= 1, M
       RTP(I,J) = R(I,J)
1      R11(I,J) = R(I,J)
C      R11 CONTAINS THE INTERCORRELATIONS OF THE PREDICTORS.
       DO 2  I= 1, M
2      R12(I) = R(I,M+1)
C      R12 CONTAINS THE CRITERION-PREDICTOR CORRELATIONS.
C
       M = 0
       WRITE (6,24)
24     FORMAT (45HOFIRST, SECOND, AND THIRD DEGREE POLYNOMIALS.)
100    M = M + 1
       ME1 = ME1 + 1
C
       CALL MATINV (R11,  M,   DET)
       DO 3   J = 1, M
       B(J) = 0.0
       DO 3   K = 1, M
3      B(J) = B(J) + R11(J,K) * R12(K)
C      B CONTAINS THE VECTOR OF BETA WEIGHTS.
C
       RSQ = 0.0
       DO 4  I=1, M
4      RSQ = RSQ + B(I) * R12(I)
       WRITE (6,90)
       WRITE (6,5) RSQ
5      FORMAT (21HOMULTIPLE R SQUARE = F8.3)
       RMULT = SQRT  (RSQ)
       WRITE (6,6) RMULT
6      FORMAT (14HOMULTIPLE R = F8.3)
       T = M + 1
       MZ = 4
       XNDF1 = T - 1.0
       WRITE (6,7) XNDF1
7      FORMAT (11HON.D.F.1 = F3.0)
       XNDF2 = EN - T
       WRITE (6,8) XNDF2
8      FORMAT (11HON.D.F.2 = F10.0)
       F = (RSQ * XNDF2) /((1.0 - RSQ) * XNDF1)
       WRITE (6,9) F
```

```
   9     FORMAT (35H0F FOR ANALYSIS OF VARIANCE ON R = F10.3)
C
         WRITE (6,10)
  10     FORMAT (13H0BETA WEIGHTS)
         WRITE (6,11) (B(I),  I = 1, M)
  11     FORMAT (10F8.3)
C
         DO 13  I= 1,M
  13     A(I) = SD(MZ) / SD(I)
         DO 14  I= 1,M

  14     B1(I) = A(I) * B(I)
         C1 = 0.0
         DO 15  I = 1,M
  15     C1 = C1 + B1(I) * XM(I)
         C = XM(MZ) - C1
         DO 16  I= 1,M
         X(I) = B(I) * R12(I)
         A(I) = R12(I) / RMULT
  16     BSQ(I) = B(I) * B(I)
C
         WRITE (6,55)
  55     FORMAT (38H0CONTRIBUTIONS TO MULTIPLE CORRELATION)
         WRITE (6,52)
  52     FORMAT (48H0   AND REGRESSION FACTOR LOADINGS,  2ND COLUMN))
         DO 53     J = 1, M
  53     WRITE (6,54) J,  VNAME(J),  X(J),   A(J)
  54     FORMAT (1H I4,3X,A6,3X,F8.3,3X,F8.3)
         WRITE (6,17)
  17     FORMAT (21H0SQUARED BETA WEIGHTS)
         WRITE (6,11) (BSQ(I),  I = 1, M)
         WRITE (6,18)
  18     FORMAT (10H0B WEIGHTS)
         WRITE (6,11) (B1(I),  I = 1, M)
         WRITE (6,19) C
  19     FORMAT (22H0INTERCEPT CONSTANT = F9.3)
C
C
C
C        ENTER POLYNOMIAL ROUTINE.
         X2(ME1) = C
         DO 27 I = 1,M
         SSD(ME1, I) = B1(I)
         R(ME1, I) = B(I) * R12(I)
  27     D(ME1, I) = B(I)
         SX(ME1) = 0.0
         DO 28 I = 1, M
  28     SX(ME1) = SX(ME1) +  R(ME1,I)
C
C        SSD(ME1, I) AND XM(ME1) CONTAIN THE RAW-SCORE
C        WEIGHTS AND INTERCEPT CONSTANT FOR THE
C        POLYNOMIAL IN ME1 DEGREE. D(ME1,I)
C        CONTAINS ITS BETA WEIGHTS.
C        SX(ME1) CONTAINS THE REDUCTION S.S. DUE TO
C        FITTING ME1 TERMS.
C
         DO 25    I = 1, MT
         DO 25    J = 1, MT
  25     R11(I,J) = RTP(I,J)
         IF(ME1 - 3) 100, 29, 29
  29     WRITE (6,47)
  47     FORMAT(29H1 ANOVA TABLE FOR POLYNOMIALS)
         WRITE (6,90)
         WRITE (6,30) SX(1)
  30     FORMAT(43H0 REDUCTION DUE TO LINEAR FIT, WITH 1 DF = F10.3)
         RESID = 1.0 - SX(1)
```

POLY (*continued*)

```
         XNDF = EN - 2.0
         XMSQU = RESID/XNDF
         WRITE (6,31) RESID,  XNDF,  XMSQU
  31     FORMAT(18H0 RESIDUAL S.S. = F10.3, 7H  DF = F7.0,18H  RESIDUAL M.S
        1. = F10.3)
         F = SX(1) / XMSQU
         WRITE (6,32) F
  32     FORMAT (20H0F FOR LINEAR FIT = F8.3)
         WRITE (6,90)
C
         WRITE (6,33) SX(2)
  33     FORMAT(54H0 REDUCTION DUE TO GENERAL QUADRATIC FIT WITH DF.2, = F1
        10.3)
         XNUM = SX(2) /2.0
         WRITE (6,34) XNUM
  34     FORMAT(18H0REDUCTION M.S. = F10.3)
         RESID2 = 1.0 - SX(2)
         XNDF = XNDF - 1.0
         XMSQ2 = RESID2/XNDF
         WRITE (6,31) RESID2,  XNDF,  XMSQ2
         F = XNUM/XMSQ2
         WRITE (6,35) F
  35     FORMAT(23H0F FOR QUADRATIC FIT = F8.3)
C
         XMSQU = SX(2) - SX(1)
         WRITE (6,36) XMSQU
  36     FORMAT(53H0REDUCTION DUE TO QUADRATIC TERM ALONE, WITH 1 DF, = F10
        1.3)
         F = XMSQU / XMSQ2
         WRITE (6,37) F
  37     FORMAT(29H0F FOR QUADRATIC TERM ALONE = F8.3)
C
         XNUM = SX(3) / 3.0
         WRITE (6,90)
         WRITE (6,38) SX(3)
  38     FORMAT(47H0REDUCTION DUE TO GENERAL CUBIC FIT WITH DF 3, F10.3)
         WRITE (6,34) XNUM
         XNDF = XNDF - 1.0
         RESID3 = 1.0 - SX(3)
         XMSQ3 = RESID3 / XNDF
         WRITE (6,31) RESID3,  XNDF,  XMSQT
         F = XNUM / XMSQ3
         WRITE (6,39) F
  39     FORMAT(27H0F FOR GENERAL CUBIC FIT = F8.3)
C
         XMSQU = SX(3) - SX(2)
         F = XMSQU / XMSQ3
         WRITE (6,40) XMSQU
  40     FORMAT(47H0REDUCTION DUE TO CUBIC TERM ALONE WITH 1 DF = F10.3)
         WRITE (6,41) F

  41     FORMAT(25H0F FOR CUBIC TERM ALONE = F8.3)
         WRITE (6,90)
         WRITE (6,90)
C
C
         READ  (5,42) START,  XINCR,  N
  42     FORMAT(2F5.0, I5)


C
C
         WRITE (6,43)
```

POLY (*continued*)

```
43      FORMAT(48H1 Y VALUES FOR STEPS ON X, FOR THREE POLYNOMIALS)
        WRITE (6,90)
        WRITE (6,44)
44      FORMAT(41H0      X      LINEAR     QUADRATIC        CUBIC)
45      FORMAT(F8.2,1X,F10.3,2X,F10.2,2X,F10.1)
        DO 46 J = 1,N
        Y1=(SSD(1,1) * START)+X2(1)
        Y2=(SSD(2,1) * START)+(SSD(2,2)* START*START) + X2(2)
        Y3 = (SSD(3,1) *START) +(SSD(3,2)*START*START)+(SSD(3,3)*START*STA
       CRT*START) + X2(3)
        WRITE (6,45) START, Y1, Y2, Y3
46      START = START + XINCR
        GO TO 101
        END
>
```

3.8 EXAMPLES OF POLYNOMIAL CURVE FITTING

The 100 pairs of X, Y data points for the POLY check problem, as presented in Table 3.11, were contrived to be sprinkled more or less around a J-shaped portion of the hyperbola described by the equation

$$Y = X + .5 X^2$$

This section of the curve is shown by the solid curve in Figure 3.3, and the 100 data points are plotted. Note that the absence of an intercept constant forces the curve through the origin. The linear fit to the data produced by POLY is

$$\hat{Y}_i = 1.1 X_i + 1.7$$

and the correlation of .79 indicates that this fit has some utility. The linear regression is shown as the straight dashed line in Figure 3.3. The quadratic fitted by POLY is

$$\hat{Y}_i = .92 X_i + .45 X_i^2 + .22$$

and is shown as the J-shaped dashed curve in Figure 3.3. The value of R_m for the quadratic fit is .94, indicating that 88 percent of the variance in Y is accounted for by the quadratic function of X. This is a very nice fit. The POLY output (Table 3.12) clearly indicates that nothing is to be gained by fitting a cubic term. Given these data, the researcher who wanted to transform X in order to linearize its relation with Y would simply substitute values of the quadratic $f(X)$ for values of X in his data set.

Incidentally, POLY could be used to transform X for another purpose, namely to normalize it. The trick would be to sort the 100 sample values of X,

TABLE 3.11 Data for POLY Check Problem

CHECK PROBLEM FOR POLY.

POINTS HAVE BEEN SELECTED MORE OR LESS AROUND THE
FOLLOWING CURVE.
HYPOTHETICAL CURVE IS Y=X+.5 X**2

100
```
   X        Y
(5X, 2F5.0)
```

#	X	Y	#	X	Y	#	X	Y
1	-3.0	1.8	35	-0.2	0.2	69	1.8	2.0
2	-3.0	1.5	36	-0.2	-0.4	70	2.0	5.4
3	-3.0	1.2	37	0.0	0.0	71	2.0	5.0
4	-2.8	0.8	38	0.2	1.2	72	2.0	4.6
5	-2.6	1.2	39	0.2	0.6	73	2.0	2.8
6	-2.6	0.4	40	0.2	-0.4	74	2.0	1.8
7	-2.5	0.6	41	0.4	1.4	75	2.0	4.0
8	-2.4	1.0	42	0.4	1.0	76	2.2	6.2
9	-2.4	0.2	43	0.4	0.8	77	2.2	3.8
10	-2.4	0.0	44	0.4	0.2	78	2.2	3.2
11	-2.2	0.8	45	0.4	-0.2	79	2.2	2.2
12	-2.2	0.4	46	0.5	0.6	80	2.4	4.6
13	-2.0	0.0	47	0.6	2.2	81	2.5	5.6
14	-1.8	0.4	48	0.8	1.8	82	2.6	7.4
15	-1.8	0.2	49	0.8	0.6	83	2.6	7.0
16	-1.8	-0.6	50	0.8	0.2	84	2.6	6.6
17	-1.6	0.0	51	1.0	1.5	85	2.6	5.0
18	-1.4	0.2	52	1.0	0.8	86	2.6	4.0
19	-1.4	-0.2	53	1.0	2.0	87	2.6	4.0
20	-1.4	-0.6	54	1.0	2.8	88	2.8	7.6
21	-1.2	-0.8	55	1.2	2.4	89	2.8	7.4
22	-1.0	0.2	56	1.2	1.2	90	2.8	7.2
23	-1.0	0.0	57	1.4	4.0	91	2.8	6.2
24	-1.0	0.5	58	1.4	3.6	92	2.8	5.8
25	-1.0	0.6	59	1.4	3.2	93	2.8	5.4
26	-0.8	-0.6	60	1.4	1.8	94	2.8	4.8
27	-0.8	-1.0	61	1.4	0.8	95	3.0	8.0
28	-0.6	0.2	62	1.5	2.6	96	3.0	7.8
29	-0.6	0.0	63	1.6	4.6	97	3.0	7.5
30	-0.6	-0.6	64	1.6	3.6	98	3.0	7.0
31	-0.5	-0.4	65	1.6	1.6	99	3.0	6.6
32	-0.4	0.4	66	1.6	1.2	100	3.0	6.2
33	-0.4	-0.6	67	1.8	4.2			
34	-0.2	0.6	68	1.8	2.4			

ranking them from smallest to largest. The ranks divided by $N + 1$ would then be used as Y scores, and since this Y would be uniformly distributed, the fitted $f(X)$ would transform X to a uniform distribution with the range $0.0 < (fX) < 1.0$. These values of $f(X)$ would then be treated as $P(z)$, or cumulative probability, values, and corresponding z, or standard score, values would be looked up in a table of the unit normal distribution. Alternatively, computer subroutines exist for evaluating z for given $P(z)$.

Our real-data example employs Project TALENT data from Appendix B,

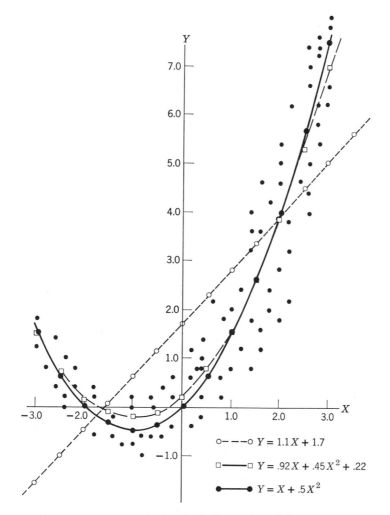

Figure 3.3. Graph of a fitted polynomial.

TABLE 3.12 Printout from Polynomial Fitting

```
CHECK PROBLEM FOR POLY.

  POINTS HAVE BEEN SELECTED MORE OR LESS AROUND
THE FOLLOWING CURVE.
HYPOTHETICAL CURVE IS Y=X+.5 X**2

POLYNOMIAL FITTING FOR 2 TESTS, 100 SUBJECTS.
TEST MEANS AND STANDARD DEVIATIONS
  1     X          .569      1.817
  2    XSQUAR     3.594      3.027
  3    X CUBE     3.398     12.099
  4     Y         2.346      2.554
CORRELATION MATRIX
  1     X         1.000      .249      .921      .788
  2    XSQUAR      .249     1.000      .323      .691
  3    X CUBE      .921      .323     1.000      .783
  4     Y          .788      .691      .783     1.000
```

```
FIRST, SECOND, AND THIRD DEGREE POLYNOMIALS.
```

```
MULTIPLE R SQUARE=.622
MULTIPLE R=.788
N.D.F.1=1
N.D.F.2=98
F FOR ANALYSIS OF VARIANCE ON R=161.020
BETA WEIGHTS
  .788
CONTRIBUTIONS TO MULTIPLE CORRELATION
  AND REGRESSION FACTOR LOADINGS, 2ND COLUMN
  1     X          .622      1.000
SQUARED BETA WEIGHTS
  .622
B WEIGHTS
  1.108
INTERCEPT CONSTANT=1.715
```

```
MULTIPLE R SQUARE=.882
MULTIPLE R=.939
N.D.F.1=2
N.D.F.2=97
F FOR ANALYSIS OF VARIANCE ON R=363.794
```

TABLE 3.12 (Continued)

```
BETA WEIGHTS
  .657    .527
CONTRIBUTIONS TO MULTIPLE CORRELATION
  AND REGRESSION FACTOR LOADINGS, 2ND COLUMN
  1    X         .518       .839
  2    XSQUAR      .364       .735
SQUARED BETA WEIGHTS
  .432    .278
B WEIGHTS
  .924    .445
INTERCEPT CONSTANT=.221
```

```
MULTIPLE R SQUARE=.883
MULTIPLE R=.940
N.D.F.1=3
N.D.F.2=96
F FOR ANALYSIS OF VARIANCE ON R=240.827
BETA WEIGHTS
  .613    .522    .049
CONTRIBUTIONS TO MULTIPLE CORRELATION
  AND REGRESSION FACTOR LOADINGS, 2ND COLUMN)
  1    X         .484       .839
  2    XSQUAR      .361       .735
  3    X CUBE      .039       .833
SQUARED BETA WEIGHTS
  .376    .273    .002
B WEIGHTS
  .862    .441    .010
INTERCEPT CONSTANT=.236
```

```
ANOVA TABLE FOR POLYNOMIALS
```

```
REDUCTION DUE TO LINEAR FIT, WITH 1 DF=.622
RESIDUAL S.S.=.378    DF=98    RESIDUAL M.S.=.004
F FOR LINEAR FIT=161.020
```

```
REDUCTION DUE TO GENERAL QUADRATIC FIT WITH
  DF.2,= .882
REDUCTION M.S.=.441
```

TABLE 3.12   (Concluded)

RESIDUAL S.S.=1.18   DF=97   RESIDUAL M.S.=.001
F FOR QUADRATIC FIT=363.794
REDUCTION DUE TO QUADRATIC TERM ALONE, WITH
  1 DF,=.261
F FOR QUADRATIC TERM ALONE=214.982

REDUCTION DUE TO GENERAL CUBIC FIT WITH DF 3,=.883
REDUCTION M.S.=.294
RESIDUAL S.S.=.117   DF=96   RESIDUAL M.S.=.001
F FOR GENERAL CUBIC FIT=240.827
REDUCTION DUE TO CUBIC TERM ALONE WITH 1 DF=.000
F FOR CUBIC TERM ALONE=.282

Y VALUES FOR STEPS ON X, FOR THREE POLYNOMIALS

X	LINEAR	QUADRATIC	CUBIC
-4.00	-2.717	3.64	3.2
-3.50	-2.163	2.44	2.2
-3.00	-1.609	1.45	1.3
-2.50	-1.055	.69	.7
-2.00	-.501	.15	.2
-1.50	.053	-.16	-.1
-1.00	.607	-.26	-.2
-.50	1.161	-.13	-.1
-.00	1.715	.22	.2
.50	2.270	.79	.8
1.00	2.824	1.59	1.5
1.50	3.378	2.61	2.6
2.00	3.932	3.85	3.8
2.50	4.486	5.31	5.3
3.00	5.040	7.00	7.1
3.50	5.594	8.91	9.1
4.00	6.148	11.04	11.4

the 12th grade male file. The dependent variable is a socio-economic environment scale (SEE) and the independent variable is a reading comprehension test score (RDG). The reason for trying a curvilinear regression was the hunch that good readers might tend more often than a linear regression would suggest to be from middle class homes where the achievement press would be higher than in upper class homes. Actually, the data analysis (Table 3.13) reveals that a statistically significant but very small gain in regression

TABLE 3.13   Printout from Polynomial Fitting

```
ONE PERCENT SAMPLE OF 12TH GRADE MALES FILE
PROJECT TALENT
   P. R. LOHNES, 1-5-67
      X=READING COMPREHENSION TEST SCORE
      Y=SOCIO-ECONOMIC ENVIRONMENT SCORE
POLYNOMIAL FITTING FOR 2 TESTS, 234 SUBJECTS.
TEST MEANS AND STANDARD DEVIATIONS
   1    RDGCOM        33.585           9.513
   2    XSQUAR      1218.098         589.273
   3    X CUBE     46404.431       29958.760
   4    SOCIOE        98.543           9.444
CORRELATION MATRIX
   1    RDGCOM    1.000     .984     .955     .368
   2    XSQUAR     .984    1.000     .992     .389
   3    X CUBE     .955     .992    1.000     .392
   4    SOCIOE     .368     .389     .392    1.000
```

```
FIRST, SECOND, AND THIRD DEGREE POLYNOMIALS.
```

```
MULTIPLE R SQUARE=.136
MULTIPLE R=.368
N.D.F.1=1
N.D.F.2=232
F FOR ANALYSIS OF VARIANCE ON R=36.394
BETA WEIGHTS
   .368
CONTRIBUTIONS TO MULTIPLE CORRELATION
   AND REGRESSION FACTOR LOADINGS, 2ND COLUMN
   1    RDGCOM         .136        1.000
SQUARED BETA WEIGHTS
   .136
```

TABLE 3.13   (Continued)

```
B WEIGHTS
  .366
INTERCEPT CONSTANT=86.265
```

```
MULTIPLE R SQUARE=.158
MULTIPLE R=.398
N.D.F.1=2
N.D.F.2=231
F FOR ANALYSIS OF VARIANCE ON R=21.681
BETA WEIGHTS
  -.466    .848
CONTRIBUTIONS TO MULTIPLE CORRELATION
  AND REGRESSION FACTOR LOADINGS, 2ND COLUMN
  1    RDGCOM    -.172    .926
  2    XSQUAR     .330    .978
SQUARED BETA WEIGHTS
   .217    .718
B WEIGHTS
  -.463    .014
INTERCEPT CONSTANT=97.533
```

```
MULTIPLE R SQUARE=.166
MULTIPLE R=.407
N.D.F.1=3
N.D.F.2=230
F FOR ANALYSIS OF VARIANCE ON R=15.256
BETA WEIGHTS
  -1.895    4.337    -2.100
CONTRIBUTIONS TO MULTIPLE CORRELATION
  AND REGRESSION FACTOR LOADINGS, 2ND COLUMN
  1    RDGCOM    -.698    .904
  2    XSQUAR    1.687    .955
  3    X CUBE    -.823    .962
SQUARED BETA WEIGHTS
  3.590    18.810    4.412
B WEIGHTS
  -1.881     .070    -.001
INTERCEPT CONSTANT=107.772
```

```
ANOVA TABLE FOR POLYNOMIALS
```

TABLE 3.13 (Concluded)

REDUCTION DUE TO LINEAR FIT, WITH 1 DF=.136
RESIDUAL S.S.=.864   DF=232   RESIDUAL M.S.=.004
F FOR LINEAR FIT=36.394   P<.01

REDUCTION DUE TO GENERAL QUADRATIC FIT WITH
   DF.2, =.158
REDUCTION M.S.=.079
RESIDUAL S.S.=.841   DF=231   RESIDUAL M.S.=.004
F FOR QUADRATIC FIT=21.681   P<.01
REDUCTION DUE TO QUADRATIC TERM ALONE, WITH
   1 DF, =.022
F FOR QUADRATIC TERM ALONE=6.159   .01<P<.05

REDUCTION DUE TO GENERAL CUBIC FIT WITH DF 3,=.166
REDUCTION M.S.=.055
RESIDUAL S.S.=.834   DF=230   RESIDUAL M.S.=.004
F FOR GENERAL CUBIC FIT=15.256, p<.01
REDUCTION DUE TO CUBIC TERM ALONE WITH 1 DF=.008
F FOR CUBIC TERM ALONE=2.185   P>.05

Y VALUES FOR STEPS ON X, FOR THREE POLYNOMIALS

X	LINEAR	QUADRATIC	CUBIC
20.00	93.576	93.71	92.7
22.00	94.307	93.93	93.0
24.00	95.039	94.25	93.5
26.00	95.770	94.69	94.2
28.00	96.501	95.23	95.1
30.00	97.232	95.88	96.0
32.00	97.963	96.64	97.1
34.00	98.694	97.51	98.1
36.00	99.425	98.48	99.2
38.00	100.157	99.57	100.3
40.00	100.888	100.76	101.4
42.00	101.619	102.07	102.3
44.00	102.350	103.48	103.2
46.00	103.081	105.00	103.9
48.00	103.812	106.63	104.4
50.00	104.543	108.36	104.7
52.00	105.274	110.21	104.8
54.00	106.006	121.16	104.6

effect is garnered by fitting a quadratic term. The regression coefficient changes from .37 for the linear fit to .40 for the quadratic fit. The cubic term is not significant. The actual curvilinear trend in the data, while very slight, runs against the hypothesis, since the results are that:

(1) The poorest readers (RDG score of 20) are predicted slightly higher on SEE by the quadratic fit than by the linear fit;

(2) The poor and middling readers (RDG scores of 22–40) are predicted lower on SEE by the quadratic fit than by the linear fit;

(3) The best readers (RDG scores of 42–54) are predicted higher on SEE by the quadratic fit than by the linear fit.

This outcome is typical of outcomes in trait and factor psychology, where quadratic regression seldom makes an important improvement over linear regression.

3.9 EXERCISES

1. Multiple correlation can often be used to develop interesting comparisons of the predictability of a criterion from two different predictor bases. Using the one percent sample of TALENT 12th grade males of the Appendix, compute (1) the multiple regression of variable 20, SEE (socioeconomic environment) on the space of variables 9–16, the abilities variables, and (2) SEE on the space of variables 18–19, the interests variables, and write an interpretation of the results.

2. Our example for polynomial fitting employed RDG as the independent variable and SEE as the dependent variable. It would be interesting to run the analysis in the other direction. Using the Appendix TALENT data, compute polynomials on the SEE variable for variable 12, RDG as the criterion. How will you get POLY to take the first variable read from the data deck as Y and the second variable read as X?

3. If A is a rectangular matrix of order 6×3, and B is the transpose of A, is it possible to obtain the inverse of the product [AB]?

4. Given:

$$R = \begin{bmatrix} 1.00 & .50 & .30 \\ .50 & 1.00 & .20 \\ .30 & .20 & 1.00 \end{bmatrix}$$

$$|R| = .68 \qquad N = 50$$

   (a) What is the minor of the element r_{23} in R?

   (b) What is the cofactor of r_{23}?

(c)   What value does r_{23} assume in $\mathbf{R}^{-1}$?

(d)   What is the rank of $\mathbf{R}$?

(e)   Partitioning $\mathbf{R}$, the multiple correlation $R_{3.21}$ was found to be .31. What is the probability that the null hypothesis $\rho_{3.21} = 0$ is true?

5.   In the POLY program, notice how the accumulations loop ending on statement 908 forms values of X^2 and X^3 and accumulations in terms of these constructed values, so that only two scores are read in for each subject (X and Y), yet a fourth order correlation matrix is developed. Can you trace the path through the program that brings control to statement 100 three times, so that first, second, and third degree polynomials get fitted by multiple regression and tested by analysis of variance?

||||||||||||||||||||||||||||||||||||||||||||||||||||||||||||||||

CHAPTER FOUR

Principal Components

4.1 MATHEMATICS OF PRINCIPAL COMPONENTS

Intercorrelations among the elements of a vector variable are the bane of the multivariate researcher's struggle for meaning. The number of parameters to be estimated for a p-element vector is

p	means
p	variances
$[(p^2 - p)/2]$	covariances
$[(p^2 - p)/2 + 2p$	parameters

If p is two, there are only five parameters. If p is ten, there are 65 parameters. If p is 30, there are 495 parameters to be estimated *and to interpret*. None of these may be ignored. Project TALENT, for example, has 60 different abilities test scores for each subject, and this vector is only part of the measurement base. The multivariate normal population underlying each sex-grade sample (e.g., ninth grade males) has 1890 parameters to be considered in any statement about these abilities. Since there are eight sex-grade samples, you can see that it is quite a challenge to make a general statement from the 60 Project TALENT scales about the abilities of the American high school student.

We have already seen in Chapter Three that there is a technical problem in small sample research using prediction schemes such as multiple regression when the predictor battery is composed of correlated elements. All of our multivariate prediction models involve the inverse of some form of the predictor dispersion, and this has as many elements that are subject to sampling fluctuations as the dispersion itself. The **b** weights of a multivariate prediction function appear to be a simple representation of the predictive values of the elements of the predictor variable, but actually the **b**'s are conditioned by all the covariance estimates in the system and must be interpreted in the light of them. Practically, this means that the **b**'s are almost uninterpretable.

Consider the virtues of a vector variable composed of uncorrelated elements. Now there are only $2p$ parameters (means plus variances) to be estimated and interpreted. Moreover, the relation of any single element of the vector variable to any external criterion variable is independent of the relation of any other element to that criterion, and can be interpreted separately and unambiguously. Given the correlations of the predictor elements with a univariate criterion arranged in the column vector $\mathbf{R}_{12}$, we have

$$\mathbf{b} = \mathbf{R}_{11}^{-1}\mathbf{R}_{12} = \mathbf{I}^{-1}\mathbf{R}_{12} = \mathbf{I}\mathbf{R}_{12} = \mathbf{R}_{12}$$

since the relationship

$$\mathbf{R}_{11} = \mathbf{I}$$

expresses the uncorrelatedness of the elements of the predictor variable. Furthermore, the squared multiple correlation coefficient is simply the sum of squares of the predictor element-criterion correlations:

$$R^2 = \mathbf{b}'\mathbf{R}_{12} = \mathbf{R}_{12}'\mathbf{R}_{12}$$

or

$$R^2 = r_{1_p}^{\ 2} + r_{2_p}^{\ 2} + \cdots + r_{p-1,\,p}^2$$

Thus the total variance explained by the multiple regression is partitioned into a set of independent components of variance, each due to one of the predictor elements. This is a truly elegant state of affairs.

Unfortunately, data seldom arrive in uncorrelated packages. When several or many different observation schemes are used to collect data on human subjects it turns out that the elements of the observation or score vector are intercorrelated. A living human being is a highly integrated system, all the overt characteristics and behaviors of which are interrelated. *If we want an uncorrelated vector variable we have to construct it by transforming the data.*

In Chapter Two we studied a general linear transformation of the standardized vector variable $\mathbf{z}$ of the form

$$\mathbf{y} = \mathbf{V}'\mathbf{z}$$

where $\mathbf{V}$ is a $p \times n$ coefficients matrix that carries the p-element variable $\mathbf{z}$ into the derived n-element variable $\mathbf{y}$. We saw that the centroid of $\mathbf{y}$ is

$$\mathbf{m}_y = \mathbf{V}'\mathbf{m}_z = \mathbf{0}$$

because $\mathbf{m}_z = \mathbf{0}$, and the dispersion of $\mathbf{y}$ is

$$\mathbf{D}_y = \mathbf{V}'\mathbf{D}_z\mathbf{V} = \mathbf{V}'\mathbf{R}\mathbf{V}$$

where $\mathbf{R}$ is the correlation matrix for $\mathbf{z}$, since the dispersion of a standardized variable is a correlation matrix.

To produce a transformation vector for $\mathbf{y}$ for which the elements are un-correlated is the same as saying that we want a $\mathbf{V}$ such that $\mathbf{D}_y$ is a diagonal matrix. That is, all the off-diagonal elements of $\mathbf{D}_y$ must be zero. Mathematicians call this an orthogonalizing transformation, and they know that there is an infinity of values for $\mathbf{V}$ that will yield a diagonal $\mathbf{D}_y$ for any given correlation matrix $\mathbf{R}$. Thus the mathematical problem "find a unique $\mathbf{V}$ such that $\mathbf{D}_y$ is diagonal" is insoluble as it stands. Some further restriction must be imposed on the problem.

Karl Pearson had a notion of a good solution to this problem as early as 1902, when he had just given Galton's correlation measure its appropriate and enduring algebraic definition as the product-moment correlation coefficient. He thought that the variances of the leading elements of $\mathbf{y}$ should be maximized. Some years later Truman Lee Kelley pondered the problem and mustered a variety of arguments for a variance-maximizing solution in psychological and educational research. His views are recapitulated by Lohnes (1966, Ch. 1). Kelley persuaded Harold Hotelling to work on the matter, and Hotelling (1933) derived the "principal components" solution which we now present. The matrix formulation of the calculus problem we borrow from T. W. Anderson (1958, Ch. 11).

For the first principal component, which will be the first element of $\mathbf{y}$, and will be defined by the coefficients in the first column of $\mathbf{V}$ (here denoted $\mathbf{v}_1$), we want a solution such that the variance of $\mathbf{y}_1$ will be maximized. Of course, this requirement has no meaning unless we place some restriction on the numbers in $\mathbf{v}_1$, so we will require that the sum of squares of the coefficients be equal to unity. What we want is

$$\frac{1}{N} \sum_{i=1}^{N} y_{1i}^2 \bigg|_{\text{max}}$$

where

$$y_{1i} = \mathbf{v}_1'\mathbf{z}_i$$

and

$$\mathbf{v}_1'\mathbf{v}_1 = 1 \ (\text{"normalizing"} \ \mathbf{v}_1)$$

But

$$\frac{1}{N} \sum_{i=1}^{N} y_{1i}{}^{2} = \frac{1}{N} \sum_{i=1}^{N} (\mathbf{v}_1{'}\mathbf{z}_i)^2$$

$$= \frac{1}{N} \sum (\mathbf{v}_1{'}\mathbf{z}_i)(\mathbf{z}_i{'}\mathbf{v}_1)$$

$$= \mathbf{v}_1{'} \frac{1}{N} \sum (\mathbf{z}_i \mathbf{z}_i{'})\mathbf{v}_1$$

$$= \mathbf{v}_1{'}\mathbf{R}\mathbf{v}_1$$

Therefore, we want to maximize $\mathbf{v}_1{'}\mathbf{R}\mathbf{v}_1$ subject to $\mathbf{v}_1{'}\mathbf{v}_1 = 1$. Let

$$\phi_1 = \mathbf{v}_1{'}\mathbf{R}\mathbf{v}_1 - \lambda_1(\mathbf{v}_1{'}\mathbf{v}_1 - 1)$$

introducing the restriction on $\mathbf{v}_1$ to the function to be maximized via the Lagrange multiplier λ_1. The vector of partial derivatives is

$$\frac{\partial \phi_1}{\partial \mathbf{v}_1} = 2\mathbf{R}\mathbf{v}_1 - 2\lambda_1\mathbf{v}_1$$

(see T. W. Anderson, 1958, page 347, Theorem 8) and, setting equal to zero, dividing out 2, and factoring gives

$$(\mathbf{R} - \lambda_1\mathbf{I})\mathbf{v}_1 = 0 \qquad (4.1.1)$$

This is recognizable as the problem of the eigenstructure of $\mathbf{R}$.

Since equation 4.1.1 is so basic it might be useful to spell it out in more detail. The partial differentiation resulted in a set of p homogeneous equations. The resulting p equations may be written

$$
\begin{aligned}
v_1(1 - \lambda) + \quad & v_2 r_{12} + \quad && v_3 r_{13} && + \ldots + \quad v_p r_{1p} && = 0 \\
v_1 r_{21} + \quad & v_2(1 - \lambda) + \quad && v_3 r_{23} && + \ldots + \quad v_p r_{2p} && = 0 \\
v_1 r_{31} + \quad & v_2 r_{32} + \quad && v_3(1 - \lambda) + && \ldots + \quad v_p r_{3p} && = 0 \\
\ \ \vdots \quad & \qquad \vdots \qquad && \quad \vdots && && \\
v_1 r_{p1} + \quad & v_2 r_{p2} + \quad && v_3 r_{p3} && + \ldots + v_p(1 - \lambda) && = 0
\end{aligned}
$$

These equations can also be written as matrices as follows:

$$
\begin{bmatrix}
(1 - \lambda_i) & r_{12} & r_{13} & \cdots & r_{1p} \\
r_{21} & (1 - \lambda_i) & r_{23} & \cdots & r_{2p} \\
r_{31} & r_{32} & (1 - \lambda_i) & \cdots & r_{3p} \\
\vdots & \vdots & \vdots & & \vdots \\
r_{p1} & r_{p2} & r_{p3} & \cdots & (1 - \lambda_i)
\end{bmatrix}
\cdot
\begin{bmatrix}
v_{1i} \\
v_{2i} \\
v_{3i} \\
\vdots \\
v_{pi}
\end{bmatrix}
=
\begin{bmatrix}
0 \\
0 \\
0 \\
\vdots \\
0
\end{bmatrix}
$$

Finally, in matrix notation this becomes equation 4.1.1.

The characteristic equation of $\mathbf{R}$ is a polynomial of degree p gotten by expanding the determinant of

$$|\mathbf{R} - \lambda\mathbf{I}| = 0 \tag{4.1.2}$$

and solving for the roots λ. Specifically, the largest eigenvalue λ_1 and its associated vector $\mathbf{v}_1$ are required. Letting λ_1 be the largest root of the characteristic equation for $\mathbf{R}$, we substitute it and its vector in equation 4.1.1 which is expanded and then multiplied by $\mathbf{v}_1'$ on the left

$$\mathbf{R}\mathbf{v}_1 - \mathbf{v}_1\lambda_1 = 0$$
$$\mathbf{R}\mathbf{v}_1 = \mathbf{v}_1\lambda_1$$
$$\mathbf{v}_1'\mathbf{R}\mathbf{v}_1 = \mathbf{v}_1'\mathbf{v}_1\lambda_1 = \lambda_1$$

so that λ_1 is clearly the variance of the normalized linear component of $\mathbf{z}$ that has maximum variance.

Next we want to find the normalized linear combination

$$y_2 = \mathbf{v}_2'\mathbf{z}$$

which, out of all the components uncorrelated with y_1, has maximum variance. The statistical restriction of uncorrelatedness of y_1 and y_2 is stated as

$$\frac{1}{N}\sum_{i=1}^{N} y_{2i}y_{1i} = 0$$

where

$$\frac{1}{N}\sum y_{2i}y_{1i} = \frac{1}{N}\sum (\mathbf{v}_2'\mathbf{z}_i)(\mathbf{v}_1'\mathbf{z}_i)$$

$$= \frac{1}{N}\sum (\mathbf{v}_2'\mathbf{z}_i)(\mathbf{z}_i'\mathbf{v}_1)$$

$$= \mathbf{v}_2'\frac{1}{N}\sum (\mathbf{z}_i\mathbf{z}_i')\mathbf{v}_1$$

$$= \mathbf{v}_2'\mathbf{R}\mathbf{v}_1$$

$$= \mathbf{v}_2'\mathbf{v}_1\lambda_1$$

But this can be zero only if

$$\mathbf{v}_2'\mathbf{v}_1 = 0$$

which states the requirement of geometric orthogonality of the eigenvectors.

Introducing a second Lagrange multiplier v_1 to get the orthogonality restriction into the calculus problem, we want to maximize

$$\phi_2 = \mathbf{v_2}'\mathbf{R}\mathbf{v_2} - \lambda_2(\mathbf{v_2}'\mathbf{v_2} - 1) - v_1\mathbf{v_2}'\mathbf{R}\mathbf{v_1}$$

$$\frac{\partial \phi_2}{\partial \mathbf{v_2}} = 2\mathbf{R}\mathbf{v_2} - 2\lambda_2 \mathbf{v_2} - 2v_1\mathbf{R}\mathbf{v_1}$$

and setting equal to zero, dividing out 2, and multiplying on the left by $\mathbf{v_1}'$ gives

$$\mathbf{v_1}'\mathbf{R}\mathbf{v_2} - \lambda_2\mathbf{v_1}'\mathbf{v_2} - v_1\mathbf{v_1}'\mathbf{R}\mathbf{v_1} = 0$$

which reduces to

$$- v_1\mathbf{v_1}'\mathbf{R}\mathbf{v_1} = 0$$

showing that v_1 must equal zero and λ_2 and $\mathbf{v_2}$ must be the second root of the characteristic equation 4.1.2 and its associated vector. This process may be continued to get $p - 1$ maximal variances λ_j for components

$$y_j = \mathbf{v_j}'\mathbf{z}$$

The last principal component, associated with the smallest eigenvalue, has no degrees of freedom, since

$$\sum_{j=1}^{p} \lambda_j = \text{trace}(\mathbf{R}) = p$$

Another fixed relationship is

$$\prod_{j=1}^{p} \lambda_j = |\mathbf{R}|$$

Letting $\mathbf{L}$ be a diagonal matrix with λ_j in the jth position on the diagonal, the full eigenstructure of $\mathbf{R}$ is given as

$$\mathbf{RV} = \mathbf{VL}$$

where

$$\mathbf{V}'\mathbf{V} = \mathbf{VV}' = \mathbf{I}$$

and

$$\mathbf{V}'\mathbf{RV} = \mathbf{L} = \mathbf{D}_y$$

In PRINCO we compute

$$\mathbf{RV} = \mathbf{VL}$$

so that λ_1 is the maximum value the variance of a component of $\mathbf{z}$ can have, subject to the normalization of the vector $\mathbf{v}_1$. Then λ_2 is a maximum given λ_1, and so on for the remaining eigenvalues. Each component has maximum variance out of all possible normalized linear functions statistically uncorrelated and geometrically orthogonal to the ones preceding it. This is the remarkable property of the principal components. T. L. Kelley observed that it is a valuable property to have in factors, if there are no more pressing considerations. In Kelley's day there were no more pressing considerations, so he accepted the principal components as research solutions. In the next chapter, we discuss the considerations that lead many researchers today to rotate the principal components into other factor solutions.

One other remarkable feature of principal components is that while the complete set of p components will exactly reproduce the correlation matrix,

$$\mathbf{R} = \mathbf{VLV'}$$

and thus accounts for all the variance in the vector variable $\mathbf{z}$, it is possible to retain in a research solution only the first n of the components, with confidence that these n factors extract more of the variance of $\mathbf{z}$ than any other set of n orthogonal factors would. This is of immense importance. If for reasons of parsimony we want to reduce the number of variables in our research from p correlated variables in $\mathbf{z}$ to $n < p$ uncorrelated variables in y, the n derived variables that will retain as much of the variance of $\mathbf{z}$ as possible are the first n principal components (or any rigid orthogonal rotation of them).

When y has fewer elements than $\mathbf{z}$, then the *rank* of the measurement space has been reduced from p to n, and we have to characterize the n principal components as *a reduced rank model for the data*. This n-rank model has discarded some of the available variance of $\mathbf{z}$, since the discarded roots from λ_{n+1} to λ_p are all nonzero. Note too that for $n < p$ the $p \times n$ coefficients matrix $\mathbf{V}$ has the property that $\mathbf{V'V} = \mathbf{I}$, but does *not* have the property that $\mathbf{VV'} = \mathbf{I}$. For the reduced rank model, we say that the coefficients matrix is column orthogonal but is not orthonormal.

By using the eigenstructure notation $\mathbf{L}$ and $\mathbf{V}$ for the sample space model we have been hiding from the issue of errors in estimates of $\mathbf{L}$ and $\mathbf{V}$ as computed from $\mathbf{R}$. Our general approach is to plan to get a very good estimate of $\mathbf{R}$ from a large and representative sample of subjects so as to minimize errors of estimation in the principal components. Most of the statistical literature on testing hypotheses about elements of $\mathbf{L}$ and $\mathbf{V}$ is beyond the level of this text (see T. W. Anderson and H. Rubin, 1956, for a start). One rudimentary precaution we do urge researchers using small samples to take to is apply the *sphericity test* to $\mathbf{R}$. This is a test of the null hypothesis that the population correlation matrix is an identity matrix. That is, the null hypothesis states that the elements of z are already uncorrelated in the population from which the sample was drawn, and the observed correlations in **R** differ from zero only by chance. If so, the population swarm in the space of the standardized variates is spherical, thus the name of the test. For any equi-density surface in a mapping of points in a space of uncorrelated variates of unit variance, the shape is a sphere.

Bartlett (1950) originated the sphericity test. It is perhaps the simplest example of a test based on Wilks' (1932) notion of taking the determinant of a dispersion matrix as the multivariate analog of the variance. Wilks called his determinant index *the generalized variance*. If we let **P** stand for the population correlation matrix, for standardized variates z, $|\mathbf{P}|$ is the generalized variance and $|\mathbf{R}|$ is its estimator. What Bartlett has contrived is an approximate chi-square test of $H_0 : |\mathbf{P}| = 1$. Of course, $|\mathbf{P}| = 1$ only if $\mathbf{P} = \mathbf{I}$. The test criterion is

$$\chi^2_{.5(p^2 - p)} = -[(N - 1) - \tfrac{1}{6}(2p + 5)]\log_e |\mathbf{R}| \qquad (4.1.3)$$

To see the "reasonableness" of Bartlett's test criterion, consider the following while examining the equation 4.1.3:

(1) The $|\mathbf{R}|$ formed from $Z'Z(1/N)$ will always be in the range 1.00 to 0.00; as $|\mathbf{R}| \to 1$, $\mathbf{R} \to I$; as $R \to I$, the dispersion approaches sphericity ($0 \le |\mathbf{R}| \le 1.00$).

(2) $\log_e$ of decimal fractions (in the range 0.00 to 1.00) are negative and have the range $-\infty$ to -0.00; as $|\mathbf{R}| \to 1.0$ then $\log_e |\mathbf{R}| \to 0.0$ and $\chi^2 \to 0.0$.

(3) a large N with respect to p will increase the χ^2 for a given $|\mathbf{R}|$; therefore a high N/p ratio increases power of test.

A nonsignificant χ^2 at some reasonable α level should lead to a decision not to factor the matrix, since the vector variable may already be treated as a set of uncorrelated elements. Knapp and Swoyer (1967) have reported on a Monte Carlo study of the power of this test. They find the power to be quite high, so that for $N = 20$, $p = 10$, and $\alpha = .05$ one is virtually certain to reject H_0 when the elements of z are intercorrelated .36 or more, and for $N = 200$, $p = 10$, and $\alpha = .05$ one is virtually certain to reject H_0 when the elements of z intercorrelate .09 or more. Their paper is a nice example of how operating characteristics of multivariate inference tests can be researched empirically.

A helpful way of conceptualizing the principal components is to think of them as factoring the correlation matrix into orthogonal and additive matrices. Each such matrix is of the form

$$\mathbf{R}_j = \lambda_j \mathbf{v}_j \mathbf{v}_j{'}$$

and

$$\mathbf{R} = \lambda_1 \mathbf{v}_1 \mathbf{v}_1' + \lambda_2 \mathbf{v}_2 \mathbf{v}_2' + \cdots + \lambda_p \mathbf{v}_p \mathbf{v}_p'$$
$$= \mathbf{R}_1 + \mathbf{R}_2 + \cdots + \mathbf{R}_p$$

This is the sense in which the full set of principal components completely "explains" the correlation matrix. Since

$$\mathbf{R} = \sum_{j=1}^{p} \lambda_j \mathbf{v}_j \mathbf{v}_j' = \sum_{j=1}^{p} \mathbf{R}_j$$

corresponds to $\mathbf{R} = \mathbf{VLV}'$ and provides a complete "theory" for $\mathbf{R}$. Each layer or matrix in the summation may be thought of as a *component theory matrix*, as it is the part of $\mathbf{R}$ accounted for by one principal component. When we decide upon a reduced rank model for $\mathbf{R}$, we actually select the first n of the $\mathbf{R}_j$ as component theory matrices and relegate the last $p - n$ of the $\mathbf{R}_j$ as error matrices. Defining our *theory* matrix as

$$\hat{\mathbf{R}} = \mathbf{R}_1 + \mathbf{R}_2 + \cdots + \mathbf{R}_n$$

and our *error* or *residual* matrix as

$$\tilde{\mathbf{R}} = \mathbf{R}_{n+1} + \mathbf{R}_{n+2} + \cdots + \mathbf{R}_p$$

we have the following theory plus error partition of $\mathbf{R}$:

$$\mathbf{R} = \hat{\mathbf{R}} + \tilde{\mathbf{R}}$$

The rank of $\hat{\mathbf{R}}$ is n and the rank of $\tilde{\mathbf{R}}$ is $p - n$. We call $\hat{\mathbf{R}}$ the theory matrix because it is what the intercorrelations among the elements of $\mathbf{z}$ would be if $\mathbf{z}$ were entirely explained by n principal components.

How do we select a value of n in actual research? One way is to look at the distribution of residual correlations in $\tilde{\mathbf{R}}$ for a given n and see if the values are suitably small and nicely distributed. We would like "errors" to be approximately normally distributed around a mean of zero. How much of a standard deviation can be tolerated for such errors is a matter of individual judgment. The larger the sample size, N, the tighter the theory for the data should be, presumably. Kaiser (1966) has expressed a variety of compelling arguments for a rule of thumb that takes as a value of n the number of eigenvalues larger than unity, and his rule seems to work well when N is small or moderate. For very large samples it may be worthwhile to take n larger than Kaiser's rule would suggest. We employ a routine called RESID to display the distribution of residual correlations following Kaiser's rule. If we feel that we want a tighter fit to the data we then specify a larger n and recompute the residuals. The solution that makes us happiest usually has to be found by trial.

In small sample research, Bartlett's sphericity test can be extended to protect against extracting too many components. Bartlett shows how to test the null hypothesis that the determinant of the population residual matrix after extraction of n components is zero, indicating that factoring should stop. After components corresponding to roots $\lambda_1, \lambda_2, \ldots, \lambda_n$ have been extracted

$$\chi^2_{.5(p-n)(p-n-1)} = -[(N-1) - \tfrac{1}{6}(2p+5) - \tfrac{2}{3}n]\log_e X_{p-n}$$

where

$$X_{p-n} = \frac{|\mathbf{R}|}{\{[\prod_{j=1}^{n} \lambda_j[(p - \sum_{j=1}^{n} \lambda_j)/(p-n)]^{p-n}\}}$$

Our PRINCO program computes this chi-square for all values of n from 0 to $p - 1$.

We pointed out in Chapter Two that it is useful to scale any transformation $\mathbf{y}$ of a vector variable $\mathbf{z}$ so that its elements have zero means and unit variances. Such a standardized transformation is called a factoring of $\mathbf{z}$, or of $\mathbf{R}$, and each linear component of the transformation is called a *factor*. The principal components already have zero means, but their nonunit variances are arrayed in the diagonal elements of $\mathbf{L}$. Since y_j has variance λ_j the principal factor may be obtained as

$$f_j = \frac{y_j}{\sqrt{\lambda_j}}$$

or for all factors

$$\mathbf{f} = \mathbf{L}^{-1/2}\mathbf{y}$$

But since

$$y_j = \mathbf{v}_j'\mathbf{z}$$

we have

$$f_j = \frac{\mathbf{v}_j'\mathbf{z}}{\sqrt{\lambda_j}}$$

and since

$$\mathbf{y} = \mathbf{V}'\mathbf{z}$$

we have

$$\mathbf{f} = \mathbf{L}^{-1/2}\mathbf{V}'\mathbf{z} = \mathbf{B}'\mathbf{z}$$

where

$$\mathbf{B} = \mathbf{VL}^{-1/2} \tag{4.1.4}$$

The matrix $\mathbf{B}$ is then the matrix of *factor score coefficients* for principal factors.

The greatest interest in any factor solution centers on the correlations between the original variables and the factors. The matrix of such test-factor correlations is called the *factor structure*. It is the primary interpretative device in principal components analysis, just as it is in any multivariate analysis which results in a factoring of a measurement battery. We have already seen its usefulness in regression analysis as a way of displaying the regression functions as factors of the predictor battery. In the factor structure the element r_{jk} gives the correlation of the jth test with the kth factor. Assuming that the content of the observation variables is well known, the correlations in the kth column of the structure help in interpreting, and perhaps naming, the kth factor. Also, the coefficients in the jth row give the best view of the factor composition of the jth test.

The derivation of the factor structure $\mathbf{S}$ is as follows:

$$\mathbf{S} = \frac{1}{N} \sum_{i=1}^{N} (\mathbf{z}_i - \mathbf{m}_z)(\mathbf{f}_i - \mathbf{m}_f)'$$

$$= \frac{1}{N} \sum \mathbf{z}_i \mathbf{f}_i'$$

$$= \frac{1}{N} \sum \mathbf{z}_i (\mathbf{L}^{-1/2}\mathbf{V}'\mathbf{z}_i)'$$

$$= \frac{1}{N} \sum (\mathbf{z}_i \mathbf{z}_i')\mathbf{VL}^{-1/2}$$

$$= \mathbf{RVL}^{-1/2}$$

and since

$$\mathbf{RV} = \mathbf{VL}$$

$$\mathbf{S} = \mathbf{VLL}^{-1/2} = \mathbf{VL}^{1/2}$$

Another set of coefficients of interest in factor analysis is the weights that compound predicted observations $\hat{\mathbf{z}}$ from factor scores $\mathbf{f}$. These regression coefficients for the multiple regression of each element of the observation vector $\mathbf{z}$ on the factor f are called factor loadings, and the matrix $\mathbf{A}$ that contains them as its rows is called the *factor pattern*. The multiple regression for the jth test and the ith subject is

$$\hat{z}_{ji} = \mathbf{a}_j * f_i$$

Note that the somewhat strained notation $\mathbf{a}_j*$ is intended to denote the jth row of the factor pattern matrix $\mathbf{A}$. Since the intercorrelations among the factors are all zero and the factors are standardized, the predictor correlation matrix is an identity matrix. Given the correlations of the n factors with the jth test in the row vector $\mathbf{r}^*_{z_{jf}}$ we have

$$\mathbf{a}_j* = \mathbf{r}^*_{z_{jf}} \mathbf{I}^{-1} = \mathbf{r}^*_{z_{jf}}$$

Putting the p row vectors $r^*_{z_{jf}}$ together in the matrix $\mathbf{S}$ of test-factor correlations, or the structure matrix, we get the full matrix of factor loadings

$$\mathbf{A} = \mathbf{SI}^{-1} = \mathbf{S} = \mathbf{VL}^{1/2}$$

Thus we see that for principal components the factor pattern $\mathbf{A}$ is the same as the factor structure $\mathbf{S}$. As a matter of fact, the same result holds for any orthogonal factor solution, although the general proof is different. It is remarkably parsimonious that the same matrix $\mathbf{S}$ should express both the theoretical composition of the tests ("explaining" the tests) and the correlations of the factors with the tests ("explaining" the factors). We now show that this same matrix is also the source of the "explanation" of the correlations among the tests.

Let us see what the theory partition of the correlation matrix is.

$$\hat{\mathbf{R}} = \frac{1}{N} \sum_{i=1}^{N} (\hat{z}_i - \mathbf{m}_z)(\hat{z}_i - \mathbf{m}_z)'$$

$$= \frac{1}{N} \sum \hat{z}_i \hat{z}_i'$$

$$= \frac{1}{N} \sum (\mathbf{Sf}_i)(\mathbf{Sf}_i)'$$

$$= \frac{1}{N} \sum (\mathbf{Sf}_i \mathbf{f}_i' \mathbf{S}')$$

$$= \mathbf{S} \frac{1}{N} \sum (\mathbf{f}_i \mathbf{f}_i') \mathbf{S}'$$

$$= \mathbf{SIS}' = \mathbf{SS}' \tag{4.1.5}$$

It is important to note that $\hat{\mathbf{R}}$ is a theoretical partition of $\mathbf{R}$, and not really a correlation matrix itself, since it does not have unities in its diagonal. In the derivation above,

$$\frac{1}{N} \sum_{i=1}^{N} \hat{z}_{ji}^{2} < 1 \qquad \text{for} \quad j = 1, 2, \ldots, p$$

assuming the reduced rank model. Technically, we partition a correlation matrix into two or more orthogonal, additive variance-covariance matrices. The elements of $\hat{\mathbf{R}}$ and $\tilde{\mathbf{R}}$ are technically partial variances and covariances, and we speak of the theory matrix and the error or residual matrix only to be descriptive. Any such orthogonal partition of $\mathbf{R}$ can be scaled into a legitimate correlation matrix by dividing the j,kth element by the square root of the product of the jth and the kth diagonal elements, but then the partitions will not be additive. Note, too, that if the full rank model is used, so that all p principal components are retained and $\mathbf{S}$ is p-square, then and only then will $\hat{\mathbf{R}} = \mathbf{R}$, with unities in the diagonal of the theory matrix, and then $\tilde{\mathbf{R}} = 0$. However, the full rank model will almost never be employed by the researcher.

Combining the results we have demonstrated we see that

$$\hat{\mathbf{R}} = \mathbf{VLV}' = \mathbf{SS}' = \sum_{j=1}^{n} \mathbf{R}_{j}$$

where

$$\mathbf{R}_{j} = \lambda_{j} \mathbf{v}_{j} \mathbf{v}_{j}' = \mathbf{s}_{j} \mathbf{s}_{j}'$$

and

$$\tilde{\mathbf{R}} = \mathbf{R} - \hat{\mathbf{R}} = \mathbf{R} - \mathbf{SS}'$$

(Remember that in our notation $\mathbf{v}_{j}$ and $\mathbf{s}_{j}$ are the jth column vectors from the matrices $\mathbf{V}$ and $\mathbf{S}$ respectively.)

How much of the variance in the p-element test battery is accounted for by this jth layer of the factor theory? It is useful to observe that p, the order and trace of $\mathbf{R}$, is the maximum variance a factor could extract, and use p as a yardstick for evaluating λ_{j}. (Yes, λ_{1} could equal p, in which extreme case all the other eigenvalues would equal zero. This would mean that $\mathbf{z}$ was already a pure unifactor measure with no error component, however, and it is not to be expected.) What we do is to divide λ_j by p to express the proportion of the total available variance that is extracted by the jth principal factor, and call this ratio P_{v_j}. This is a nicely scaled index since it sums to unity for the full rank model:

$$\sum_{j=1}^{p} \frac{\lambda_j}{p} = 1$$

For the reduced rank model

$$P_v = \sum_{j=1}^{n} P_{v_j} = \sum_{j=1}^{n} \frac{\lambda_j}{p}$$

gives the proportion of the total available variance in the test battery that is retained by the model for the data. Since

$$\lambda_j = \sum_{k=1}^{p} s_{kj}^{2}$$

the proportion of variance extracted by a factor can be computed by summing the squares of a column of the structure matrix S, and the overall proportion of battery variance retained by the reduced rank model can be computed from the sum of these column sums of squares for S

$$P_v = \sum_{j=1}^{n} \left(\frac{\sum_{k=1}^{p} s_{kj}^{2}}{p} \right)$$

The row sum of squares for S is also interesting because it expresses the proportion of the variance of a test (or element of the vector variable z) that is extracted by all n factors. This value is called the *achieved communality* for the test, and is defined as

$$h_k^{2} = \sum_{j=1}^{n} s_{kj}^{2}$$

The square of the correlation of test k with factor j gives the part of the variance of the test accounted for by that factor, and the sum of these squares for n factors is the communality, or explained variance, for the test.

4.2 NUMERICAL EXAMPLE OF PRINCIPAL COMPONENTS

Starting with the data matrix $\mathbf{X}$, we first form $\mathbf{R}$ following the procedures outlined in Section 2.4.

$$\mathbf{X} = \begin{bmatrix} 7 & 4 & 3 \\ 4 & 1 & 8 \\ 6 & 3 & 5 \\ 8 & 6 & 1 \\ 8 & 5 & 7 \\ 7 & 2 & 9 \\ 5 & 3 & 3 \\ 9 & 5 & 8 \\ 7 & 4 & 5 \\ 8 & 2 & 2 \end{bmatrix} \rightarrow \mathbf{R} = \begin{bmatrix} 1.00 & .67 & -.10 \\ .67 & 1.00 & -.29 \\ -.10 & -.29 & 1.00 \end{bmatrix}$$

Solving for the roots[1] of $\mathbf{R}$, we obtain $\lambda_1 = 1.769$, $\lambda_2 = .927$ and $\lambda_3 = .304$. Each of these roots satisfies the equation 4.1.2 $|\mathbf{R} - \lambda_i \mathbf{I}| = 0$. The sum of the roots (3.000) is equal to the trace of $\mathbf{R}$, and the continued product, $\prod_{i=1}^{3} \lambda_i = .499$, is the determinant of $\mathbf{R}$.

To test if there are significant relationships among these three variables, Bartlett's sphericity test is applied. Starting with equation 4.1.3

$$\chi^2 = -(9 - 1 - \tfrac{5}{6})\log_e .499$$
$$= -(7.16)(-.695)$$
$$= 4.99 \quad \text{where} \quad ndf = \tfrac{1}{2}(3^2 - 3) = 3$$

Since a χ^2 of 4.99 with 3 degrees of freedom is not even significant at the .10 level, interest in accounting for the relations among these three variables would normally cease, since we cannot reject the null hypothesis that $\mathbf{P} = \mathbf{I}$. However, for the purpose of this numerical example, we proceed to describe the other calculations involved in principal components analysis.

Substituting $\lambda_1 = 1.769$ and $\mathbf{R}$ into equation 4.1.1 we obtain,

$$\begin{bmatrix} -.769 & .67 & -.10 \\ .67 & -.769 & -.29 \\ -.10 & -.29 & -.769 \end{bmatrix} \cdot \begin{bmatrix} v_{11} \\ v_{21} \\ v_{31} \end{bmatrix} = \begin{bmatrix} 0 \\ 0 \\ 0 \end{bmatrix}$$

[1] The actual solution will be described in Section 4.3.

which is the matrix expression for three homogeneous equations with three unknowns. This yields a nontrivial solution for v since $[\mathbf{R} - \lambda\mathbf{I}]$ is not of full rank. Repeating this operation for the second and third roots then yields the matrix $\mathbf{V}$, containing the resulting three normalized vectors as columns:

$$\mathbf{V} = \begin{bmatrix} .64 & .38 & -.66 \\ .69 & .10 & .72 \\ -.34 & .91 & .20 \end{bmatrix}$$

Notice that if you premultiply $\mathbf{V}$ by its transpose, an identity matrix results. That is, $\mathbf{V}'\mathbf{V} = \mathbf{I}$.

Forming the matrix $\mathbf{L}^{1/2}$, a diagonal whose matrix elements are the square roots of the eigenvalues of $\mathbf{R}$, we can then obtain $\mathbf{S}$, the factor structure, from $\mathbf{V}$, using $\mathbf{L}^{1/2}\mathbf{V} = \mathbf{S}$.

$$\begin{bmatrix} 1.33 & 0 & 0 \\ 0 & .96 & 0 \\ 0 & 0 & .55 \end{bmatrix} \cdot \begin{bmatrix} .64 & .38 & -.66 \\ .69 & .10 & .72 \\ -.34 & .91 & .20 \end{bmatrix} = \begin{bmatrix} .85 & .37 & -.37 \\ .91 & .09 & .40 \\ -.45 & .88 & 11 \end{bmatrix}$$

So, for example, .91 is the correlation between variable 2 and the first principal component. Similarly, the coefficients matrix $\mathbf{B}$ is formed using the reciprocals in the diagonals of $\mathbf{L}^{1/2}$, following (4.1.4).

$$\mathbf{B} = \mathbf{L}^{-1/2}\mathbf{V} = \begin{bmatrix} .48 & .40 & -1.20 \\ .52 & .10 & 1.31 \\ -.26 & .95 & .37 \end{bmatrix}$$

The principal components program reports the square of the multiple correlation between each variable and the others in the set. These are useful in considering how many factors to use in subsequent analyses. Also, they are easily derived from the diagonal of $\mathbf{R}^{-1}$ which is also needed for later computations.

$$\mathbf{R}^{-1} = \begin{bmatrix} 1.84 & -1.28 & -.18 \\ -1.28 & 1.98 & .44 \\ -.18 & .44 & 1.11 \end{bmatrix}$$

$$R^2_{1 \cdot 23} = 1 - \frac{1}{1.84} = .456$$

$$R^2_{2 \cdot 13} = 1 - \frac{1}{1.98} = .496$$

$$R^2_{3 \cdot 12} = 1 - \frac{1}{1.11} = .098$$

TABLE 4.1   Printout of PRINCO Numerical Example

```
R INVERSE
  SECTION 1
  ROW             1              2              3
   1             1.84          -1.28          -0.18
   2            -1.28           1.98           0.44
   3            -0.18           0.44           1.11
```

DETERMINANT OF R=0.4987441E 00

FOR SPHERICITY TEST, CHI SQUARE=4.99 AND NDF=3

FACTOR	EIGEN—VALUE	PERCENT TRACE	CUM PERCENT	N.D.F.	CHI—SQUARE
1	1.7688	59.0	59.0	3	4.99
2	0.9271	30.9	89.9	1.00	0.47
3	0.3041	10.1	100.0	0	0.0

FACTOR PATTERN. FACTORS ARE COLUMNS, TESTS ARE ROWS.

```
  SECTION 1
  ROW          1      2       3
   1          0.85   0.37   -0.37
   2          0.91   0.09    0.40
   3         -0.45   0.88    0.11
```

TEST	COMMUNALITY	MULT R SQUARE
1	1.000	0.456
2	1.000	0.496
3	1.000	0.098

FACTOR SCORE COEFFICIENTS. FACTORS ARE COLUMNS, TESTS ARE ROWS.

```
  SECTION 1
  ROW          1      2       3
   1          0.48   0.40   -1.20
   2          0.52   0.10    1.31
   3         -0.26   0.95    0.37
```

The theory matrix of rank 1 is formed from the first column of **S** following (4.1.5)

$$
\begin{bmatrix} .85 \\ .91 \\ -.45 \end{bmatrix} \cdot [.85 \quad .91 \quad -.45] = \begin{bmatrix} .73 & .78 & -.39 \\ .78 & .83 & -.41 \\ -.39 & -.41 & .21 \end{bmatrix} = \hat{\mathbf{R}}
$$

This estimate of **R** is the three-dimensional variance-covariance matrix for the N points after projecting them from their original position in that 3-space on to the first principal component. Converting $\hat{\mathbf{R}}$ to a correlation matrix

$$
\begin{bmatrix} \dfrac{1}{\sqrt{.73}} & 0 & 0 \\ 0 & \dfrac{1}{\sqrt{.83}} & 0 \\ 0 & 0 & \dfrac{1}{\sqrt{.21}} \end{bmatrix} \cdot \begin{bmatrix} .73 & .78 & -.39 \\ .78 & .83 & -.41 \\ -.39 & -.41 & .21 \end{bmatrix} \cdot \begin{bmatrix} \dfrac{1}{\sqrt{.73}} & 0 & 0 \\ 0 & \dfrac{1}{\sqrt{.83}} & 0 \\ 0 & 0 & \dfrac{1}{\sqrt{.21}} \end{bmatrix}
$$

we obtain

$$
\begin{bmatrix} 1.0 & 1.0 & 1.0 \\ 1.0 & 1.0 & 1.0 \\ 1.0 & 1.0 & 1.0 \end{bmatrix}
$$

which is what one would expect since all the points now lie on a straight line (the principal axis).

The roster of factor scores, **F**, can be obtained from **ZB**, where **Z** is **X** converted to standard score form.

$$
\mathbf{F} = \begin{bmatrix}
.41 & -.69 & .06 \\
-2.11 & .07 & .63 \\
-.46 & -.32 & .30 \\
1.62 & -1.00 & .70 \\
.70 & 1.09 & .65 \\
-.86 & 1.32 & -.85 \\
-.60 & -1.31 & .86 \\
.94 & 1.72 & -.04 \\
.22 & .03 & .34 \\
.15 & -.91 & -2.65
\end{bmatrix}
$$

For example, the first factor score for the first student is .41 and can be formed by:

$$.48\left(\frac{7.0 - 6.9}{1.45}\right) + .52\left(\frac{4.0 - 3.5}{1.50}\right) - .26\left(\frac{3 - 5.1}{2.66}\right) = .41$$

since the mean vector is [6.9 3.5 5.1], the standard deviations are [1.45 1.50 2.66], and $\mathbf{b_1}' = [.48 \quad .52 \quad -.26]$.

Since the three variables represented in matrix $\mathbf{F}$ have zero means and unit variance, $1/N$ times $\mathbf{F'F}$ yields a correlation matrix equal to I.

$$\frac{1}{N}\mathbf{F'F} = \begin{bmatrix} 1.00 & .00 & .00 \\ .00 & 1.00 & .00 \\ .00 & .00 & 1.00 \end{bmatrix}$$

This is expected since the principal components are orthogonal. The computer results for this numerical example are summarized in Table 4.1.

4.3 RESEARCH EXAMPLE

In a study titled "Redundancy in Student Records" Lohnes and Marshall (1965) employed the first two principal components of a 21-element junior high school records vector, composed of 13 standardized test scores and 8 teacher-assigned course marks, to demonstrate that "most of the important information . . . could be included in one general factor." The correlation matrix for the 230 subjects was striking in that all the correlations were positive and most were high. The lowest correlation was .48 between PGAT Number score and eighth-grade science mark. The highest correlation was .81 between MAT language score and seventh-grade English mark. As reported in Table 4.2, the first principal component absorbed 68 percent of the total variance. Every one of the 21 variables correlated positively and above .75 with this g-type factor. The second factor was a bipolar one on which all 13 tests loaded positively and all 8 course marks loaded negatively. This second component accounted for only 6 percent of the total variance, but its obvious interpretation is intriguing. Some students are test pleasers and others are teacher pleasers, relatively. The authors believe that their results suggest that school records could be cleaned up by replacing test scores and course marks with one or a few factor scores.

In the next chapter it will be argued that rotation of the reference axes in the subspace established by principal components analysis is usually desirable.

TABLE 4.2   Principal Component Pattern and Structure Matrix

Test Score	Loading for PC1	Loading for PC2	$h^{2\,a}$	Course Grade	Loading for PC1	Loading for PC2	$h^{2\,a}$
PGAT "Verbal"	.829	.289	.77	Seventh grade			
PGAT "Reasoning	.849	.212	.77	English	.877	−.165	.80
PGAT "Number"	.766	.287	.67	Eighth-grade			
MAT "Word				English	.826	−.397	.84
Knowledge"	.846	.245	.78	Seventh-grade			
MAT "Reading"	.868	.239	.81	arithmetic	.809	−.195	.69
MAT "Spelling"	.835	.025	.70	Eighth-grade			
MAT "Language"	.881	.064	.82	arithmetic	.788	−.174	.65
MAT "Study Skills—				Seventh-grade			
Language"	.825	.071	.69	social studies	.855	−.316	.83
MAT "Arithmetic				Eighth-grade			
Computation"	.864	.035	.75	social studies	.770	−.409	.76
MAT "Arithmetic				Seventh-grade			
Problems"	.852	.153	.75	science	.789	−.281	.70
MAT "Social Studies"	.810	.162	.68	Eighth-grade			
MAT "Study Skills—				science	.757	−.406	.74
Social Studies"	.822	.166	.70				
MAT "Science"	.811	.326	.76				

[a] The communality for the variable, representing the proportion of its variance preserved in the two factors.

However, in this research example the principal components analysis not only established that the rank of the model should be two and the location of the model's rank two subspace of the 21-dimensional measurement space, but also clearly specified in the first two principal factors practically ideal locations for the reference vectors.

4.4   THE PRINCO PROGRAM

This program can perform a complete principal components analysis of the data. That the p factors reported for the pth order input correlation matrix represent a full-rank model for the data can be seen from the communalities of 1.000 reported for all variables in the check problem. The punchout of the full factor pattern and full score coefficients matrices has been arranged so that unwanted factors can be stripped off the decks and discarded easily. The punched **R**-inverse should be saved if there is likely to be interest in computing factor score coefficients for rotated factors (see Chapter Five). Users of smaller, slower computers may use the program to compute and report a reduced-rank model for a number of factors specified on the control card.

```
C         PRINCIPAL COMPONENTS.    A COOLEY-LOHNES PROGRAM.
C
C     THIS PROGRAM COMPUTES A PRINCIPAL COMPONENTS ANALYSIS OF AN
C   M-SQUARE CORRELATION MATRIX.
C
C   INPUT
C
C   1) FIRST TEN CARDS OF DATA DECK DESCRIBE THE PROBLEM IN A TEXT
C   THAT WILL BE REPRODUCED ON THE OUTPUT.   DO NOT USE COLUMN 1.
C   2) CONTROL CARD (CARD 11),     COLS 1-2   M = NUMBER OF VARIABLES
C                                  COLS 3-7   N = NUMBER OF SUBJECTS
C                                  COLS 8-9   L = NUMBER OF FACTORS
C          TO BE EXTRACTED(I.E., DESIRED RANK FOR MODEL)
C                                  COL 10 L1 = 0 IF R IS FULL RANK,
C                                         L1 = 1 IF R IS REDUCED
C          RANK (AS FROM FACTOR PROGRAM).
C   3) FORMAT CARD (CARD 12),  FOR R MATRIX.  AS PUNCHED BY CORREL,
C          FORMAT IS (10X, 7F10.7 / (10X, 7F10.7)).
C   4) UPPER-TRIANGULAR R MATRIX, FROM CORREL OR OTHER SOURCE.
C
C   PUNCHED OUTPUT
C
C   1) R-INVERSE, UPPER-TRIANGULAR, FORMAT (10X, 5E14.7)
C   2) TRANSPOSED FACTOR PATTERN, A-PRIME, FORMAT (10X,7F10.7)
C   3) SCORE COEFFICIENTS, C-PRIME, FORMAT (10X,5E14.7).
C
C   REQUIRED SUBROUTINES ARE MPRINT, MATINV, AND HOW.
C
C
      DIMENSION  TIT(16), R(100,100), A(100,100), W(100), X(100),
     C   Y(100), Z(100), U(100), V(100), T(100), S(100)
C
   1  WRITE (6,2)
   2  FORMAT(55H1PRINCIPAL COMPONENTS.   A COOLEY-LOHNES PROGRAM.        )
      DO 3   J = 1, 10
      READ  (5,4) (TIT(K),    K = 1, 16)
   3  WRITE (6,4) (TIT(K),    K = 1, 16)
   4  FORMAT (16A5)
      READ(5,5) M, N, L
   5  FORMAT(I2,I5,I2)
      WRITE(6,6) M, N, L
   6  FORMAT(3H0  I3, 8H TESTS, I6, 16H SUBJECTS.   RANK I3,8H  MODEL.)
      EM = M
      EL = L
      EN = N
      READ  (5,4) (TIT(J),   J = 1, 16)
      DO 19   J = 1, M
  19  READ  (5,TIT) (R(J,K),   K = J, M)
  18  FORMAT (10X, 7F10.7 / (10X, 7F10.7))
      DO 20   J = 1, M
      DO 20   K = J, M
      A(J,K) = R(J,K)
   7  FORMAT(20H0CORRELATION MATRIX )
      A(K,J) = R(J,K)
  20  R(K,J) = R(J,K)
      WRITE(6,7)
      CALL MPRINT (R, M)
C
      IF (L1) 21,21,271
  21  CALL MATINV (A, M, DET)
      DO 23   J = 1, M
  23  WRITE (7,24) J,   (A(J,K),   K = J, M)
  24  FORMAT(4H ROW I3,3X, 5E14.7 / (10X, 5E14.7))
      WRITE (6,25) DET
  25  FORMAT(20H0DETERMINANT OF R =   E14.7)
C
```

```
          NDF = (M * M - M)/ 2
          CHISQ = -((EN-1.0)-((1.0/6.0)*(2.0*EM+5.0))) * ALOG (DET)
          WRITE (6,26) CHISQ,   NDF
26        FORMAT(36H0FOR SPHERICITY TEST,   CHI-SQUARE =   F9.2,16H    AND N.D.
          CF. =  16)
          DO 27   J = 1, M
27        W(J) = 1.0 - (1.0 / A(J,J))
271       CONTINUE
          CALL HOW (M, 100, L, R, X, A, U, V, Y, Z)
          TEMP = 0.0
          TEMP1 = 0.0
          TEMP2 = 1.0
          DO 28 J = 1, L
          U(J) = (X(J) / EM) * 100.0
          TEMP = TEMP + U(J)
          V(J) = TEMP
          TEMP1 = TEMP1 + X(J)
          S(J) = TEMP1
          TEMP2 = TEMP2 * X(J)
          T(J) = TEMP2
          V(J) = TEMP
          Y(J) = SQRT  (X(J))
          Z(J) = 1.0 / Y(J)
          DO 28   K = 1, M
          R(K,J) = A(K,J) * Z(J)
28        A(K,J) = A(K,J) * Y(J)
          IF (L1) 282,282,280
280       DO   281 J= 1,M
          W(J) = 0.0
281       Z(J) = 0.0
          NDF = 0.0
          GO TO 30
282       CONTINUE
          Y(1) = NDF
          Y(M) = 0.0
          Z(1) = CHISQ
          Z(M) = 0.0
          EK = EL - 1.0
          K = L - 1
          J = L
29        Y(J) = (EM*EM - 2.0*EM*EK - EM + EK*EK + EK) / 2.0
          TEMP = DET/(T(K) * (((EM - S(K)) / (EM - EK)) **(EM - EK)))
          Z(J) = -(EN - 1.0 - ((2.0 * EM + 5.0)/ 6.0) - (2.0 * EK /3.0))
         C   * ALOG(TEMP)
          K = K - 1
          EK = EK - 1.0
          J = J - 1
          IF (J - 1)   30, 30,   29
30        WRITE (6,31)
31        FORMAT(72H0FACTOR EIGENVALUE   PERCENT TRACE   CUM PERCENT   N.D.F.
         C   CHI-SQUARE    )
          DO 16 J = 1, L
          NDF = Y(J)
16        WRITE (6,32) J,   X(J),   U(J),   V(J),   NDF,   Z(J)
32        FORMAT(3H0  I3,F9.4,6X,F8.1,8X,F8.1,4X,I4,3X,F9.2)
          WRITE (6,33)
33        FORMAT(55H0FACTOR PATTERN. FACTORS ARE COLUMNS, TESTS ARE ROWS.  )
          DO 8 J = 1, M
8         WRITE(6,17) J, (A(J,K), K = 1, L)
          DO 40 J = 1, L
40        WRITE(7,15) J, (A(K,J),K=1,M)
15        FORMAT(4H ROW,I3,3X,7F10.7/(10X,7F10.7))
17        FORMAT (4H ROW,I3,3X, 10F8.3/(10X, 10F8.3))
          WRITE (6,34)
34        FORMAT(55H0TEST    COMMUNALITY    MULT R SQUARE                    )
          DO 35 J = 1, M
```

PRINCO (*continued*)

```
        TEMP = 0.0
        DO 36 K = 1, L
36      TEMP = TEMP + A(J,K) * A(J,K)
35      WRITE (6,37) J,     TEMP,    W(J)
37      FORMAT (3H0   I3,6X,F6.3,6X,F6.3)
        WRITE (6,38)
38      FORMAT(65H0FACTOR SCORE COEFFICIENTS. FACTORS ARE COLUMNS, TESTS A
        CRE ROWS.                       )
        DO 9 J = 1, M
9       WRITE(6,17) J,  (R(J,K), K=1,L)
        DO 39 J = 1, L
39      WRITE (7,24) J,     (R(K,J),    K = 1, M)
C    NOTE THAT IN PUNCHED SCORE COEFFICIENTS MATRIX, ROWS ARE
C    FACTORS AND COLUMNS ARE TESTS, ALLOWING THE DISCARDING OF
C    PART OF THE DECK IF A REDUCED RANK MODEL IS TO BE SCORED.
C
        END
>
```

4.5 EIGENSTRUCTURE OF A SYMMETRIC MATRIX

The correlation matrix is a real, symmetric matrix of full rank, which means that all its eigenvalues are real, positive numbers and the corresponding eigenvectors are real. It is conventional to scale the eigenvectors so that the sum of squares of elements of any vector is one, in order that

$$\mathbf{V}\mathbf{V}' = \mathbf{V}'\mathbf{V} = I$$

where $\mathbf{V}$ is the matrix of column eigenvectors. There are several ways to compute the eigenstructure of a positive definite, symmetric matrix. One scheme involves iterating for an eigenvalue and its eigenvector, then exhausting the matrix of these and iterating on the residual matrix for the next eigenvalue and vector, and so on. This is a practical scheme for doing small matrices on small computers, although we will propose a more useful noniterative scheme for large matrices and large computers. Our main interest in the iteration and exhaustion method is that it demonstrates that the matrix is composed of layers of eigenstructure components of the form

$$\lambda_j \mathbf{v}_j \mathbf{v}_j'$$

and that the largest eigenvalue of the residual after extraction and exhaustion of k layers is the $k + 1$ eigenvalue of the original matrix. That is, for

$$\tilde{\mathbf{A}} = \mathbf{A} - \sum_{j=1}^{k} \lambda_j \mathbf{v}_j \mathbf{v}_j'$$

the solution to the determinantal equation

$$|\bar{\mathbf{A}} - \lambda_{max}| = 0$$

yields for λ_{max} the eigenvalue λ_{k+1} of $\mathbf{A}$. Each layer removed is of rank one (meaning that the matrix $\lambda_j \mathbf{v}_j \mathbf{v}_j'$ has only one nonzero eigenvalue), and the rank of $\bar{\mathbf{A}}$ after k layers have been removed is $m - k$.

The iteration for an eigenvalue and its vector begins with an arbitrary trial vector $\mathbf{y}_1 = \mathbf{1}$, which is used as a postmultiplier of the matrix $\mathbf{A}$. If $\mathbf{y}_1$ is an eigenvector, $\mathbf{A}\mathbf{v}$ will be a multiple of $\mathbf{A}\mathbf{y}_1$, the multiplier being the eigenvalue λ_1. In general, $\mathbf{y}_1$ will not be an eigenvector. By iteration, successive vectors may be produced from the multiplication

$$\mathbf{x}_r = \mathbf{A}\mathbf{v}_{r-1}$$

where r is the number of the current iteration. The approximation to the eigenvalue after each iteration is taken to be the leading (first) element of the previous product vector $\mathbf{x}$, and this trial eigenvalue is divided out of all the elements of that product vector to produce the trial eigenvector. Thus

$$\mathbf{v}_r = \mathbf{x}_r \left(\frac{1}{e_r}\right)$$

where e_r is the first element of $\mathbf{x}_r$. Figure 4.1 diagrams this iteration procedure, and a little study of the figure will clarify what is done. This iteration scheme will produce convergence to the largest eigenvalue, provided that the second largest eigenvalue is not equal to the first. The iteration process continues until the eigenvector stabilizes to a preset test tolerance, as we have programed the method. Then the eigenvector is scaled to unit length, the matrix is exhausted of this layer of its eigenstructure, and iteration begins for the next layer.

Convergence can be quite slow when two roots are close to each other in value. As a practical matter, it is helpful to power the matrix before beginning the procedure, in order to separate the roots more. The relationship that is capitalized on is that given

$$\mathbf{A}\mathbf{v} = \lambda\mathbf{v}$$

then

$$\mathbf{A}^k\mathbf{v} = \lambda^k\mathbf{v}$$

Simply squaring the matrix before entering ITER-S will speed things up considerably. Of course, you must then remember to take the square roots of the eigenvalues reported by the subroutine.

As it stands, the ITER-S subroutine will compute the complete eigenstructure of the matrix. Researchers using small computers would find it desirable

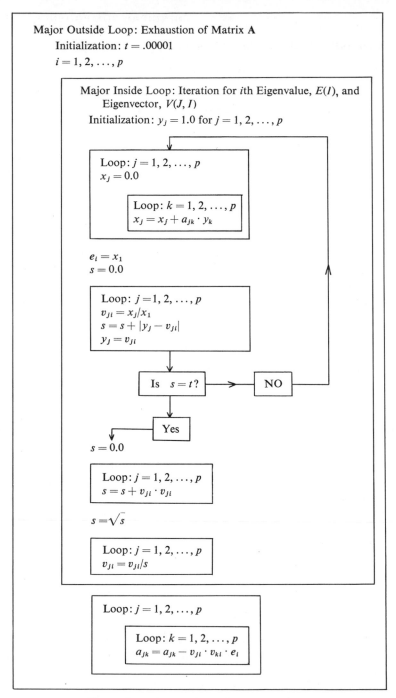

Figure 4.1. Diagram of simplified ITER-S algorithm.

to make the index L, which controls the number of layers extracted, an argument of the subroutine call, so that only a preset number of the largest eigenvalues would be extracted. Another approach would be to test each eigenvalue against 1.00, and extract only one eigenvalue smaller than 1.00 automatically. An interesting relationship between the eigenstructures of $\mathbf{A}$ and of $\mathbf{A}^{-1}$ is that the smallest eigenvalue of $\mathbf{A}$ is the largest eigenvalue of $\mathbf{A}^{-1}$, so that if the

ITER S

```
      SUBROUTINE ITER S (A, M, E, V)
C     A COOLEY-LOHNES TEACHING SUBROUTINE.
C
C     THIS PROGRAM COMPUTES THE EIGENVALUES AND EIGENVECTORS OF A REAL
C     SYMMETRIC MATRIX BY THE METHODS OF ITERATION AND EXHAUSTION.
C     ARGUMENTS ARE (1) A = INPUT MATRIX, (2) M = ORDER OF A,
C     (3) E = EIGENVALUES, (4) V = MATRIX OF COLUMN EIGENVECTORS.
C
      DIMENSION  A(50,50), E(50), V(50,50),  X(50), Y(50), Z(50)
      L = 1
      TEST = .00001
1     IT = 0
      DO 2    J = 1, M
2     Y(J) = 1.0
3     IT = IT + 1
      DO 4    J = 1, M
      X(J) = 0.0
      DO 4    K = 1, M
4     X(J) = X(J) + A(J,K) * Y(K)
      E(L) = X(1)
      SUM1 = 0.0
      DO 5    J = 1, M
      V(J,L) = X(J) / X(1)
      SUM1 = SUM1 + ABSF(Y(J) - V(J,L))
5     Y(J) = V(J,L)
      IF (IT - 10) 9,  6,  9
6     IF (SUM2 - SUM1)   7, 7,  9
7     WRITE(6, 8)
8     FORMAT (34H0LROOT NOT CONVERGING. ERROR STOP.)
      CALL EXIT
9     SUM2 = SUM1
      IF (SUM1 - TEST)  10, 10,  3
10    SUM1 = 0.0
      DO 11    J = 1, M
11    SUM1 = SUM1 + V(J,L) * V(J,L)
      SUM1 = SQRTF (SUM1)
      DO 12    J = 1, M
12    V(J,L) = V(J,L) / SUM1
C     VECTOR HAS BEEN NORMALIZED TO UNIT LENGTH.
C     ITERIATION FOR L-TH ROOT AND VECTOR IS COMPLETED.
C     ENTER MATRIX EXHAUSTION PROCEDURE.
      DO 13    J = 1, M
      DO 13    K = 1, M
13    A(J,K) = A(J,K) - V(J,L) * V(K,L) * E(L)
      IF (M - L)   15, 15,  14
14    L = L + 1
      GO TO 1
15    RETURN
      END
```

smallest eigenvalue only of **A** is desired, $\mathbf{A}^{-1}$ can be submitted to ITER-S and the routine can be stopped after it produces one eigenvalue and vector.

A program titled TEST ITERS has been listed to provide a means of demonstrating that ITER-S produces an eigenstructure solution. The test program reports the input matrix, the computed eigenstructure, the matrix **V′V**, which should approximate an identity matrix, and the matrix **VΛV′**, which should approximate the input matrix.

TEST ITER S

```
C       TEST ITER S.  A COOLEY-LOHNES ROUTINE.
        DIMENSION R(50,50),E(50),V(50,50),A(50),B(50),C(50),D(50),X(50,50)
        C, FMT(16)
   1    READ (5, 2)      M
   2    FORMAT (I2)
        READ (5, 3)      (FMT(J),   J = 1, 16)
   3    FORMAT (16A5)
        DO 4   J = 1, M
   4    READ (5, FMT)    (R(J,K),   K = J, M)
        DO 5   J = 1, M
        DO 5   K = 1, M
   5    R(K,J) = R(J,K)
        WRITE(6, 6)
   6    FORMAT (16H1INPUT MATRIX, R)
        DO 7   J = 1, M
   7    WRITE(6, 10)     (J, (R(J,K),   K = J, M))
  10    FORMAT (7H   ROW I3, 7F10.3 / (10X, 7F10.3))
C
        CALL ITER S (R, M, E, V)
        WRITE(6, 8)
   8    FORMAT (12HOEIGENVALUES)
        WRITE(6, 9)      (J, E(J),   J = 1, M)
   9    FORMAT (10X, 10(I3, F7.3))
        WRITE(6, 11)
  11    FORMAT (13HOEIGENVECTORS)
        DO 12   J = 1, M
  12    WRITE(6, 13)     (J, (V(K,J),   K = 1, M))
  13    FORMAT(7HOVECTORI3,7F10.5/(10X,7F10.5))
        DO 14   J = 1, M
        DO 14   K = 1, M
        R(J,K) = 0.0
        DO 14   L = 1, M
  14    R(J,K) = R(J,K) + V(L,J) * V(L,K)
        WRITE(6, 15)
  15    FORMAT (11HOVPRIME * V)
        DO 16   J = 1, M
  16    WRITE(6, 10)     (J, (R(J,K),   K = 1, M))
        DO 17   J = 1, M
        DO 17   K = 1, M
  17    R(J,K) = V(J,K) * E(K)
        DO 18   J = 1, M
        DO 18   K = 1, M
        X(J,K) = 0.0
        DO 18   L = 1, M
  18    X(J,K) = X(J,K) + R(J,L) * V(K,L)
        WRITE(6, 19)
  19    FORMAT (20HOV * LAMBDA * VPRIME)
        DO 20   J = 1, M
  20    WRITE(6, 10)     (J, (X(J,K),   K = 1, M))
        GO TO 1
        END
>
```

4.6 THE HOW SUBROUTINE

Fairly recently the creative efforts of three American mathematicians have converged to yield a set of procedures of very high quality for the numerical analysis of eigenstructures. Since the names of these innovators are Householder, Ortega, and Wilkinson, the subroutine that incorporates their inventions has been named SUBROUTINE HOW. The routine was originally programed by David W. Matula, under the direction of William Meredith at the University of California at Berkeley Computation Center. We have recoded Matula's excellent routine only in hopes of increasing its transparency somewhat for students.

We can give a thumbnail sketch of the methods of HOW, but the really interested reader should go to the chapters by the originators of the methods in Ralston and Wilf, Volume II (1967) for a complete account. First, HOW tri-diagonalizes the input matrix, using Householder's method of finding a similarity orthogonal transformation (Wilkinson, 1960). Then the routine uses Ortega's method of Sturm sequences to compute the eigenvalues of the tri-diagonal form, which are also the eigenvalues of the input matrix (Ortega, 1960). Finally, Wilkinson's procedure is used to get the eigenvectors of the tri-diagonal form, one at a time (if they are requested), and these are transformed to vectors of the input matrix. Since the algorithms are not iterative, time requirement becomes a function of matrix size. On the IBM 7090 HOW takes about seven seconds to get the complete eigenstructure of a 20th order matrix, and about five minutes to get a complete solution for a 100th order matrix. The separation of the roots influences precision, but in general the solutions for statistical matrices are of excellent quality. Repeating, zero, and negative roots do not upset the routine, but the user will find that exactly equal roots are assigned equal eigenvectors, and he must remember that vectors associated with negative roots are imaginary. Fortunately, matrices arising in statistical data analysis do not have negative roots and usually will not have equal or zero roots. Great credit is due all the people who have contributed to making these procedures available to us, as they have expedited the analysis of large-scale multivariate problems perhaps more than any other single set of numerical methods has.

The argument list of HOW is very long, primarily to specify work areas to be shared in memory with the main program. In the argument, the variables are

(1) order of input matrix
(2) dimensioned size of input matrix in main program
(3) number of eigenvectors required

```
      SUBROUTINE HOW (MVAR, MDIM, NVECT, R, E, V, A, B, C, D)
C     EIGENVALUES AND EIGENVECTORS OF A SYMMETRIC MATRIX, ALGORITHM
C         BY HOUSEHOLDER, ORTEGA, AND WILKINSON.  ORIGINAL PROGRAM BY
C     DAVID W. MATULA UNDER THE DIRECTION OF WILLIAM MEREDITH,UNIVERSITY
C         OF CALIFORNIA, BERKELEY, 1962.
C     MODIFIED BY P. R. LOHNES, PROJECT TALENT, 1966.
C
C     M IS THE ORDER OF THE INPUT MATRIX, R.
C     MD IS THE DIMENSIONED SIZE OF R IN THE MAIN PROGRAM.
C     NV IS THE NUMBER OF EIGENVECTORS TO BE COMPUTED.
C     E IS THE VECTOR IN WHICH THE EIGENVALUES ARE RETURNED.
C     V IS THE MATRIX IN WHICH THE EIGENVECTORS ARE RETURNED.
C         THE EIGENVECTORS ARE STORED AS COLUMNS IN V.
C     A, B, C, AND D ARE WORKSPACE VECTORS.
C
      DIMENSION  R(1), E(1), V(1), A(1), B(1), C(1), D(1)
      EQUIVALENCE (S1,IS1),(S2,IS2)
C
      M = MVAR
      MD = MDIM
      NV = NVECT
      IF (M - 1)   100, 97, 96
 96   M1 = M - 1
C     TRI-DIAGONALIZE THE MATRIX.
      M2 = M1 * MD + M
      M3 = M2 - MD
      M4 = MD + 1
      L = 0
      DO 1   I = 1, M2, M4
      L = L + 1
 1    A(L) = R(I)
      B(1) = 0.0
      IF (M - 2)  13,  2,  3
 3    KK = 0
      DO 15   K = 2, M1
      KL = KK + K
      KU = KK + M
      KJ = K + 1
      SUM = 0.0
      DO 4   J = KL, KU
 4    SUM = SUM + R(J) **2
      S = SQRT(SUM)
      Z = R(KL)
      B(K) = SIGN (S, -Z)
      S = 1.0 / S
      C(K) = SQRT (ABS (Z) * S + 1.0)
      X = SIGN  (S / C(K),  Z)
      R(KL) = C(K)
      DO 5   I = KJ, M
      JJ = I + KK
      C(I) = X * R(JJ)
 5    R(JJ) = C(I)
      DO 8   J = K, M
      JJ = J + 1
      D(J) = 0.0
      L = KK + J
      DO 6   I = K, J
      L = L + MD
 6    D(J) = D(J) + R(L) * C(I)
      IF (JJ - M)  7,  7,  9
 7    DO 8   I = JJ, M
      L = L + 1
 8    D(J) = D(J) + R(L) * C(I)
 9    X = 0.0
      DO 10   J = K, M
 10   X = X + C(J) * D(J)
```

```
         X = .50 * X
         DO 11   I = K, M
  11     D(I) = X * C(I) - D(I)
         LL = KK
         KK = KK + MD
         DO 15   I = K, M
         LL = LL + MD
         DO 15   J = I, M
         L = LL + J
  15     R(L) = R(L) + D(I) * C(J) + D(J) * C(I)
         L = 1
         DO 12   I = 1, M
         X = A(I)
         A(I) = R(L)
         R(L) = X
  12     L = L + M4
   2     B(M) = R(M3)
C        COMPUTE EIGENVALUES.
  13     BD = ABS (A(1))
         DO 14   I = 2, M
  14     BD = AMAX1(BD, ABS (A(I)) + B(I)**2)
         BD = BD + 1.0
         DO 16   I = 1, M
         A(I) = A(I) / BD
         B(I) = B(I) / BD
         D(I) = 1.0
  16     E(I) = -1.0
         DO 37   K = 1, M
  17     IF((D(K)-E(K))/AMAX1(ABS (D(K)),ABS (E(K)),1.0E-9)-1.0E-6)37,37,18
  18     X = (D(K) + E(K)) * .50
         IS2 = 1
         C(1) = A(1) - X
         IF (C(1))   19,  20,  20
  19     IS1 = -1
         N = 0
         GO TO 21
  20     IS1 = 1
         N = 1
  21     DO 31   I = 2, M
         IF (B(I))   22,  26,  22
  22     IF (B(I-1))   23,  27,  23
  23     IF(ABS (C(I-1))+ABS (C(I-2))-1.0E-15)   24,  25,  25
  24     C(I-1) = C(I-1) * 1.0E15
         C(I-2) = C(I-2) * 1.0E15
  25     C(I) = (A(I)-X)*C(I-1)-B(I)**2 * C(I-2)
         GO TO 28
  26     C(I) = (A(I)-X) * SIGN (1.0, S1)
         GO TO 28
  27     C(I) = (A(I)-X) * C(I-1) - SIGN (B(I)**2, S2)
  28     S2 = S1
         IF (C(I))   29,  30,  29
  29     S1 = SIGN (S1,C(I))
         IF (IS2+IS1)   30,  31,  30
  30     N = N + 1
  31     CONTINUE
         N = M - N
         IF (N - K)  34,  32,  32
  32     DO 33   J = K, N
  33     D(J) = X
  34     N = N + 1
         IF (M - N)   17,  35,  35
  35     DO 36   J = N, M
         IF (X - E(J))   17,  17,  36
  36     E(J) = X
         GO TO 17
  37     CONTINUE
```

```
        DO 38    I = 1, M
        A(I) = A(I) * BD
        B(I) = B(I) * BD
38      C(I) = (D(I) + E(I)) * BD * .50
        M1 = M
        K = 1
39      I = 1
40      DO 43    J = 1, M1
        IF (I - J)  41,  43,  41
41      IF (C(I) - C(J))   43,  43,  42
42      I = J
        GO TO 40
43      CONTINUE
        E(K) = C(I)
        K = K + 1
        M1 = M1 - 1
        IF (I - M1 - 1)   44,  46,  46
44      DO 45    M2 = 1, M1
45      C(M2) = C(M2+1)
46      IF (M1 - 1)   47,  47,  39
47      E(K) = C(1)
        IF (ISIGN  (1, NV))   79,  76,  76
76      DO 77    I = 1, M
77      C(I) = E(I)
        J = M
        DO 78    I = 1, M
        E(I) = C(J)
78      J = J - 1
79      CONTINUE
C
C
C       DECIDE WHETHER TO COMPUTE EIGENVECTORS, AND IF SO, HOW MAN
C
        IF (NV)   48,  99,  48
48      KX = IABS(NV)
        J = 1
        DO 98    INV = 1, KX
        X = A(1) - E(INV)
        Y = B(2)
        M1 = M - 1
        DO 54    I = 1, M1
        IJ = J + I - 1
        IF (ABS (X) - ABS (B(I+1)))   49,  51,  53
49      C(I) = B(I+1)
        D(I) = A(I+1) - E(INV)
        V(IJ) = B(I+2)
        Z = -X / C(I)
        X = Z * D(I) + Y
        IF (M1 - I)  50,  54,  50
50      Y = Z * V(IJ)
        GO TO 54
51      IF (X)   53,  52,  53
52      X = 1.0E-10
53      C(I) = X
        D(I) = Y
        V(IJ) = 0.0
        X = A(I+1) - (B(I+1) / X * Y + E(INV))
        Y = B(I+2)
54      CONTINUE
        MJ = M + J - 1
        IF (X)   56,  60,  56
56      V(MJ) = 1.0 / X
57      I = M1
        IJ = J + I - 1
        V(IJ) = (1.0 - D(I) * V(MJ))/ C(I)
        X = V(MJ)**2 + V(IJ)**2
58      I = I - 1
```

```
        IJ = J + I - 1
        IF (I)    59,   61,   59
59      V(IJ) = (1.0 - (D(I) * V(IJ+1) + V(IJ) * V(IJ+2))) / C(I)
        X = X + V(IJ)**2
        GO TO 58
60      V(MJ) = 1.0E10
        GO TO 57
61      X = SQRT(X)
        DO 62    I = 1, M
        IJ = J + I - 1
62      V(IJ) = V(IJ) / X
        J1 = M1 * MD - MD
        K = M
        GO TO 66
63      K = K - 1
        J1 = J1 - MD
        Y = 0.0
        DO 64    I = K, M
        IJ = J + I - 1
        L = J1 + I
64      Y = Y + V(IJ) * R(L)
        DO 65    I = K, M
        IJ = J + I - 1
        L = J1 + I
65      V(IJ) = V(IJ) - Y * R(L)
66      IF (J1)    63,   67,   63
67      NPLUS = 0
        NMIN = 0
        DO 70    I = 1, M
        IJ = J + I - 1
        IF (V(IJ))    68,   69,   69
68      NMIN = NMIN + 1
        GO TO 70
69      NPLUS = NPLUS + 1
70      CONTINUE
        IF (NPLUS - NMIN)    71,   73,   73
71      DO 72    I = 1, M
        IJ = J + I - 1
72      V(IJ) = -V(IJ)
73      CONTINUE
98      J = J + MD
C       RESTORE THE INPUT MATRIX.
99      MD1 = MD + 1
        JJ = MD1
        M1 = M * MD
        DO 75    I = 2, M1, MD1
        K = I
        DO 74    J = JJ, M1, MD
        R(K) = R(J)
74      K = K + 1
75      JJ = JJ + MD1
        GO TO 100
97      E(1) = R(1)
        V(1) = 1.0
100     RETURN
        END
>
```

(4) the input matrix
(5) vector to contain the eigenvalues
(6) matrix to contain the column eigenvectors
(7), (8), (9), (10) four work-space vectors.

A feature of Matula's coding of the routine is that the matrices and vectors are not dimensioned in the subroutine, so the source or object programs can be combined with any main program without adjustment. However, the second argument must represent the main program dimensioned size of all matrices and vectors used by the subroutine (i.e., listed in the argument).

The program TEST ITERS may be used to demonstrate HOW if the following call statement replaces the call on ITER-S:

```
CALL HOW   (M, 50, M, R, E, V, A, B, C, D)
```

4.7 EXERCISES

1. Given

$$R = \begin{bmatrix} 1.00 & .50 & .30 \\ .50 & 1.00 & .20 \\ .30 & .20 & 1.00 \end{bmatrix} \quad \text{and} \quad N = 50;$$

(a) If the largest two roots of R are 1.68 and .83, what does the third root equal?

(b) What is the probability that the population correlation matrix from which R was sampled, is an identity matrix?

(c) The structure of the first two factors is:

$$S = \begin{bmatrix} .84 & .17 \\ .79 & .42 \\ .60 & -.79 \end{bmatrix}$$

What is the correlation between the second variable and the first factor? What proportion of the variance of variable three is explained by these first two factors?

(d) What is the rank of the matrix product formed by multiplying SS'?

2. Using either the TALENT males or females of Appendix B, do a principal components analysis of the eight mental abilities measures.

III

Selected Factor Analysis Procedures

5.1 INTRODUCTION TO FACTOR ANALYSIS

Factor analysis has become the generic term for a variety of procedures developed for the purpose of analyzing the intercorrelations within a set of variables. The variables may be test scores, test items, questionnaire responses, etc. Some procedures of factor analysis have been developed to explore a specific hypothesis regarding the basic structure of mental abilities. On the other hand, principal components analysis is a generally useful procedure whenever the task is to determine the minimum number of independent dimensions needed to account for most of the variance in the original set of variables.

The historical development of this field illustrates the frequent controversies that occur as a new science or scientific technique emerges. Each pioneer seemed to feel his procedure was *the* method of factor analysis. Only recently have students of factor analysis begun to see that the different procedures are suitable for different purposes and usually involve different assumptions regarding the nature of human attributes. Harman (1960), Chapter 6, has done a comprehensive and thorough job of comparing in detail the factor solutions that have received the most attention. A brief general discussion of several aspects of factor analysis is presented here, followed by selected computational procedures.

One situation in which factor analytic techniques could be applied would be that of scaling a set of responses that sample a particular psychological or sociological domain. For instance, the investigator might have several responses to questionnaire items concerned with socioeconomic status (SES). What this set of items measures in common defines the basic dimension for SES.

Figure 5.1 may clarify this point. Here X and Y represent two variables

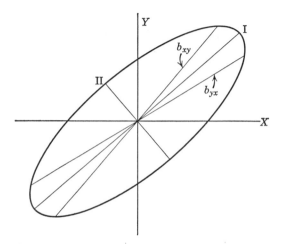

Figure 5.1.

such as family income and prestige rating of father's occupation. In this test space model, individuals (or families) are plotted according to their reported values for X and Y (or some transformation of the reported values). The ellipse suggests the general outline of the bivariate swarm resulting from the plot of N individuals. The X and Y exhibit a high positive correlation in this case. The task is to produce a composite score measuring what these variables have in common and producing a maximum variance among individuals. This is accomplished by projecting all points perpendicularly onto the principal axis of the ellipse (I). This procedure is called principal-components analysis. The principal axis, or component, defines the factor or basic dimension the variables are measuring in common. The principal axis should not be confused with regression lines, as indicated by the regression coefficients b_{xy} and b_{yx} in the diagram.

If a third variable is introduced, the test space is three-dimensional and the ellipse becomes an ellipsoid. In our example a third variable might be the assessed valuation of the family home, and the principal axis of the ellipsoid is again the best measure of what these variables have in common, socio-economic status in this case. As more and more variables are included, additional significant components of SES may appear, indicating that the domain is not unidimensional. Thus principal-components analysis not only reveals how several measures of a domain can be combined to produce maximum discrimination among individuals along a single dimension, but often reveals that several independent dimensions are required to define the domain under investigation adequately.

Another situation in which factor analysis may be used is that in which we wish to reduce the dimensionality of a set of variables by taking advantage of their intercorrelations. For instance, in an example presented in Chapter Four, 21 variables from school records were reduced to a more manageable two dimensions. Principal-components analysis accomplished this by finding the principal axis of the hyperellipsoid in the 21-space (similar to the ellipse in the two-space) along which there was maximum variance. Then a second axis was constructed, orthogonal to the first, along which the remaining variance was maximized (similar to Axis *II* in Figure 5.1). At this point in the sequence of locating new axes in the ellipsoid, an insignificant amount of variance was found in the remaining dimensions, thus reducing the number of dimensions required, from 21 to two.

The last use of factor analysis to be considered here is to find ways of identifying fundamental and meaningful dimensions of a multivariate domain. The investigation of mental abilities has received the most attention in this regard. Psychologists have identified and have agreed on a few of these primary mental abilities, or differential aptitudes, as they are sometimes called. This "construct-seeking" task of factor analysis is most frequently accomplished today by first conducting a principal-components analysis, and by then using the resulting principal factors as a set of reference axes for determining the simplest structure, or most easily interpretable set of factors, for the domain in question. This whole process, which Harman (1960) calls multiple-factor analysis, is most easily understood from the person space model.

The person space, or sample as it is often called, is N-dimensional, where N is the sample size. The p tests are converted to standard score units and are plotted in the N space as points or vectors. That is, the scores of N subjects on test j locate a point for that test in the N space, and a line from the origin to that point becomes the vector representation of test j. The cosine of the angle between pairs of normalized test vectors is the correlation between the two tests. Figure 5.2 illustrates the sample space for the data in Table 5.1. Sample spaces of two dimensions are not very interesting because there are only two possible locations for a test vector, if standard scores are based on that particular sample of two. This is another demonstration that the correlation must be either $+1.00$ or -1.00 for the case where $N = 2$ ($\cos 0° = 1.00$ and $\cos 180° = -1.00$).

Now N dimenstions are not needed to describe p vectors, where $p < N$. Think, for instance, of *two* test vectors in a *three*-dimensional space. The two test vectors define a plane, and only a plane is needed to describe the relationship between the two tests. The problem is to decide among the

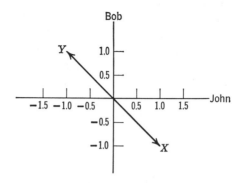

Figure 5.2.

TABLE 5.1   Standard Score Roster

Name	X	Y
John	1	−1
Bob	−1	1

infinite sets of reference axes capable of adequately describing a particular configuration of tests. Figure 5.3 illustrates this problem by showing two of the pairs of axes which could serve as reference axes (factors) for describing the plane in which tests X and Y lie. Here axes A and B are rotated to positions C and D, respectively.

Principal-components analysis defines a unique set of reference axes for a given combination of p variables using the maximum variance criterion as described. However, that procedure does not usually result in a satisfactory set of reference axes for psychological interpretation. A new set of axes is formed by rotating the derived principal-component axes. This rotation process can best be described by a fabricated example.

In order to illustrate the frequent advantages of rotation, one of the authors devised a set of eight scores for each of a hundred rectangles. First, an arbitrary length L and width W were selected for each rectangle; then the eight scores were produced from the formulas of Table 5.2, where e_{ji} is a

TABLE 5.2   Formulas for Scores on Eight Tests

Test	Formula
1	$X_{1i} = L_i$
2	$X_{2i} = W_i$
3	$X_{3i} = 10L_i + e_{3i}$
4	$X_{4i} = 10W_i + e_{4i}$
5	$X_{5i} = 20L_i + 10W_i + e_{5i}$
6	$X_{6i} = 20L_i + 20W_i + e_{6i}$
7	$X_{7i} = 10L_i + 20W_i + e_{7i}$
8	$X_{8i} = 40L_i + 10W_i + e_{8i}$

one-digit random number introduced as the error component on test j for individual i. The eight scores for the 100 rectangles are found in Appendix C.

The intercorrelations among the eight scores were computed and the **R** matrix (Table 5.3) was subjected to a principal-components analysis, with

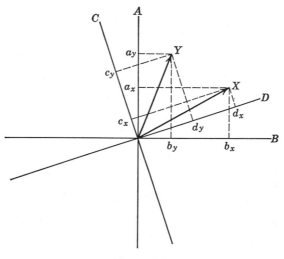

Figure 5.3.

TABLE 5.3   Means, Standard Deviations, and Correlations for Eight Variables on One Hundred Rectangles

Variables	1	2	3	4	5	6	7	8
Means	17.8	11.7	190.6	119.1	476.3	591.5	416.1	831.3
S.D.	6.5	5.4	66.5	54.3	148.5	181.0	134.2	274.1
Intercorrelations								
1		.140	.987	.168	.931	.804	.597	.980
2			.160	.930	.491	.693	.877	.331
3				.185	.927	.807	.608	.972
4					.489	.671	.835	.347
5						.962	.848	.984
6							.950	.903
7								.743
8								

the results displayed in Table 5.4. Notice that two components account for 98.5 percent of the total variance in the battery of eight tests and that the first component is a general factor, whereas the second is a bipolar factor in which the tests dominated by length have negative loadings and those dominated by width have positive loadings.

A rotation was then performed on the matrix of adjusted factor loadings

TABLE 5.4   Two Principal Components Accounting for 98.5 Percent of the Trace of **R** from Eight Rectangle Tests on One Hundred Subjects

	First Component	Second Component
Latent Root	5.96	1.93
Percent of Trace	74.4	24.1
Variable	Factor Loadings	
1	.8534	−.5188
2	.6337	.7623
3	.8578	−.4998
4	.6332	.7356
5	.9837	−.1761
6	.9923	.0776
7	.9264	.3664
8	.9391	−.3418

of the eight tests on the two principal components selected. The result (Table 5.5) was that the general factor was destroyed and two group factors were produced. The Varimax criterion was used in rotation here. This is described in Section 5.3.

In our example eight test vectors were located in an N-dimensional sample space ($N = 100$). The cosines of the angles between the eight test vectors, that is, the intercorrelations, cause the vectors to tend to lie in a common plane within the N space. Principal-components analysis was used to define the relative locations of the tests in the plane. The resulting reference axes are labeled I and II in Figure 5.4.

The coordinates of the termini of the eight test vectors in this two-dimensional factor space are listed in Table 5.4. The relative locations in the figure

TABLE 5.5   Two Rotated
Factors from Varimax

Test	L Factor	W Factor
1	.997	.058
2	.089	.987
3	.989	.076
4	.103	.965
5	.909	.414
6	.772	.627
7	.554	.828
8	.967	.252

of the test vector termini are indicated by circled numbers, the numbers referring to the eight tests.

Axes I and II are not the only axes that can define this two-space. The axes resulting from principal-components analysis were rotated to positions A and B, and a new set of test coordinates were obtained. These are listed in Table 5.5. Figure 5.4 shows the location of axes L and W. These new, rotated axes might be preferable for purposes of interpreting the basic dimensions of the domain measured by the eight tests. This is because the new coordinates are more " simple " in the sense that a given variable tends to have a high coefficient for only one new axis, and each factor has zero, or near zero, coefficients for at least some of the variables.

The rotated solution consists of two distinct factors, length (variables 1, 3, 5, 6, and 8) and width (variables 2, 4, 7). The components solution, on the other hand, produced one general factor and one bipolar factor. One general and $p - 1$ bipolar factors are typically the results of principal-components analysis. (Technically, p common factors are produced although not all of them are usually useful.)

Which of the infinite number of possible factor solutions is preferable is generally a theoretical question. We are familiar with rectangles, and so we tend to prefer the simple structure solution produced by rotation. Length and width are the two basic dimensions for rectangles; hence factors L and W "make sense."

We could argue, however, that the principal-components solution is also interpretable. The first factor, I, is a general size or "bigness" factor and

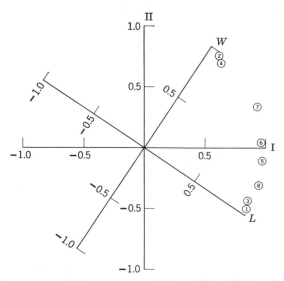

Figure 5.4. Test vector termini in two-space using two different factor solutions. Here I and II are the principal-components axes, and L and W are the Varimax-rotated axes.

the second, bipolar factor, II, is length ($-$) versus width ($+$). The second factor would be dependent only on the relative lengths and widths of rectangles and independent of overall size.

The new coordinates for the eight variables listed in Table 5.5 are the loadings for the corresponding factor. Both individuals and tests can be plotted in the factor space. The relative location of neither tests nor individuals is affected by the orthogonal rotation of factors. The interpretation of the factor solution alone is affected.

The brief description of factor analysis presented in this section is intended only as an introduction for the reader who has not been previously exposed to the procedures.

5.2 EXTRACTION OF ARBITRARY ORTHOGONAL FACTORS

Factors are derived measurement constructs that may have the virtues of parsimony, orthogonality, increased reliability, and increased normality over the observation measures from which they are derived. With the aid of the computer it is easy to transform a p-element observation vector into an n-element factor vector, but there are some hard decisions to be made by the researcher before he turns the computer loose in this fashion. The major problem is the indeterminancy of a factor solution. Even after the rank of the model for the data is decided upon there is an infinity of possible n-rank factor solutions for a given correlation matrix and, of course, rank is a very arbitrary decision in the first place. In fact, the choice of a solution for a given rank can be even more arbitrary than the convention of adopting "varimaxed principal components" suggests. Also, we should not overlook the many arbitrary decisions about variables and sample(s) that lead to the correlation matrix itself. For example, the decision on how to handle missing data is infrequently discussed in research reports, but the problem has been swept under the rug only after some decision was made. (Our approach is to eliminate from the sample all subjects with incomplete score vectors.) Data analysis is a science in which the quality of judgments that are made has much to do with the quality of outcomes. The mere possession of automated data processing procedures does not guarantee worthy outcomes of research.

We are advocating principal components as a primary factor analysis procedure, but we want to acknowledge that there are many worthy schemes for factoring a matrix. To illustrate this point, we now describe what we term extraction of arbitrary factors, which is a method that allows the researcher to specify the loadings he would like to have on each factor in succession. The algorithm causes the computer to respect the prescription for each factor as far as it can, subject to the restriction that the factor must be placed orthogonal to all previously extracted factors. In this scheme the first factor can be quite arbitrary, as the computer needs only to adjust the requested loadings to make the obtained loadings consistent with the relations in the correlation structure. The requirement of mutual orthogonality imposes increasingly severe restraints on the location of the successive factors as factoring continues.

We have taken the method of extracting arbitrary factors from Overall (1962), which also presents an elegant proof of the orthogonality of the extracted factors to each other and to any factors of the residual matrix after their extraction. This last point is crucial, because we intend to propose an occasional use of a mixed-methods strategy, in which the researcher would first extract one or more arbitrary factors from his correlation matrix in order to remove certain sources of variance he wants to control initially, and then would apply principal components analysis to the residual covariance matrix. What Overall shows is that the arbitrary factors and the principal components of the residual will all be mutually orthogonal.

Starting with an arbitrary vector of factor coefficients $\bar{\mathbf{h}}_1$ that has zeros in every position except the positions of variables we want to "load," where it has numbers indicating the relative weights desired, this arbitrary vector is scaled so that

$$\mathbf{h}_1'\mathbf{R}\mathbf{h}_1 = 1$$

This is accomplished by computing

$$\mathbf{h}_1 = \left\{\frac{1}{\sqrt{\bar{\mathbf{h}}_1'\mathbf{R}\bar{\mathbf{h}}_1}}\right\}\bar{\mathbf{h}}_1$$

The factor structure correlations ($\mathbf{s}_1$) for the first factor are computed as

$$\mathbf{s}_1 = \mathbf{R}\mathbf{h}_1$$

This is a perfectly general method for defining an arbitrary factor on a vector variable with dispersion $\mathbf{R}$. This first factor is now extracted, or we sometimes say "exhausted," from $\mathbf{R}$, leaving the first residual matrix $\tilde{\mathbf{R}}_1$

$$\tilde{\mathbf{R}}_1 = \mathbf{R} - \hat{\mathbf{R}}_1 = \mathbf{R} - \mathbf{s}_1\mathbf{s}_1'$$

Specifying a second arbitrary factor vector $\bar{\mathbf{h}}_2$, again we transform to unit variance:

$$\mathbf{h}_2'\tilde{\mathbf{R}}_1\mathbf{h}_2 = 1$$

where

$$\mathbf{h}_2 = \left\{\frac{1}{\sqrt{\bar{\mathbf{h}}_2'\tilde{\mathbf{R}}_1\bar{\mathbf{h}}_2}}\right\}\bar{\mathbf{h}}_2$$

The loadings for the second factor are

$$\mathbf{s}_2 = \tilde{\mathbf{R}}_1\mathbf{h}_2$$

The matrix is now exhausted of this second factor

$$\tilde{\mathbf{R}}_2 = \tilde{\mathbf{R}}_1 - \hat{\mathbf{R}}_2 = \mathbf{R} - \hat{\mathbf{R}}_1 - \hat{\mathbf{R}}_2 = \tilde{\mathbf{R}}_1 - \mathbf{s}_2\mathbf{s}_2'$$

The residual matrix $\tilde{\mathbf{R}}_2$ is the variance-covariance matrix from which we may then extract a third arbitrary factor by continuing the method. Of course, after $p - 1$ arbitrary factors have been extracted, the last factor has no degrees of freedom and its placement will not be influenced by the specification vector in any way. Alternately, we could now compute the principal components of $\tilde{\mathbf{R}}_2$. We would find that $\tilde{\mathbf{R}}_2$ has two zero eigenvalues, indicating that we have reduced the rank of the correlation matrix by two in the process of extracting two arbitrary factors.

To see that the first two arbitrary factors are uncorrelated, consider that the factors are

$$f_1 = \mathbf{b}_1'\mathbf{z} \quad \text{and} \quad f_2 = \mathbf{b}_2'\mathbf{z}$$

where

$$\mathbf{b}_1 = \mathbf{R}^{-1}\mathbf{s}_1 \quad \text{and} \quad \mathbf{b}_2 = \mathbf{R}^{-1}\mathbf{s}_2$$

by multiple regression theory. That is, we get factor score coefficients by regressing the factors on the vector variable $\mathbf{z}$. The uncorrelatedness requirement is

$$r_{f_1 f_2} = 0$$

where

$$
\begin{aligned}
r_{f_1 f_2} &= \frac{1}{N} \sum_{i=1}^{N} f_{1i} f_{2i} \\
&= \frac{1}{N} \sum (\mathbf{b}_1'\mathbf{z}_i)(\mathbf{b}_2'\mathbf{z}_i) \\
&= \frac{1}{N} \sum (\mathbf{b}_1'\mathbf{z}_i)(\mathbf{z}_i'\mathbf{b}_2) \\
&= \mathbf{b}_1' \frac{1}{N} \sum (\mathbf{z}_i \mathbf{z}_i')\mathbf{b}_2 \\
&= \mathbf{b}_1'\mathbf{R}\mathbf{b}_2 \\
&= \mathbf{b}_1'\mathbf{R}\mathbf{R}^{-1}\mathbf{s}_2 \\
&= \mathbf{b}_1'\mathbf{s}_2 \\
&= \mathbf{b}_1'(\mathbf{R} - \mathbf{s}_1\mathbf{s}_1')\mathbf{h}_2 \quad [\text{since } \mathbf{s}_2 = \tilde{\mathbf{R}}_1\mathbf{h}_2] \\
&= \mathbf{b}_1'(\mathbf{R} - \mathbf{R}\mathbf{b}_1\mathbf{b}_1'\mathbf{R})\mathbf{h}_2 \quad [\text{since } \mathbf{s}_1 = \mathbf{R}\mathbf{b}_1] \\
&= \mathbf{b}_1'\mathbf{R}\mathbf{h}_2 - \mathbf{b}_1'\mathbf{R}\mathbf{b}_1\mathbf{b}_1'\mathbf{R}\mathbf{h}_2
\end{aligned}
$$

but

$$\mathbf{b}_1'\mathbf{R}\mathbf{b}_1 = (\mathbf{h}_1'\mathbf{R}\mathbf{R}^{-1})\mathbf{R}(\mathbf{R}^{-1}\mathbf{R}\mathbf{h}_1) = \mathbf{h}_1'\mathbf{R}\mathbf{h}_1 = 1$$

therefore

$$r_{f_1 f_2} = \mathbf{b}_1'\mathbf{Rh}_2 - 1(\mathbf{b}_1'\mathbf{Rh}_2) = 0$$

This method may be used to prove the uncorrelatedness of each pair of arbitrary factors.

The Project TALENT data can provide a numerical example of FACTOR, our program which extracts arbitrary orthogonal factors. For the 234 TALENT males, the (1) Reading Comprehension, (2) Creativity and (3) Mechanical Reasoning tests yields the following means, standard deviations, and correlations:

	1	2	3
Means	33.59	10.41	13.57
Std. Dev.	9.49	3.88	3.57
1	1.00	.60	.44
2	.60	1.00	.53
3	.44	.53	1.00

If we wanted to extract two orthogonal factors, one heavily loaded on Reading Comprehension, the other on Mechanical Reasoning, then two hypothesis vectors would be input to FACTOR along with the $\mathbf{R}$ matrix from CORREL.

Hypothesis Vectors

$\bar{\mathbf{h}}_1$	$\bar{\mathbf{h}}_2$
1.00	.00
.00	.00
.00	1.00

In this case $\bar{\mathbf{h}}_1$ is already scaled so that $\bar{\mathbf{h}}_1'\mathbf{R}\bar{\mathbf{h}}_1 = 1.00$, therefore $\mathbf{h}_1 = \bar{\mathbf{h}}_1$, and the factor structure for the first factor ($\mathbf{s}_1$) is found:

$$\mathbf{s}_1 = \mathbf{Rh}_1 = \begin{bmatrix} 1.00 & .60 & .44 \\ .60 & 1.00 & .53 \\ .44 & .53 & 1.00 \end{bmatrix} \cdot \begin{bmatrix} 1.00 \\ .00 \\ .00 \end{bmatrix} = \begin{bmatrix} 1.00 \\ .60 \\ .44 \end{bmatrix}$$

This makes sense since, if we use the coefficients in h_1 to produce factor scores ($f_i = h_1'z_i$), those factor scores will have the same correlations with variables (1), (2) and (3) as does (1). Having extracted this first factor, the residual matrix ($\tilde{\mathbf{R}}_1$) is found by first forming $s_1 s_1'$:

$$\tilde{\mathbf{R}}_1 = s_1 s_1' = \begin{bmatrix} 1.000 \\ .600 \\ .443 \end{bmatrix} \cdot [1.000 \quad .600 \quad .443] = \begin{bmatrix} 1.00 & .60 & .44 \\ .60 & .36 & .26 \\ .44 & .26 & .20 \end{bmatrix}$$

Then the residual matrix

$$\tilde{\mathbf{R}}_1 = \mathbf{R} - \hat{\mathbf{R}}_1 = \begin{bmatrix} 0.0 & 0.0 & 0.0 \\ 0.0 & .64 & .27 \\ 0.0 & .27 & .80 \end{bmatrix}$$

To obtain a second factor orthogonal to the first one, the second hypothesis vector is scaled to produce unti variance on the residuals of $\mathbf{h}_1$.

$$\bar{\mathbf{h}}_2 \, \tilde{\mathbf{R}}_1 \bar{\mathbf{h}}_2 = [.0 \quad .0 \quad 1.0] \cdot \begin{bmatrix} .0 & .0 & .0 \\ .0 & .64 & .27 \\ .0 & .27 & .80 \end{bmatrix} \cdot \begin{bmatrix} .0 \\ .0 \\ 1.0 \end{bmatrix} = .80$$

Therefore

$$\mathbf{h}_2 = \bar{\mathbf{h}}_2 \, \frac{1}{\sqrt{.80}} = \begin{bmatrix} .0 \\ .0 \\ 1.12 \end{bmatrix}$$

and

$$s_2 = \tilde{\mathbf{R}}_1 \mathbf{h}_2 = \begin{bmatrix} .0 & .0 & .0 \\ .0 & .64 & .27 \\ .0 & .27 & .80 \end{bmatrix} \cdot \begin{bmatrix} .00 \\ .00 \\ 1.12 \end{bmatrix} = \begin{bmatrix} .00 \\ .30 \\ .90 \end{bmatrix}.$$

For the complete factor pattern then,

$$\mathbf{S} = \begin{bmatrix} 1.00 & .00 \\ .60 & .30 \\ .44 & .90 \end{bmatrix}$$

the communalities are the sums of squares of the rows of $\mathbf{S}$, or

$$\begin{bmatrix} 1.00 \\ .44 \\ 1.00 \end{bmatrix}$$

and the residual variance-covariance matrix contains only one nonzero element ($\tilde{r}_{22} = 0.56$), which is the variance of the Creativity Test that is not explained by the two-factor model. All other variances and covariances are completely accounted for.

```
C        EXTRACTION OF ARBITRARY ORTHOGONAL FACTORS.
C        A COOLEY - LOHNES PROGRAM.
C
C        THIS PROGRAM PERMITS THE USER TO DESCRIBE, IN TERMS OF HYPOTHESIS
C     VECTORS, THE SEQUENTIAL ARBITRARY FACTORS HE DESIRES TO HAVE
C     APPROXIMATED BY ORTHOGONAL FACTORS.  THE METHOD OF EXTRACTION AND
C     EXHAUSTION IS THAT DESCRIBED BY J. E. OVERALL (1962).  THE FACTOR
C     PATTERN FOR THE EXTRACTED FACTORS IS PUNCHED OUT, AS IS THE
C     RESIDUAL VARIANCE-COVARIANCE MATRIX AFTER THE EXTRACTION OF THE
C     LAST ARBITRARY FACTOR.  SINCE THIS METHOD WILL OFTEN BE USED TO
C     REMOVE CONTROL FACTORS, THE RESIDUAL MATRIX MAY BE ENTERED INTO
C     PRINCO FOR FURTHER FACTORING.
C
C        INPUT
C     1) TEN CARDS OF TEXT DESCRIBING THE PROBLEM FOR THE OUTPUT.
C     2) CONTROL CARD (CARD 11),    COLS 1-2   M = ORDER OF THE R MATRIX,
C                                   COLS 3-4   N = NUMBER OF ARBITRARY
C                                                  FACTORS TO BE EXTRACTED.
C     3) FORMAT CARD (CARD 12) FOR THE R MATRIX.
C     4) UPPER-TRIANGULAR CORRELATION, OR R MATRIX.
C     5) A SET OF N HYPOTHESIS VECTORS, PUNCHED TO FORMAT (40F2.1).
C          EACH VECTOR STARTS ON A NEW CARD.
C
C
C        PUNCHED OUTPUT
C     1) TRANSPOSED FACTOR PATTERN, A-PRIME, N X M, TO FORMAT (10X,7F10.7)
C     2) RESIDUAL R MATRIX, UPPER-TRIANGULAR, RANK M-N, FORMAT
C        (10X, 7F10.7).
C
C        SUBROUTINE MPRINT IS REQUIRED.
C
C
       DIMENSION  TIT(16), R(100,100),  A(100,100),  V(100), W(100)
C
    1  WRITE (6,2)
    2  FORMAT(54H1ARBITRARY FACTORING.  A COOLEY-LOHNES PROGRAM.          )
       DO 3   J = 1, 10
       READ  (5,4) (TIT(K),    K = 1, 16)
    3  WRITE (6,4) (TIT(K),    K = 1, 16)
    4  FORMAT (16A5)
       READ  (5,5) M, N
    5  FORMAT (2I2)
       WRITE (6,6) N, M
    6  FORMAT(9H0EXTRACT I3,14H FACTORS FROM I3,7H TESTS.)
       READ  (5,4) (TIT(J),    J = 1, 16)
       DO 7   J = 1, M
    7  READ  (5,TIT) (R(J,K),    K = J, M)
       DO 8   J = 1, M
       DO 8   K = J, M
    8  R(K,J) = R(J,K)
C
       DO 10   L = 1, N
       READ (5,9) (V(J),    J = 1, M)
    9  FORMAT (40F2.1)
       WRITE (6,11)
   11  FORMAT (18H0HYPOTHESIS VECTOR)
       WRITE (6,12) (V(J),    J = 1, M)
   12  FORMAT (20(1X, F4.2, 1X))
       DO 13   J = 1, M
       W(J) = 0.0
       DO 13   K = 1, M
   13  W(J) = W(J) + V(K) * R(K,J)
       C = 0.0
       DO 14   J = 1, M
   14  C = C + W(J) * V(J)
       C = SQRT  (C)
```

```
        DO 15    J = 1, M
  15    W(J) = V(J) / C
        DO 16    J = 1, M
        A(J,L) = 0.0
        DO 16    K = 1, M
  16    A(J,L) = A(J,L) + R(J,K) * W(K)
        DO 17    J = 1, M
        DO 17    K = 1, M
  17    R(J,K) = R(J,K)  -  A(J,L) * A(K,L)
C    R IS NOW A RESIDUAL MATRIX WITH L FACTORS REMOVED, AND IS THUS
C    OF RANK (M - L).  LOADINGS FOR FACTOR L ARE IN COLUMN L OF A.
  10    CONTINUE
C
        DO 18    J = 1, N
        V(J) = 0.0
        DO 18    K = 1, M
  18    V(J) = V(J) + A(K,J) * A(K,J)
        WRITE (6,19)
  19    FORMAT(55H0PROPORTION OF VARIANCE EXTRACTED BY EACH FACTOR          )
        EM = M
        C = 0.0
        DO 20    J = 1, N
        V(J) = V(J) / EM
        C = C + V(J)
  20    W(J) = C
        DO 21    J = 1, N
  21    WRITE (6,22) J,  V(J),  W(J)
  22    FORMAT(7H0FACTORI3,14H   PROP. VAR. F5.3,15H,    CUM. PROP. F6.3)
        DO 23    J = 1, M
        V(J) = 0.0
        DO 23    K = 1, N
  23    V(J) = V(J) + A(J,K) * A(J,K)
        WRITE (6,24) N
  24    FORMAT(62H0 ACTUAL COMMUNALITIES FOR TESTS FOR ARBITRARY MODEL OF
        CRANK      I3)
        WRITE (6,25) (J,   V(J),   J = 1, M)
  25    FORMAT (1H0, 6(6H TEST , I3,1X, F4.2, 5X))
        WRITE (6,26)
  26    FORMAT (15H0FACTOR PATTERN    )
        DO 27    J = 1, M
  27    WRITE (6,28) J,  (A(J,K),   K = 1, N)
  28    FORMAT(6H0TEST I3,1X,10F10.3 / (10X, 10F10.3))
        WRITE (6,29) N
  29    FORMAT(50H0RESIDUAL VARIANCE-COVARIANCE MATRIX FOR RANK       I3)
        CALL MPRINT (R, M)
        DO 30    J = 1, N
  30    WRITE (7,31) J,   (A(K,J),   K = 1, M)
  31    FORMAT(6HFACTORI3,1X,7F10.7 / (10X, 7F10.7))
C    NOTE THAT PUNCHED OUTPUT HAS FACTORS IN ROWS, TESTS IN COLUMNS.
        DO 33    J = 1, M
  33    WRITE (7,32) J,   (R(J,K),   K = J, M)
  32    FORMAT(3HROWI3,4X,   7F10.7 / (10X,7F10.7))
        GO TO 1
        END
>
```

5.3 ANALYTIC ORTHOGONAL ROTATION

Since Thurstone laid down the principles of simple structure, factor analysts have been interested in schemes for improving on the solution offered by n principal components by rotating the components to positions in which the factor pattern comes closer to Thurstone's criteria. The idea of analytic rotation schemes is to have the computer further transform the principal components in ways which preserve the elegance, tractability, and utility of the set of components, while garnering the additional virtue of a closer approximation to simple structure, so that the new variables may be more readily named and understood by researchers. It is very important to recognize that all the value of a set of n components *as a set* is preserved under analytic rotation. That is, $\hat{\mathbf{R}}$, $\bar{\mathbf{R}}$, and the communalities are undisturbed. What does change are the specifications for the separate factors given by the elements of $\mathbf{S}$. The row sums of squares for $\mathbf{S}$ do not change, but the elements of each row are modified toward simple structure. This means that an effort is made to have all s_{jk} approach either zero or unity, on the grounds that very high and very low factor loadings are easily interpreted, whereas middle-sized loadings give trouble. Henry Kaiser, the inventor of the varimax rotation scheme said, "Since a factor is a vector of correlation coefficients, the most interpretable factor is one based upon correlation coefficients which are maximally interpretable" (Kaiser, 1958).

Thurstone's (1947) criteria for simple structure are:

1. Each row of the factor matrix should have at least one zero.
2. If there are m common factors, each column of the factor matrix should have at least m zeros.
3. For every pair of columns of the factor matrix there should be several variables whose entries vanish in one column but not in the other.
4. For every pair of columns of the factor matrix, a large proportion of the variables should have vanishing entries in both columns when there are four or more factors.
5. For every pair of columns of the factor matrix there should be only a small number of variables with nonvanishing entries in both columns.

Although Thurstone's criteria for simple structure guided graphic rotation for many years, attempts to develop objective rotation methods directly from his principles proved fruitless. However, a major breakthrough was made by Carroll (1953) when he proposed the first practical analytical solution for the rotation problem. The high-speed computer, which became available to factorists at about the same time, made such procedures feasible. Soon several different computer approaches were available to the investigator who wished to make rigid orthogonal rotations of multiple factors in the effort to improve the structures of solutions.

The first analytical schemes for rotation were designed to simplify the rows of the factor matrix S. The loadings for each variable were modified so that each variable had high loadings on the fewest possible factors and zero or near-zero loadings on the remaining factors. Among the creators of early row-simplifying methods were Carroll (1953), Saunders (1953), Neuhaus and Wrigley (1954), and Ferguson (1954). Subsequently, these procedures were all shown to be algebraically identical and were classified as the quartimax method (Harman, 1960, p. 301). The quartimax method maximizes the variance of the squared loadings

$$\sum_{j=1}^{p} \sum_{k=1}^{n} s_{jk}^{4} \bigg|_{\max}$$

Since each row sum of squares must be left undisturbed, what the quartimax method does is to simplify each row, or test, by maximizing within-row variances of squared loadings.

An alternative to quartimax is the varimax solution, which involves the simplification of columns rather than of rows of the factor matrix. The varimax criterion has become the most widely accepted and employed standard for orthogonal rotations of factors since its development by Kaiser (1958). It has several advantages over the quartimax method. The emphasis in varimax is on cleaning up factors rather than variables. For each factor varimax rotation tends to yield high loadings for a few variables. The rest of the loadings in the factor will be near zero. In the varimax criterion, Kaiser defines the simplicity of a factor as the variance of its squared loadings:

$$V_k = \frac{\{p \sum_{j=1}^{p} (s_{jk}^{2})^2 - (\sum_{j=1}^{p} s_{jk}^{2})^2\}}{p^2}$$

where s_{jk} is the new factor loading for variable j on factor k; $j = 1, 2, \ldots, p$, and $k = 1, 2, \ldots, n$. Then for the entire factor matrix the varimax criterion is

$$V = \sum_{k=1}^{n} V_k = \sum_{k=1}^{n} \left\{ \frac{[p \sum_{j=1}^{p} (s_{jk}^{2})^2 - (\sum_{j=1}^{p} s_{jk}^{2})^2]}{p^2} \right\} \bigg|_{\max}$$

Experience with the varimax criterion as defined above soon revealed a slight bias in the resulting solutions. The rotation tended to produce factors

```
C          VARIMAX (OR QUARTIMAX) ROTATION OF ORTHOGONAL FACTORS.
C          A COOLEY - LOHNES PROGRAM.
C
C          THIS PROGRAM READS AN N (FACTORS) BY M (TESTS) TRANSPOSED
C     ORTHOGONAL FACTOR PATTERN AND PERFORMS EITHER A VARIMAX OR A
C     QUARTIMAX ROTATION, BY METHODS DESCRIBED BY H. F. KAISER (1958).
C     TEST COMMUNALITIES ARE NOT ALTERED BY THE ROTATION. PROPORTIONS
C     OF VARIANCE EXTRACTED BY THE ROTATED FACTORS ARE REPORTED.
C     THE NEW FACTOR PATTERN IS PUNCHED FOR TRANSFER TO THE COEFF PROGRAM
C     SO THAT FACTOR SCORE COEFFICIENTS MAY BE COMPUTED, AND FOR TRANSFER
C     TO THE RESID PROGRAM SO THAT VARIOUS REDUCED RANK MODELS BASED
C     ON IT MAY BE TESTED FOR FIT.
C
C     INPUT
C     1) A TEN CARD TEXT DESCRIBING THE PROBLEM TO BE PRINTED OUT.
C     2) CONTROL CARD (CARD 11),    COLS 1-2   N = NUMBER OF FACTORS,
C                                   COLS 3-4   M = NUMBER OF TESTS,
C                                   COL 5    METHOD = 0 FOR VARIMAX OR
C                                            METHOD = 1 FOR QUARTIMAX.
C     3) FACTOR PATTERN, AS N ROWS, WITH M ELEMENTS PER ROW, PUNCHED
C        TO FORMAT (10X, 7F10.7), AS FROM THE PRINCO PROGRAM.
C
C
C
C     PUNCHED OUTPUT
C     1) TRANSPOSED FACTOR PATTERN, A-PRIME, N X M, TO FORMAT
C        (10X, 7F10.7).
C
C
          DIMENSION   A(100,100),  C(100),    TIT(16)
C
   1      WRITE (6,2)
   2      FORMAT(55H1ROTATE ORTHOGONAL FACTORS.  A COOLEY-LOHNES PROGRAM.  )
          DO 3   J = 1, 10
          READ   (5,4) (TIT(K),    K = 1, 16)
   3      WRITE (6,4) (TIT(K),    K = 1, 16)
   4      FORMAT (16A5)
          READ  (5,5) N,  M,  METHOD
   5      FORMAT (2I2, I1)
C     PLEASE DO NOT TRY TO ROTATE ONE FACTOR.  MINIMUM VALUE FOR N IS 2.
          EP = .00116
          IF (METHOD)   6, 6,  8
   6      WRITE (6,7) N, M
   7      FORMAT (20H0VARIMAX ROTATION OF I3,17H FACTORS BASED ONI3,7H TESTS
         C.  )
          GO TO 10
   8      WRITE (6,9) N,  M
   9      FORMAT(22H0QUARTIMAX ROTATION OFI3,17H FACTORS BASED ONI3,7H TESTS
         C.  )
  10      DO 50   J = 1, N
  50      READ  (5,51) (A(K,J),    K = 1, M)
  51      FORMAT (10X, 7F10.7 / (10X, 7F10.7))
          DO 13   J = 1, M
          C(J) = 0.0
          DO 11   K = 1, N
  11      C(J) = C(J) + A(J,K) ** 2
          WRITE (6,12) J,  C(J)
  12      FORMAT (7H0TEST  I3,15H     COMMUNALITY  F6.3)
          C(J) = SQRT (C(J))
          DO 13   K = 1, N
  13      A(J,K) = A(J,K) / C(J)
C         NORMALIZES ROWS OF A.
          N1 = N - 1
C
  14      NR = 0
          DO 23   I = 1, N1
```

```
        I1 = I + 1
        DO 23    J = I1, N
        A1 = 0.0
        B1 = 0.0
        C1 = 0.0
        D1 = 0.0
        DO 15    K = 1, M
        U = A(K,I) ** 2  -  A(K,J) ** 2
        V = A(K,I)  *  A(K,J)  *  2.0
        A1 = A1 + U
        B1 = B1 + V
        C1 = C1 + U ** 2  - V ** 2
15      D1 = D1 + U * V * 2.0
        IF (METHOD)    16, 16,  17
16      EM = M
        QN = D1 - 2.0 * A1 * B1 / EM
        QD = C1 - (A1 ** 2 - B1 ** 2) / EM
        GO TO 18
17      QN = D1
        QD = C1
18      IF (ABS (QN) + ABS (QD))    20, 20,  19
19      IF (ABS (QN) - ABS (QD))    21,  22,  35
21      EM = ABS  (QN / QD)
        IF (EM - EP)    25,  24,  24
24      CS = COS  (ATAN  (EM))
        SN = SIN  (ATAN  (EM))
        GO TO 26
25      IF (QD)    27,  20,  20
27      SP = .70710678
        CP = SP
        GO TO 29
35      EM = ABS  (QD / QN)
        IF (EM - EP)    31,  30, 30
30      SN = 1.0 / SQRT  (1.0 + EM ** 2)
        CS = SN * EM
        GO TO 26
31      CS = 0.0
        SN = 1.0
        GO TO 26
22      CS = .70710678
        SN = CS
26      EM = SQRT  ((1.0 + CS) * 0.5)
        CS1 = SQRT  ((1.0 + EM) * 0.5)
        SN1 = SN / (4.0 * CS1 * EM)
        IF (QD)    32, 33, 33
32      CP = .70710678 * (CS1 + SN1)
        SP = .70710678 * (CS1 - SN1)
        GO TO 34
33      CP = CS1
        SP = SN1
34      IF (QN)  28,  29, 29
28      SP = -SP
        GO TO 29
20      NR = NR + 1
        GO TO 23
29      DO 36    K = 1, M
        EM = A(K,I) * CP + A(K,J) * SP
        A(K,J) = A(K,J) * CP - A(K,I) * SP
36      A(K,I) = EM
23      CONTINUE
        IF (NR - (N * N1) / 2)    14,  37,  14
C
37      DO 38    K = 1, M
        DO 38    L = 1, N
38      A(K,L) = A(K,L) * C(K)
C       DENORMALIZES ROWS OF A.
```

ROTATE (*continued*)

```
         WRITE (6,39)
 39      FORMAT(55HONEW FACTOR PATTERN. COLUMNS ARE FACTORS.                    )
         DO 40    J = 1, M
 40      WRITE (6,41) J,   (A(J,K),    K = 1, N)
 41      FORMAT(6HOTEST I3,1X, 10F10.3  / (10X, 10F10.3))
         DO 42    J = 1, N
 42      WRITE (7,43) J,   (A(K,J),    K = 1, M)
 43      FORMAT(6HFACTORI3,1X, 7F10.7 / (10X, 7F10.7))
C    NOTE PUNCHED PATTERN HAS FACTORS AS ROWS, TESTS AS COLUMNS.
         DO 44    J = 1, N
         C(J) = 0.0
         DO 44    K = 1, M
 44      C(J) = C(J) + A(K,J) ** 2
         EM = M
         DO 45    J = 1, N
 45      C(J) = C(J) / EM
         WRITE (6,46)
 46      FORMAT(55HOPROPORTION OF VARIANCE ACCOUNTED FOR BY EACH FACTOR.  )
         DO 47    J = 1, N
 47      WRITE (6,48) J,   C(J)
 48      FORMAT(8H-FACTOR I3,15H  VAR. PROP. = F5.3)
         GO TO 1
         END
 >
```

---

which had disparate values for the column sums $\sum_{j=1}^{p} s_{jk}^2$. This bias was removed by redefining the varimax criterion to provide what Kaiser calls the *normal varimax* criterion:

$$V = \sum_{k=1}^{n} \left\{ \frac{[p \sum_{j=1}^{p} (s_{jk}^2/h_j^2)^2 - (\sum_{j=1}^{p} s_{jk}^2/h_j^2)^2]}{p^2} \right\} \Bigg|_{max}$$

where h_j^2 is the communality of test j. After rotation to a maximum value of V, each vector is readjusted by multiplying each row of the resulting factor matrix by the square root of the test's communality. Rotation does not affect the values of the communalities of the tests.

The criterion V is maximized by the iterative application of a set of trigonometric functions. Kaiser (1958) delineates the method fully. The ROTATE program will perform either a quartimax or a varimax rotation, depending on the value of the argument METHOD in the control card. Notice that the iteration procedures for quartimax and varimax are identical except in the trigonometric functions applied in rotating a pair of rows and columns. Note too that although in principle orthogonal rotation involves an orthogonal transformation matrix **T** such that the new pattern **S** is produced by postmultiplying the original pattern by **T**, this method of computing rotations does not build up or report **T**.

We have incorporated the rotation procedures in a separate program to allow maximum flexibility to the user in experimenting with rotation of models of varying rank. Since a rotation performed on a subset of a selected set of orthogonal factors modifies only the locations of the reference axes within the subspace spanned by the subset of factors, without any influence on the other factors in the selected model and without interference with the orthogonality between the factors of the rotated subset and those of the remainder of the selected set, it is quite legitimate to withhold from rotation any factors of the original factoring which are deemed to be already properly placed, and to submit for rotation only a subset deemed improvable via rotation.

5.4 THE DISTRIBUTION OF ERRORS FOR A FACTOR MODEL

A factor analytic solution usually partitions a rank p observed correlation matrix $\mathbf{R}$ into two orthogonal, additive matrices such that

$$\mathbf{R} = \hat{\mathbf{R}} + \tilde{\mathbf{R}}$$

where $\hat{\mathbf{R}}$ is the reduced-rank factor theory matrix, of rank n for n factors, and $\tilde{\mathbf{R}}$ is the residual matrix of rank $p - n$. Remember that

$$\hat{\mathbf{R}} = \mathbf{SS}'$$

One source of information about the goodness of the theory for the data given by the n factors is the distribution of the errors for the theory given by the elements of $\tilde{\mathbf{R}}$. There are two types of error entries in $\tilde{\mathbf{R}}$ which require differentiation. On the diagonal are found the unexplained proportions of variance for the elements of the vector variable $\mathbf{z}$. These may be thought of as the complements of the obtained communalities,

$$\tilde{r}_{jj} = 1 - h_j^2 \qquad j = 1, 2, \ldots, p$$

Since the communalities are usually computed directly from the factor pattern, their complements are readily available. The analyst is interested in the proportion of observed battery variance accounted for by n factors, but he looks at that directly as the average h^2 for the factor pattern.

How large should a communality be? To answer this requires an estimate of the amount of variance a test has in common with other tests in the system. The most straightforward approach to such an estimate is to compute the squared multiple correlation of the test with a best linear function of the other $p - 1$ tests. These squared multiple correlations are lower bounds for the theoretical communalities of the tests. They may be thought of as redundancy measures for the tests. It seems to be good practice to require that the achieved communalities equal or exceed these lower bounds, and to choose n so that for every test, $j = 1, 2, \ldots, p$,

$$h_j{}^2 \geq R_{M_j}{}^2$$

Given the inverse of $\mathbf{R}$, the required squared multiple correlations are directly obtainable. Let r^{jj} be the jth diagonal element of $\mathbf{R}^{-1}$. Then

$$R_{M_j}{}^2 = 1 - \frac{1}{r^{jj}}$$

There are several ways to compute $\mathbf{R}^{-1}$, but if a complete eigensolution for $\mathbf{R}$ has been computed, as we are assuming it has, then

$$\mathbf{R}^{-1} = \mathbf{VL}^{-1}\mathbf{V}'$$

so that the inverse follows very simply from the eigenstructure. (Parenthetically, it can be seen that $\mathbf{R}$ must be of full rank if it is to have an inverse, for if its rank is less than its order, then for at least one eigenvalue the computation of $\mathbf{L}^{-1}$ involves trying to divide by zero.) We have provided for the computation of the multiple R^2 of each element of $\mathbf{z}$ with all the other elements in the PRINCO program, in order to make these lower bounds for obtained communalities available.

The other type of error entry in $\tilde{\mathbf{R}}$ is provided by the off-diagonal elements, and is errors in predicting or fitting observed correlations between elements of $\mathbf{z}$. Each error of this type occurs twice in $\tilde{\mathbf{R}}$, so we look only at the entries above the main diagonal. We would like these errors to be normally distributed around a mean of zero with a small variance. The program RESID reads the $\mathbf{R}$ matrix and a factor pattern $\mathbf{S}$, computes and reports the $\hat{\mathbf{R}}$ and $\tilde{\mathbf{R}}$ matrices for a given model, and distributes separately as grouped-data frequency distributions the diagonal variance errors and the off-diagonal correlation errors. The mean and standard deviation of each of these types of errors are reported.

The selection of rank for a factor model will always be influenced by several considerations, including and perhaps emphasizing the interpretability of the rotated factors resulting. It is usually desirable to retain enough factors for rotation to demonstrate that all major factors have been accounted for and that some nearly unique factors (significant loadings on only one test) have been reached. It is better to take too many rather than too few factors into rotation. However, after rotation a final selection of factors to be retained can be influenced by comparison of error distributions for alternative selections, if RESID is run on each of the alternatives.

Ours is an approach to data analysis which seeks to account for observed variances and covariances to a reasonable but not a high degree of precision. We do not seek a high degree of accuracy in fitting a model to data because we know there are substantial and unspecified degrees of error in the data themselves. We are perhaps naive in our willingness to live with the unreliabilities of our measurements. There is the possibility of a research strategy which involves heavy investment in reliability estimates for observed measurements, and then uses these reliabilities to adjust observed variances and covariances before factor analysis is done. When the **R** matrix has been adjusted in both its diagonal and off-diagonal elements for presumed degrees of unreliability in the elements of **z** it makes sense to seek a much higher degree of precision in the factor model itself. For an interesting comparison of this research strategy with ours, compare Shaycoft's (1967) factor analyses of TALENT data with ours, which we describe partially in this chapter, and which is documented in Lohnes (1966). The choice of rank for the factor model is perhaps the most important decision the factor analyst makes.

5.5   SCORING FACTORS: PROGRAMS COEFF AND FSCOR

We have already seen that the scoring of principal components requires only a rescaling of the eigenvectors to standardize the factor scores,

$$\mathbf{f}_i = \mathbf{B}'\mathbf{z}_i$$

where the factor score coefficients are defined as

$$\mathbf{B} = \mathbf{V}\mathbf{L}^{-1/2}$$

This is a special case, however, and has no application to the general orthogonal factors case, which prevails after varimax rotation of n components. The solution for factor score coefficients in the general case depends on simple multiple regression of the factors, one at a time, on the space of the original variables.

If each column of the factor structure, **S**, is construed as a set of criterion (the factor)-predictor (the original tests) correlations, premultiplication of the column by the inverse of the intercorrelation matrix for predictors yields a vector of regression weights for the multiple regression of that factor on the tests. Letting the vector $\mathbf{s}_k$ be the kth column of **S**

$$\mathbf{b}_k = \mathbf{R}^{-1}\mathbf{s}_k$$

```
C        RESIDUALS FROM A FACTOR THEORY.    A COOLEY-LOHNES PROGRAM.
C
C        THIS PROGRAM READS IN A CORRELATION MATRIX AND A FACTOR PATTERN.
C     IT COMPUTES THE PARTITION OF THE CORRELATION MATRIX INTO A THEORY
C     MATRIX AND A RESIDUAL MATRIX, AND DOES A GROUPED-DATA FREQUENCY
C     DISTRIBUTION FOR THE RESIDUALS. ON ONE SUBMISSION, UP TO TEN
C     DIFFERENT RANK MODELS MAY BE TESTED. THAT IS, TEN VALUES OF N MAY
C     BE SPECIFIED, WHERE THE MODEL IS TO BE BASED ON THE FIRST N FACTORS.
C
C        INPUT
C
C     1) TEN CARDS DESCRIBING THE PROBLEM, TO BE A TEXT ON THE OUTPUT.
C     2) CONTROL CARD (CARD 11),    COLS 1-2   M = ORDER OF THE R MATRIX,
C           COLS 3-4   N = RANK OF FIRST MODEL,   COLS 5-6   N = RANK
C           OF SECOND MODEL,    ETC. (UP TO TEN VALUES OF N ARE
C           ACCEPTABLE, ALTHOUGH LESS THAN 10 MAY BE SUPPLIED).
C     3)FORMAT CARD (CARD 12) SPECIFIES INPUT FORMAT FOR R MATRIX.
C     4) R MATRIX IN UPPER-TRIANGULAR FORM.
C     5) TRANSPOSED FACTOR PATTERN, A-PRIME, SQUARE OF ORDER M.
C        NOTE THAT ROWS ARE FACTORS AND COLUMNS ARE TESTS IN A-PRIME.
C
C        SUBROUTINE MPRINT IS REQUIRED.
C
      DIMENSION   TIT(16), R(80,80), A(80,80), B(80,80),  N(10), N1(42),
     C N2(42)
C
    1 WRITE (6,2)
    2 FORMAT(55H1RESIDUALS FROM A FACTOR THEORY.   A COOLEY-LOHNES PROGR)
      DO 3    J = 1, 10
      READ   (5,4) (TIT(K),    K = 1, 16)
    3 WRITE (6,4) (TIT(K),    K = 1, 16)
    4 FORMAT (16A5)
      READ   (5,5) M,   (N(J),   J = 1, 10)
    5 FORMAT (11I2)
      READ   (5,4) (TIT(J),    J = 1, 16)
      DO 6    J = 1, M
    6 READ   (5,TIT) (R(J,K),    K = J, M)
      DO 8    J = 1, M
    8 READ   (5,9) (A(J,K),    K = 1, M)
    9 FORMAT (10X, 7F10.7 / (10X, 7F10.7))
      DO 10    J = 1, M
      DO 10    K = J, M
   10 R(K,J) = R(J,K)
      JL = 1
   11 IF (N(JL))    46, 46,  12
   12 NIT = N(JL)
      DO 13    J = 1, M
      DO 13    K = 1, M
      B(J,K) = 0.0
      DO 13    I = 1, NIT
   13 B(J,K) = B(J,K) + A(I,J) * A(I,K)
      WRITE (6,14) NIT
   14 FORMAT(24H0THEORY MATRIX FOR RANK I3)
      CALL MPRINT (B, M)
      DO 141   J = 1, M
      DO 141   K = 1, M
  141 B(J,K) = R(J,K) - B(J,K)
      WRITE (6,15) NIT
   15 FORMAT(26H0RESIDUAL MATRIX FOR RANK I3)
      CALL MPRINT (B, M)
C
      V1 = .20
      V2 = 100.0
      DO 16    J = 1, 42
      N1(J) = 0
   16 N2(J) = 0
```

```
         SUM1 = 0.0
         SUM2 = 0.0
         LIM = V1 * V2 + .000001
         V = 1.0 / V2
         SS1 = 0.0
         SS2 = B(M,M) **2
         MM = M - 1
         DO 17   J = 1, MM
         V3 = B(J,J) + .000002
         L1 = V3 * V2 - (V3 - ABS (V3)) / (2.0 * V3) * .9999
         SUM2 = SUM2 + B(J,J)
         SS2 = SS2 + B(J,J)**2
         NN = L1 + 22
         IF (L1 - LIM)  18,  19, 19
19       N2(42) = N2(42) + 1
         GO TO 22
18       IF (L1 + LIM)  20,  21, 21
20       N2(1) = N2(1) + 1
         GO TO 22
21       N2(NN) = N2(NN) + 1
22       LL = J + 1
         DO 17    K = LL, M
         V3 = B(J,K) + .000002
         L1 = V3 * V2 - (V3 - ABS (V3))/(2.0 * V3) * .9999
         NN = L1 + 22
         IF (L1 - LIM)  24,  23, 23
23       N1(42) = N1(42) + 1
         GO TO 27
24       IF (L1 + LIM)  25,  26, 26
25       N1(1) = N1(1) + 1
         GO TO 27
26       N1(NN) = N1(NN) + 1
27       SUM1 = SUM1 + B(J,K)
17       SS1 = SS1 + B(J,K) **2
         V3 = B(M,M) + .000002
         L1 = V3 * V2 - (V3 - ABS (V3)) /(2.0 * V3) * .9999
         NN = L1 + 22
         IF (L1 - LIM)  43,  42, 42
42       N2(42) = N2(42) + 1
         GO TO 47
43       IF (L1 + LIM)  44,  45, 45
44       N2(1) = N2(1) + 1
         GO TO 47
45       N2(NN) = N2(NN) + 1
47       WRITE (6,28)
28       FORMAT(19HOOFF-DIAGONAL TERMS ,41X,13HMAIN DIAGONAL)
         WRITE (6,29)
29       FORMAT(1H0,2(14HLOWER LIMIT OF, 46X))
         WRITE (6,30)
30       FORMAT(1H ,2(15HCLASS INTERVAL , 5X, 9HFREQUENCY, 31X))
         WRITE (6,31)
31       FORMAT (1H0)
         IF (LIM - 20)  33,  32,  33
32       KR = 41
         KR1 = 2
         V4 = .20
         GO TO 34
33       KR = 31
         KR1 = 11
         V4 = .01
34       WRITE (6,35)
35       FORMAT(1H+,7X,2(9HAND ABOVE, 51X))
         WRITE (6,36) V4,  N1(42),  V4,  N2(42)
36       FORMAT(1H ,2(F6.2, 14X, I4, 36X))
         DO 37    J = KR1,  KR
```

RESID (*continued*)

```
        V4 = V4 - V
        L2 = 43 - J
   37   WRITE (6,36) V4,   N1(L2),  V4,  N2(L2)
        WRITE (6,38) V1,   N1(1),   V1,  N2(1)
   38   FORMAT(1H ,2(11HLESS THAN -, F3.2, 7X, I3, 36X))
        S1 = (M * M - M) / 2
        S2 = M
        WRITE (6,39) S1, S2
   39   FORMAT(1H0,2(3H  N, 27X, F4.0, 26X))
        SUM1 = SUM1 / S1
        SUM2 = SUM2 / S2
        SS1 = SS1 / S1 - (SUM1 / S1)**2
        SS2 = SS2 / S2 - (SUM2 / S2)**2
        SS1 = SQRT  (SS1)
        SS2 = SQRT  (SS2)
        WRITE (6,40) SUM1,   SUM2
   40   FORMAT(1H0,2(4HMEAN, 26X, F10.4, 20X))
        WRITE (6,41) SS1, SS2
   41   FORMAT (1H0,2(4HS.D., 26X, F10.4, 20X))
C
        JL = JL + 1
        IF (JL - 10) 11, 11,   46
   46   GO TO 1
        END
>
```

If we assemble the n column vectors b_k in the matrix B of score coefficients, then

$$B = R^{-1}S$$

and

$$f_i = B'z_i$$

Program COEFF takes the inverse of R as punched by PRINCO and the selected factors as the transposed pattern S' (n rows and p columns) which may be assembled as any mix of orthogonal factors from PRINCO, FACTOR, and ROTATE, and computes and punches the transposed coefficients matrix B' (n rows and p columns). Program FSCOR reads test means and standard deviations as punched by CORREL, the B' matrix from COEFF or from PRINCO, and then reads each raw score vector and computes and punches the corresponding factor score vector in z-score terms. The user who desires a different scale convention (we favor two-digit scores with mean 50 and standard deviation 10) may readily modify FSCOR to suit his whim (we would multiply each factor z-score by 10 and add 50 to the product). If really large numbers of score vectors are to be processed the standardizing and scaling operations should be built into the coefficients matrix itself.

In our Project TALENT research we have converted tens of thousands of 100-element raw score vectors to 22-element factor score vectors, and have reported on a series of predictive validity studies computed in the factor score space (Cooley and Lohnes, 1968). We believe that the results of this longitudinal psychometric research into the predictable aspects of career development are much more understandable in the factor rubrics than they would have been in the original measurement rubrics.

The notion of scoring factors by regressing them on the original measures is very general, since every factoring method reports the structure correlations of the factors with the original measures. There is a numerical analysis problem that can arise in any regression operation, however, in that R^{-1} may be computed with poor precision because of near collinearities among the original measures, leading to near singularity of R. The closer the approximation to rank reduction the greater this problem becomes. The indicators are, of course, very small determinants and nearly zero eigenvalues. It can happen that R^{-1} will be so poorly computed by MATINV that the intercorrelations among the factor scores will not be very close to zero and the correlations of the factor scores with the original measures will not be very close to the structure coefficients.

The principal components coefficients give exact solutions for the principal factors, not regression estimates. The following algebra shows that after any rigid rotation such as Varimax of a subset of the principal components, coefficients giving an exact solution for the rotated components are readily available from the formula

$$B_1 = S_1(S_1'S_1)^{-1}$$

where S_1 is the structure matrix (or pattern) for the rotated factors, and B_1 is the corresponding score coefficients.[1] Let S be the full rank principal-components pattern matrix, square and of order p, such that the full set of principal factors "loaded" by their pattern weights completely reproduce the original score matrix:

$$Z = FS' \tag{5.5.1}$$

Equation 5.5.1 is the fundamental equation of factor analysis. We rewrite it with subscripts giving the orders of the elements for clarity:

$$Z_{(N \times p)} = F_{(N \times p)} S'_{(p \times p)}$$

---

[1] The proof is credited to Professor S. David Farr, SUNY at Buffalo, to whom we are indebted for his permission to publish this very useful development.

In eigenstructure terms the equivalent of (5.5.1) is

$$\mathbf{Z} = \mathbf{F}(\mathbf{V}\mathbf{L}^{1/2})'$$
$$= \mathbf{Z}(\mathbf{V}\mathbf{L}^{-1/2})(\mathbf{V}\mathbf{L}^{1/2})'$$
$$= \mathbf{Z}\mathbf{V}\mathbf{L}^{-1/2}\mathbf{L}^{1/2}\mathbf{V}'$$
$$= \mathbf{Z}\mathbf{V}\mathbf{V}' = \mathbf{Z}$$

Solving (5.5.1) for $\mathbf{F}$ gives:

$$\mathbf{Z}\mathbf{S} = \mathbf{F}\mathbf{S}'\mathbf{S}$$
$$\mathbf{Z}\mathbf{S}(\mathbf{S}'\mathbf{S})^{-1} = \mathbf{F}(\mathbf{S}'\mathbf{S})(\mathbf{S}'\mathbf{S})^{-1}$$
$$\mathbf{F} = \mathbf{Z}\mathbf{S}(\mathbf{S}'\mathbf{S})^{-1} \tag{5.5.2}$$

In eigenstructure terms we get the results of Chapter 4 by substituting in (5.5.2) that chapter's definition of $\mathbf{S}$ as

$$\mathbf{S} = \mathbf{V}\mathbf{L}^{1/2}$$
$$\mathbf{F} = \mathbf{Z}(\mathbf{V}\mathbf{L}^{1/2})[(\mathbf{V}\mathbf{L}^{1/2})'(\mathbf{V}\mathbf{L}^{1/2})]^{-1}$$
$$= \mathbf{Z}\mathbf{V}\mathbf{L}^{1/2}(\mathbf{L}^{1/2}\mathbf{V}'\mathbf{V}\mathbf{L}^{1/2})^{-1}$$
$$= \mathbf{Z}\mathbf{V}\mathbf{L}^{1/2}\mathbf{L}^{-1}$$
$$= \mathbf{Z}\mathbf{V}\mathbf{L}^{-1/2} = \mathbf{Z}\mathbf{B}$$

Any rigid rotation of all the principal components amounts to postmultiplying $\mathbf{S}$ or $\mathbf{F}$ by an orthonormal transformation matrix $\mathbf{T}$ such that

$$\mathbf{T}\mathbf{T}' = \mathbf{T}'\mathbf{T} = \mathbf{I}$$

and

$$\mathbf{S}_1 = \mathbf{S}\mathbf{T}$$

and

$$\mathbf{F}_1 = \mathbf{F}\mathbf{T}$$

Manipulating the last two equations gives:

$$\mathbf{S}_1\mathbf{T}' = \mathbf{S}\mathbf{T}\mathbf{T}' = \mathbf{S}$$

and

$$\mathbf{F}_1\mathbf{T}' = \mathbf{F}\mathbf{T}\mathbf{T}' = \mathbf{F}$$

Substituting in (5.5.2) we get

$$\mathbf{F}_1\mathbf{T}' = \mathbf{Z}(\mathbf{S}_1\mathbf{T}')[(\mathbf{S}_1\mathbf{T}')'(\mathbf{S}_1\mathbf{T}')]^{-1}$$
$$= \mathbf{Z}(\mathbf{S}_1\mathbf{T}')(\mathbf{T}\mathbf{S}_1'\mathbf{S}_1\mathbf{T}')^{-1}$$
$$= \mathbf{Z}\mathbf{S}_1\mathbf{T}'\mathbf{T}(\mathbf{S}_1'\mathbf{S}_1)^{-1}\mathbf{T}'$$

and postmultiplying both sides by $\mathbf{T}$ gives:

$$\mathbf{F}_1 = \mathbf{Z}\mathbf{S}_1\mathbf{T}'\mathbf{T}(\mathbf{S}_1'\mathbf{S}_1)^{-1}\mathbf{T}'\mathbf{T}$$

or

$$\mathbf{F}_1 = \mathbf{Z}\mathbf{S}_1(\mathbf{S}_1'\mathbf{S}_1)^{-1} \qquad (5.5.3)$$

To move to the usual case of a rigidly rotated reduced rank solution, we note that if the n rotated factors are reported in $\mathbf{S}_{1_{(p \times n)}}$ and the remaining $p - n$ unrotated factors are retained in $\mathbf{S}_{2_{(p \times p - n)}}$, we have the following partition of $\mathbf{S}$

$$\mathbf{S} = (\mathbf{S}_1 : \mathbf{S}_2)$$

Similarly, if the n rotated factor scores are reported in $\mathbf{F}_{1_{(N \times n)}}$ and the remaining $p - n$ unrotated principal factor scores are retained in $\mathbf{F}_{2_{(N \times p - n)}}$, we have the partition

$$\mathbf{F} = (\mathbf{F}_1 : \mathbf{F}_2)$$

We can now define a transformation matrix with the feature that the first n columns of $\mathbf{S}$ or $\mathbf{F}$ are operated on and the remaining $p - n$ columns are not disturbed:

$$\mathbf{T} = \begin{pmatrix} \mathbf{T}_1 & \vdots & \mathbf{N} \\ \hdotsfor{3} \\ \mathbf{N} & \vdots & \mathbf{I} \end{pmatrix}$$

where $\mathbf{N}$ is a null matrix (all elements zero) and $\mathbf{I}$ is the identity matrix. Letting $\mathbf{S}_c$ and $\mathbf{F}_c$ be the first parts of the partitioned principal factors structure and score matrices before rotation we have:

$$(\mathbf{S}_1 : \mathbf{S}_2) = (\mathbf{S}_c\mathbf{T} : \mathbf{S}_2)$$

and

$$(\mathbf{F}_1 : \mathbf{F}_2) = (\mathbf{F}_c\mathbf{T} : \mathbf{F}_2)$$

From (5.5.3)

$$(\mathbf{F}_1 : \mathbf{F}_2) = \mathbf{Z}(\mathbf{S}_1 : \mathbf{S}_2)\begin{pmatrix} \mathbf{S}_1'\mathbf{S}_1 & \vdots & \mathbf{S}_1'\mathbf{S}_2 \\ \hdotsfor{3} \\ \mathbf{S}_2'\mathbf{S}_1 & \vdots & \mathbf{S}_2'\mathbf{S}_2 \end{pmatrix}^{-1} \qquad (5.5.4)$$

but

$$\mathbf{S}_1'\mathbf{S}_2 = (\mathbf{S}_c\mathbf{T}_1)'\mathbf{S}_2 = \mathbf{T}_1'\mathbf{S}_c'\mathbf{S}_2 = \mathbf{T}_1'\mathbf{N} = \mathbf{N}$$

and $S_2'S_1$ is similarly null, so that (5.5.4) can be written:

$$(\mathbf{F}_1 : \mathbf{F}_2) = \mathbf{Z}(\mathbf{S}_1 : \mathbf{S}_2)\begin{pmatrix} \mathbf{S}_1'\mathbf{S}_2 & \mathbf{N} \\ \hline \mathbf{N} & \mathbf{S}_2'\mathbf{S}_2 \end{pmatrix}^{-1}$$

$$= \mathbf{Z}(\mathbf{S}_1 : \mathbf{S}_2)\begin{pmatrix} (\mathbf{S}_1'\mathbf{S}_1)^{-1} & \mathbf{N} \\ \hline \mathbf{N} & (\mathbf{S}_2'\mathbf{S}_2)^{-1} \end{pmatrix}$$

$$= \mathbf{Z}[\mathbf{S}_1(\mathbf{S}_1'\mathbf{S}_1)^{-1} : \mathbf{S}_2(\mathbf{S}_2'\mathbf{S}_2)^{-1}]$$

and

$$\mathbf{F}_1 = \mathbf{Z}\mathbf{S}_1(\mathbf{S}_1'\mathbf{S}_1)^{-1}$$

while $\mathbf{F}_2$ is simply the remaining unrotated principal factor scores.

Some readers will recognize that (5.5.4) has a long history in the literature on factor scoring as the least-squares solution for the regression of the original measures on the common factors. Harman (1967, Sec. 16.7) describes this as "the short method" that avoids calculating $\mathbf{R}^{-1}$ but depends for its goodness on the closeness of the approximation of $\mathbf{R}$ by $\mathbf{SS}'$, and Horst (1965, Sec. 20.6) shows that it minimizes the sum of squared errors between $\mathbf{Z}$ and $\hat{\mathbf{Z}}$ as regressed on $\mathbf{F}$. Clearly Horst and many others prefer to regress tests on factors, whereas Harman and many others including the present authors prefer to regress factors on tests, if a regression solution is needed. Farr's point is that rigid rotations of exactly defined components are just as exactly defined and don't have to be regressed at all.

By actual computation on the rectangles test problem the reader can verify that

$$\mathbf{F}_1 = \mathbf{Z}\mathbf{R}^{-1}\mathbf{S}_1$$

and

$$\mathbf{F}_1 = \mathbf{Z}\mathbf{S}_1(\mathbf{S}_1'\mathbf{S}_1)^{-1}$$

produce almost exactly the same results when $\mathbf{R}^{-1}$ is well computed. Both ways yield factor scores that are nicely orthogonal, that reproduce the structure coefficients of $\mathbf{S}_1$ nicely when actually correlated with $\mathbf{Z}$, and that correlate almost perfectly factor for factor across methods. Farr[1] provides an example from research experience involving 60 tests and 16 Varimax factors in which a poor $\mathbf{R}^{-1}$ leads to trouble that is ameliorated by working with $(\mathbf{S}_1'\mathbf{S}_1)^{-1}$. In that example the determinant of $\mathbf{R}$ had 15 leading zeroes. However, there is no need to go beyond the *prima facie* case for using an exact solution when it is available.

---

[1] S. David Farr, SUNY at Buffalo, Private communication.

COEFF

```
C       FACTOR SCORE COEFFICIENTS.    A COOLEY-LOHNES PROGRAM.
C
C       THIS PROGRAM READS IN R-INVERSE AND A, THE FACTOR PATTERN, AND
C       COMPUTES C, THE FACTOR SCORE COEFFICIENTS MATRIX ( A SYSTEM OF
C       MULTIPLE REGRESSION BETA WEIGHTS, REGRESSING THE FACTORS ON THE
C       STANDARDIZED VECTOR VARIABLE FROM WHICH THEY WERE DERIVED).
C
C       INPUT
C       1) A TEN CARD TEXT DESCRIBING THE PROBLEM FOR THE PRINTOUT.
C       2) CONTROL CARD (CARD 11),    COLS 1-2   N = NUMBER OF FACTORS,
C                                      COLS 3-4   M = NUMBER OF TESTS.
C       3) R-INVERSE, UPPER-TRIANGULAR, FORMAT (10X, 5E14.7), AS PUNCHED
C            BY PRINCO.
C       4) A, THE PATTERN, RECTANGULAR WITH N ROWS (FACTORS) AND M COLUMNS
C            (TESTS), FORMAT (10X, 7F10.7), AS PUNCHED BY PRINCO,
C            FACTOR, ROTATE, OR ANY MIX OF FACTORS FROM THESE PROGRAMS.
C
C
C       PUNCHED OUTPUT
C       1) C-PRIME, THE COEFFICIENTS MATRIX, NXM, FORMAT (10X, 5E14.7).
C
C
        DIMENSION   TIT(16),  R(100,100),  A(50,100),  C(50,100)
C
1       WRITE (6,2)
2       FORMAT(55H1FACTOR SCORE COEFFICIENTS.  A COOLEY-LOHNES PROGRAM.  )
        DO 3    J = 1, 10
        READ   (5,4) (TIT(K),   K = 1, 16)
3       WRITE  (6,4) (TIT(K),   K = 1, 16)
4       FORMAT (16A5)
        READ   (5,5) N,   M
5       FORMAT (2I2)
        WRITE  (6,6) N,   M
6       FORMAT(17H0COEFFICIENTS FOR   I3,13H FACTORS AND   I3,7H TESTS.)
        DO 7    J = 1, M
7       READ   (5,8) (R(J,K),   K = J, M)
8       FORMAT (10X, 5E14.7)
        DO 9    J = 1, M
        DO 9    K = J, M
9       R(K,J) = R(J,K)
        DO 10   J = 1, N
10      READ   (5,11) (A(J,K),   K = 1, M)
11       FORMAT (10X, 7F10.7)
        DO 12   J = 1, N
        DO 12   K = 1, M
        C(J,K) = 0.0
        DO 12   L = 1, M
12      C(J,K) = C(J,K) + A(J,L) * R(L,K)
        DO 13   J = 1, M
13      WRITE (6,14) J,  (C(K,J),   K = 1, N)
14      FORMAT(5H0TEST I3,2X, 10F10.3 / (10X, 10F10.3))
        DO 15   J = 1, N
15      WRITE (7,16) J,  (C(J,K),   K = 1, M)
16      FORMAT(7H FACTOR I3, 5E14.7 / (10X, 5E14.7))
        GO TO 1
        END
>
```

```
C      FACTOR SCORES.    A COOLEY-LOHNES PROGRAM.
C
C      THIS PROGRAM COMPUTES FACTOR SCORES FOR NS SUBJECTS ON N FACTORS
C      FROM M TESTS.
C
C      INPUT
C      1) TEN CARDS DESCRIBING THE PROBLEM IN A TEXT FOR THE OUTPUT.
C      2) CONTROL CARD (CARD 11),    COLS 1-2   M = NUMBER OF TESTS,
C                                    COLS 3-4   N = NUMBER OF FACTORS,
C                                    COLS 5-9   NS = NUMBER OF SUBJECTS.
C                                    COL   10   L = 0 TO READ PATTERN
C                                               L = 1 TO READ COEFFICIENTS
C      3) FORMAT CARD (CARD 12) FOR SCORES.
C         SUBJECT ID IS READ (INTEGER CONVENTION) BEFORE EACH SCORE VECTOR.
C      4) TEST MEANS (10X, 5E14.7)
C      5) TEST STANDARD DEVIATIONS (10X, 5E14.7)
C      6) PATTERN MATRIX TRANSPOSE(10X,7F10.7), IN WHICH ROWS ARE FACTORS
C         AND COLUMNS ARE TESTS.  OR, IF L = 1,
C         COEFFICIENTS MATRIX (10X, 5E14.7), IN WHICH ROWS ARE FACTORS
C         AND COLUMNS ARE TESTS.  4), 5), AND 6) AS PUNCHED BY
C         PRINCO OR COEFF.
C      7) SCORE VECTORS.  ID IS READ FROM FIRST FIELD.
C
C
C      PUNCHED OUTPUT
C      1) FACTOR SCORE VECTORS FOR SUBJECTS ARE PUNCHED, TO BE READ BY
C         FORMAT (20X, 10F6.2). (I.D.S ARE INCLUDED IN SKIPPED FIELD.)
C
C
C      SUBROUTINE MATINV IS REQUIRED.
C
C
       DIMENSION  TIT(16),  C(50,100),  W(100),  X(100),  Y(100),
      C    Z(50),   V(50),   D(50,100)
C
  1    WRITE (6,2)
  2    FORMAT(51H1FACTOR SCORES.      A COOLEY-LOHNES PROGRAM            )
       WRITE (7,2)
       DO 3    J = 1, 10
       READ  (5,4) (TIT(K),    K = 1, 16)
  3    WRITE (6,4) (TIT(K),    K = 1, 16)
  4    FORMAT (16A5)
       READ  (5,5) M,   N,   NS,   L
  5    FORMAT (2I2, I5, I1)
       WRITE (6,6) M,   N,   NS
  6    FORMAT(1H0I3,7H TESTS,I3,9H FACTORS,I6,10H SUBJECTS.  )
       WRITE (7,6) M,   N,   NS
       READ  (5,4) (TIT(J),    J = 1, 16)
       READ  (5,7) (W(J),    J = 1, M)
  7    FORMAT (10X, 5E14.7)
       READ  (5,7) (Y(J),    J = 1, M)
       DO 8    J = 1, N
       IF(L)   30,30,31
  30   READ  (5,27) (C(J,K),    K = 1, M)
  27   FORMAT(10X,7F10.7)
       GO TO 8
  31   READ  (5,7) (C(J,K),    K = 1, M)
  8    CONTINUE
C
       IF(L)   15,15,16
  15   DO 17   J = 1, N
       DO 17   K = 1, N
       D(J,K) = 0.0
       DO 17   I = 1, M
  17   D(J,K) = D(J,K) + C(J,I) * C(K,I)
       CALL MATINV(D,N,DET)
```

FSCOR (*continued*)

```
          WRITE(6,18)   DET
    18    FORMAT(30H0DETERMINANT OF A-PRIME * A = E14.7)
          DO  19 K = 1, M
          DO 20   J = 1, N
          Z(J) = 0.0
          DO 20 I = 1, N
    20    Z(J) = Z(J) + C(I,K) * D(I,J)
          DO 19  I = 1, N
    19    C(I,K) = Z(I)
C     C NOW CONTAINS COEFFICIENTS COMPUTED AS
C           (A * ((A-PRIME * A) INVERSE)) TRANSPOSE.
C
C
C
    16    DO 9    J = 1, N
          DO 9    K = 1, M
    9     C(J,K) = C(J,K) / Y(K)
          DO 10    J = 1, N
          Z(J) = 0.0
          DO 10    K = 1, M
    10    Z(J) = Z(J) + C(J,K) * W(K)
C     RAW SCORE WEIGHTS ARE NOW IN C, AND CORRECTIONS ARE IN Z.
C
          DO 14   I = 1, NS
          READ   (5,TIT) ID,   (X(J),    J = 1, M)
          DO 11    J = 1, N
          V(J) = 0.0
          DO 11    K = 1, M
    11    V(J) = V(J) + C(J,K) * X(K)
          DO 12    J = 1, N
    12    V(J) = V(J) - Z(J)
          WRITE (6,13) I,   ID,   (V(J),    J = 1, N)
    13    FORMAT (I7,3X,I7,3X, 10F6.2  / (20X, 10F6.2))
    14    WRITE (7,13) I,   ID,   (V(J),    J = 1, N)
          GO TO 1
          END
    >
```

5.6   RESEARCH EXAMPLE

One facet of the extensive assessments of about 440,000 high school students collected by Project TALENT in 1960 was a set of 38 typical performance scales. Lohnes (1966) correlated these 38 scales, along with sex and school grade (9th or 12th) for a random sample of about 16,000 subjects. His objective was to obtain a suitable reduced rank factor model for these "motives." Table 5.6 names the variables. The first 11 were scaled from an autobiographical inventory, the next 10 from an adjectival check list, and the last 17 from an interests inventory. In the 40th order correlation matrix point biserial correlations were used to represent the relations of the 38 scales with the two dichotomous design variables of sex and grade. The correlation

TABLE 5.6   38 Motives Domain Variables

	Mnemonic	Code	Name of Scale
1	MEM	A-001	Memberships
2	LEA	A-002	Leadership Roles
3	HOB	A-003	Hobbies
4	WOR	A-004	Work
5	SOC	A-005	Social
6	REA	A-006	Reading
7	STU	A-007	Studying
8	CUR	A-008	Curriculum
9	COU	A-009	Courses
10	GRA	A-010	Grades
11	GUI	A-001	Guidance
12	NSO	R-601	Sociability
13	NSS	R-602	Social Sensitivity
14	NIM	R-603	Impulsiveness
15	NVI	R-604	Vigor
16	NCA	R-605	Calmness
17	NTI	R-606	Tidiness
18	NCU	R-607	Culture
19	NLE	R-608	Leadership
20	NSC	R-609	Self-confidence
21	NMP	R-610	Mature Personality
22	IPS	P-701	Physical Science, Engineering, Mathematics
23	IBS	P-702	Biological Science, Medicine
24	IPU	P-703	Public Service
25	ILL	P-704	Literary, Linguistic
26	ISS	P-705	Social Service
27	IAR	P-706	Artistic
28	IMU	P-707	Musical
29	ISP	P-708	Sports
30	IHF	P-709	Hunting, Fishing
31	IBM	P-710	Business Management
32	ISA	P-711	Sales
33	ICO	P-712	Computation
34	IOW	P-713	Office Work
35	IMT	P-714	Mechanical, Technical
36	IST	P-715	Skilled Trades
37	IFA	P-716	Farming
38	ILA	P-717	Labor

of sex with grade was computed as a phi coefficient and turned out to be zero.

It was assumed that the sex and grade effects in the data were strictly resultants of constant values of group membership. That is, it was assumed that the sole influence of sex and grade was to provide a constant increment or decrement for each sex and grade group on each measurement scale. The total sample **R** matrix would be disturbed by the influence of correlated subsample means under this assumption. It was necessary to purge the correlation system of the linear effects of sex and grade before factoring for latent dimensions of individual differences.

Using the FACTOR program, arbitrary factors were passed right through the sex and grade variables, extracting all sex and grade linked variance from the total sample **R** matrix. After this extraction the communalities of the sex and grade variables were exactly one and the residual variance-covariance matrix had two rows and columns entirely zeros. Ignoring them, the remainder of the residual matrix was exactly analogous to the error covariance matrix for the restricted linear model

$$z_{jki} = b_j + c_k + e_{jki}$$

where **b** is the sex effect and **c** is the grade effect. The four populations of the design are assumed to have equal dispersions, and factoring the residual covariance matrix

$$\tilde{\mathbf{R}} = \mathbf{R} - \hat{\mathbf{R}}_{\text{Sex}} - \hat{\mathbf{R}}_{\text{Grade}} = \mathbf{R} - s_S s_S' - s_G s_G'$$

is factoring the maximum likelihood estimate of the common dispersion. The locus of orthogonality of the individual differences factors will then be within each sex-grade population.

Table 5.7 reports the names of eleven Varimaxed principal factors of the residual matrix, ranked according to percentage of generalized variance extracted. The control factors of sex and grade are ranked for comparison. The 13 factors accounted for 76 percent of the generalized variance, and there were several indications that no further useful common factors were available. Table 5.8 reports the edited factor pattern, which testifies to the almost amazing power of the Varimax algorithm to create meaningful structure for correlation data. Note that the communalities for the rank 13 model are mostly well above the squared multiple R's. Table 5.9 reports the distribution of residual covariances from the rank 13 model and seems to show a reasonably tight fit of the model to the data.

The structure for the motives domain variables established by the rank 13 model was interpreted as follows:

> The salient features of the pattern are first, that an almost general factor of the adjectival check list scales explains most of the intercorrelation among the

TABLE 5.7   Motives Domain Factors

Mnemonic	Factor Name	Variance Extracted (in percent)
CON	Conformity Needs	11.1
SEX	Sex	9.1
BUS	Business Interests	8.7
OUT	Outdoors, Shop Interests	6.8
SCH	Scholasticism	6.6
CUL	Cultural Interests	5.8
SCI	Science Interests	4.3
GRD	Grade	4.2
ACT	Activity Level	4.0
LEA	Leadership	3.1
IMP	Impulsion	2.8
SOC	Sociability	2.8
INT	Introspection	2.4

(13 factors extract 71.5% of variace)

adjectival self-concepts by means of a social desirability response set. Second, two factors representing independent dimensions of striving explain most of the intercorrelation among the autobiographical inventory scales. Third, five independent factors account for the intercorrelation among the interest scales, one of which is the sex factor. The most noteworthy feature of the hierarchy of factors is the almost complete absence of interlocking relations among the three types of indicators. Each of the seven major factors is defined by indicators drawn exclusively from one subset of scales. The grouping is:

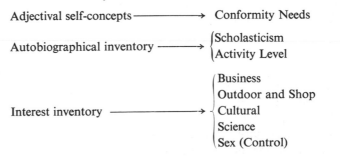

Cooley and Lohnes (1968) report on an extensive program of predictive validity studies of these factors, in which they scored the factors for thousands of TALENT subjects and related the factor scores to subsequent developmental criteria. They argue that the prediction system built on the factor

TABLE 5.8 Motives Domain Variable-Factor Correlations ≥ .35

Test	CON	SEX	BUS	OUT	SCH	CUL	SCI	GRD	ACT	LEA	IMP	SOC	INT	h^2	R^2
MEM									60					61	31
LEA										83				75	17
HOB									62					68	44
WOR									71					64	29
SOC												62		66	26
REA					39								55	66	25
STU					72									74	52
CUR					70									62	35
COU					53			44						56	40
GRA					75									66	41
GUI	63				55									54	39
NSO	72											43		68	48
NSS														66	56
NIM											87			83	16
NVI	67													61	45
NCA	74													66	52
NTI	75													68	53
NCU	72													70	58
NLE	51									44				61	39
NSC	45												66	76	30
NMP	78													75	64
IPS		47					62							82	77
IBS			51				74							75	56
IPU			39				37							64	55
ILL			46			68								82	73
ISS		−49				35								65	63
IAR						70								70	55
IMU						77								70	44
ISP		35		50										68	50
IHF		50		61										72	58
IBM			74											78	71
ISA			74											68	58
ICO		−55	79											73	62
IOW			62											74	67
IMT		63		51										80	83
IST		35	45	67										84	81
IFA				77										73	55
ILA			45	61										79	68

TABLE 5.9 Off Diagonal Terms
(Upper Triangle) of Residual
Matrix

Lower Limit of Class Interval	Frequency
.00	0
.19	0
.18	0
.17	0
.16	0
.15	0
.14	0
.13	0
.12	0
.11	0
.10	0
.09	1
.08	4
.07	7
.06	7
.05	10
.04	29
.03	27
.02	45
.01	94
—.00	168
—.01	81
—.02	78
—.03	69
—.04	46
—.05	37
—.06	19
—.07	20
—.08	18
—.09	6
—.10	2
—.11	3
—.12	1
—.13	3
—.14	1
—.15	1
—.16	1
—.17	0
—.18	1
—.19	1
—.20	0
N	780
Mean	—.0070
S.D.	.0355

variables is much more understandable than one based on the original variables would be. They demonstrate that for some criteria, at least, the predictive efficiency of the full rank system is conserved completely by the reduced rank model.

5.7 EXERCISES

1. What sequence of programs would you use to obtain varimax factor scores starting with raw data? Try it with the 234 Project TALENT male ability data. If you then ran the resulting factor scores back through CORREL, what would you expect to find? How many factors did you rotate?

2. Expand the numerical example of FACTOR in Section 5.2 by including all eight TALENT ability measures, but again using reading comprehension and mechanical reasoning as your two hypothesized "factors." How do these results compare with varimax loadings for corresponding factors from Exercise 1?

‖‖‖

CHAPTER SIX

Canonical Correlation

6.1 MATHEMATICS OF CANONICAL CORRELATION

An investigator frequently has two vector variables, z_{1i} and z_{2i}, available for a sample of subjects, each vector variable representing multiple measurements from a particular domain of traits. It might be that z_{1i} would be a set of predictor measures, or independent measures in a research, and z_{2i} would be a set of criterion measures, or dependent measures. Another possibility is that the two domains are conceptually different, although they are measured concurrently on the subjects. Thus, z_{1i} might be a set of ability tests, and z_{2i} a set of interest inventory scales. The research question then would be to display the interrelatedness of interests and abilities. Of course, the bivariate correlations between pairs of measures taken one from each domain are of interest, but there may be a great many of these. For example, if there are 10 ability tests and 10 interest scales, there are 100 bivariate correlations between pairs of interests and abilities. To try to think about all these correlations simultaneously is very difficult if one is trying to generalize about the extent and nature of interrelationships of the domains. If there are p_1 elements in z_1 and p_2 elements in z_2, and $p = p_1 + p_2$, then the square, symmetric correlation matrix R of order p is subdivided into

$$R = \left[\begin{array}{c|c} R_{11} & R_{12} \\ \hline R_{21} & R_{22} \end{array} \right]$$

such that $\mathbf{R}_{11}$ contains the intercorrelations among the elements of $\mathbf{z}_1$ and is of order p_1, $\mathbf{R}_{22}$ is the intercorrelations among the elements of $\mathbf{z}_2$ and is of order p_2, and $\mathbf{R}_{12} = \mathbf{R}_{21}'$ contains the cross-correlations between elements of $\mathbf{z}_1$ and $\mathbf{z}_2$ and is of order $p_1 \times p_2$.

The interrelationships between two sets of measurements made on the same subjects can be studied by canonical correlation method. Originally developed by Hotelling (1935, 1936), the canonical correlation is the maximum correlation between linear functions of the two vector variables. However, after that pair of linear functions that maximally correlates has been located, there may be an opportunity to locate additional pairs of functions that maximally correlate, subject to the restriction that the functions in each new pair must be uncorrelated with all previously located functions in both domains. That is, each pair of functions is so determined as to maximize the correlation between the new pair of canonical variates, subject to the restriction that they be entirely orthogonal to all previously derived linear combinations.

The canonical correlation model uses the same analytic trick to display the structure of relationships across domains of measurement that the factor model uses to display the structure of relationships within a domain. The trick is to reduce the dimensionality to a few linear functions of the measures under study. The factor model selects linear functions of tests that have maximum variances, subject to restrictions of orthogonality. The canonical model selects linear functions that have maximum covariances between domains, subject again to restrictions of orthogonality. Geometrically, the canonical model can be considered an exploration of the extent to which individuals occupy the same relative positions in one measurement space as they do in the other. We want to know to what extent the vector variable $\mathbf{z}_1$ maps subjects the same as does the vector variable $\mathbf{z}_2$.

A rigorous derivation of the canonical correlation model may be found in Anderson (1958, Chapter 12). The partial differentiation involved is complicated because it requires fitting two sets of weights simultaneously. We will not attempt to describe it here. A good approach to understanding the canonical model is to think of it as a stepwise procedure. First, the model derives a component of each vector variable such that the covariance between the components is maximized. The conditions are

$$x = \mathbf{c}'\mathbf{z}_1 \quad \text{and} \quad y = \mathbf{d}'\mathbf{z}_2$$

where

$$m_x = 0, \quad s_x^2 = 1 \quad \text{and} \quad m_y = 0, \quad s_y^2 = 1$$

and

$$R_c = \frac{1}{N} \sum_{i=1}^{N} x_i y_i \bigg|_{\text{maximum}}$$

When the vectors of factor coefficients $\mathbf{c}$ and $\mathbf{d}$ have been derived by the differential calculus, the value of R_c is the maximum correlation that can be developed between a linear function of the variable $\mathbf{z}_1$ and a linear function of the variable $\mathbf{z}_2$. We say that the *canonical factors* x and y are the factors of the two batteries that have the highest correlation.

Besides the canonical correlation coefficient (R_c), interest centers on the interpretation of the canonical factors x and y. We want to know which elements in the vector variables contribute most heavily to the maximally correlated components. Once again, a factor structure giving the correlations of the canonical factors with the elements of the vector variables on which they are defined is desired. The structure for each set is arrived at as follows:

$$\mathbf{s}_1 = \frac{1}{N} \sum \mathbf{z}_{1i} x_i \qquad \text{where} \qquad i = 1 \text{ to } N$$

$$= \frac{1}{N} \sum \mathbf{z}_{1i} (\mathbf{c}' \mathbf{z}_{1i})$$

$$= \frac{1}{N} \sum \mathbf{z}_{1i} \mathbf{z}_{1i}' \mathbf{c}$$

$$= \mathbf{R}_{11} \mathbf{c} \tag{6.1.1}$$

Similarly

$$\mathbf{s}_2 = \frac{1}{N} \sum \mathbf{z}_{2i} y_i$$

$$= \mathbf{R}_{22} \mathbf{d} \tag{6.1.2}$$

The proportion of variance extracted from the first battery by canonical factor x is $(\mathbf{s}_1' \mathbf{s}_1)/p_1$, and the *redundancy* of battery 1 given the availability of battery 2 which is displayed by canonical factor x is denoted R_{d_x} and is computed as

$$R_{d_x} = \frac{\mathbf{s}_1' \mathbf{s}_1}{p_1} R_c^2$$

That is, the redundancy is the proportion of variance extracted by the factor times the proportion of shared variance between the factor and the corresponding canonical factor of the other battery. R_{d_x} is a new analytic tool invented by Stewart and Love (1968) and explored in depth by Miller (1969). We think that in some ways it is a more interesting number than R_c, since it expresses the amount of actual overlap between the two batteries that is packaged in the first canonical relationship *as seen from the side of z_1 as added to an already available z_2*. The new coefficient R_{d_x} is intended to show what proportion of the variance in vector variable z_1 is found through the first canonical correlation to be redundant to the variance in vector variable z_2 if the latter is already available.

The proportion of variance extracted from the second battery by canonical factor y is $(s_2's_2)/p_2$, and the redundancy of battery 2 given the availability of battery 1 which is displayed by canonical factor y is denoted R_{d_y} and is computed as

$$R_{d_y} = \frac{s_2's_2}{p_2} R_c^2$$

It is important to see why $R_{d_x} \neq R_{d_y}$. The shared variance of x and y is R_c^2, but the variance extracted by x from z_1 is not the same as the variance extracted by y from z_2. It is possible, for example, that x is a major factor of z_1, perhaps very close to the first principal component of z_1, while y is a trivial factor of z_2, perhaps very close to the last principal component of z_2. If such is the case, the redundancy of z_1 given z_2 as packaged in x should be much greater than the redundancy of z_2 given z_1 as packaged in y, and indeed it will be true that $R_{d_x} > R_{d_y}$. Before we had the new coefficient of redundancy we were prone to look at R_c^2 as a measure of the overlap between the two batteries. Actually, it is only a measure of the overlap between the two canonical variates x and y, and these may or may not be important factors of their respective batteries. The Stewart and Love team and Miller, working independently and at the same time, seem to have arrived at a very useful expression for the degrees of relationship between batteries as displayed by the canonical model.

Next, the canonical model derives a pair of components that are maximally correlated subject to the restriction that they must be uncorrelated with the first pair of components. If we now subscript the first canonical factors as x_1 and y_1, the second will be x_2 and y_2, and the conditions are:

$$x_2 = c_2'z_1 \qquad \text{and} \qquad y_2 = d_2'z_2$$

where

$$m_{x_2} = 0, \quad s_{x_2}^2 = 1 \qquad \text{and} \qquad m_{y_2} = 0, \quad s_{y_2}^2 = 1$$

and

$$r_{x_2 x_1} = 0, \quad r_{x_2 y_1} = 0, \quad r_{y_2 y_1} = 0, \quad r_{y_2 x_1} = 0$$

and

$$R_{c_2} = \frac{1}{N} \sum_{i=1}^{N} x_{2i} y_{2i} \Big|_{\text{maximum}}$$

Of course this second canonical correlation will be smaller than the first. It will have a structure vector for each battery also. If this second canonical correlation is neither statistically insignificant nor trivially small, the model provides a third canonical relation for inspection, and so on up to the lesser of p_1 and p_2.

Letting n be the number of canonical relations selected for retention in a canonical model for a set of data, the numbers to be reported include the n canonical correlation coefficients, the $p_1 \times n$ matrix $\mathbf{C}$ of factor score coefficients for the first battery, the $p_2 \times n$ matrix $\mathbf{D}$ of factor score coefficients for the second battery, the $p_1 \times n$ matrix $\mathbf{S}_1$ of structure coefficients for the first battery, the $p_2 \times n$ matrix $\mathbf{S}_2$ of structure coefficients for the second battery, and the n sets of values of R_{d_x} and R_{d_y}. It should be noted that

$$\mathbf{S}_1 = R_{11}\mathbf{C} \qquad \text{and} \qquad \mathbf{S}_2 = R_{22}\mathbf{D}$$

A further feature of the redundancy analysis proposed by Stewart and Love and by Miller involves the summation of the n values of R_{d_x} to express in one index the total redundancy of battery 1 given battery 2 and the rank n canonical model for the data. That is,

$$R_{d_1} = \sum_{k=1}^{n} R_{d_{x_k}}$$

expresses the proportion of the battery 1 variance that is redundant to the battery 2 variance according to the rank n canonical model. Similarly,

$$R_{d_2} = \sum_{k=1}^{n} R_{d_{y_k}}$$

expresses the proportion of the battery 2 variance that is redundant to the battery 1 variance according to the rank n canonical model. This total redundancy measure is nonsymmetric, so that $R_{d_1} \neq R_{d_2}$. The necessity for a nonsymmetric measure of redundancy is made crystal clear if we consider the case in which z_2 contains alternate-form versions of a few of the many tests contained in z_1. Suppose z_1 consists of eight Differential Aptitude Tests, while z_2 consists of alternate-forms of the Verbal and Reasoning tests only. Now, z_2 should be almost totally redundant given the availability of z_1, but z_1 should be only moderately redundant given the availability of z_2.

Another approach to the redundancy estimate R_d is to consider the correlations of the elements of $\mathbf{z}_1$ with the canonical factors of $\mathbf{z}_2$.

$$\mathbf{r}_{z_1 y_k} = \frac{1}{N} \sum \mathbf{z}_1 (\mathbf{d}_k' \mathbf{z}_2)'$$

$$= \mathbf{R}_{12} \mathbf{d}_k$$

We are soon going to show that by definition

$$\mathbf{c}_k = \mathbf{R}_{11}{}^{-1}\mathbf{R}_{12}\,\mathbf{d}_k/R_{c_k}$$

from which we get

$$\mathbf{R}_{12}\,\mathbf{d}_k = \mathbf{R}_{11}\mathbf{c}_k\,R_{c_k}$$

and since

$$\mathbf{s}_{1_k} = \mathbf{R}_{11}\mathbf{c}_k$$

it follows that

$$\mathbf{r}_{z_1 y_k} = \mathbf{s}_{1_k}R_{c_k}$$

That is, the correlations of the elements of $\mathbf{z}_1$ with the kth canonical factor of $\mathbf{z}_2$ can be obtained by multiplying $\mathbf{s}_{1_k}$, the kth column of the structure matrix $\mathbf{S}_1$, by R_{c_k}, the kth canonical correlation coefficient. Similarly

$$\mathbf{r}_{z_2 x_k} = \mathbf{R}_{21}\mathbf{c}_k = \mathbf{s}_{2_k}\,R_{c_k}$$

Now we can get

$$R_{d_{x_k}} = \frac{1}{p_1}\sum_{j=1}^{p_1} r^2_{z_1 y_{k_j}}$$

and

$$R_{d_{y_k}} = \frac{1}{p_2}\sum_{j=1}^{p_2} r^2_{z_2 x_{k_j}}$$

Finally, R_{d_1} is the sum of the k values in the vector $\mathbf{R}_{d_x}$ and R_{d_2} is the sum of the k values in the vector $\mathbf{R}_{d_y}$.

The actual computation of the canonical correlation analysis involves the solution of a complicated eigenstructure problem which can be expressed in terms of the partitions of the correlation matrix for $\mathbf{z}_1$ and $\mathbf{z}_2$ together as:

$$(\mathbf{R}_{22}{}^{-1}\mathbf{R}_{21}\mathbf{R}_{11}{}^{-1}\mathbf{R}_{12} - \lambda_j\mathbf{I})\mathbf{d}_j = 0 \qquad (6.1.3)$$

with the restriction equation

$$\mathbf{d}_j'\mathbf{R}_{22}\,\mathbf{d}_j = 1$$

In assigning $\mathbf{z}_1$ and $\mathbf{z}_2$ it will be helpful to arrange matters so that $p_2 \le p_1$. We will assume that this convention has been followed. Section 6.2 of this chapter discusses methods for solving the eigenstructure of a nonsymmetric matrix such as $\mathbf{R}_{22}{}^{-1}\mathbf{R}_{21}\mathbf{R}_{11}{}^{-1}\mathbf{R}_{12}$. However, those methods will report eigenvectors $\mathbf{v}_j$ arranged as columns in a $p_2 \times n$ matrix $\mathbf{V}$ such that

$$\mathbf{V}'\mathbf{R}_{22}\,\mathbf{V} = \boldsymbol{\theta}$$

where θ is a diagonal matrix. In order to rescale the eigenvectors so that the restriction equation will be satisfied and the coefficients in $\mathbf{D}$ will define canonical factors with unit variances, we observe that

$$\theta^{-1/2}\mathbf{V}'\mathbf{R}_{22}\mathbf{V}\theta^{-1/2} = \theta^{-1/2}\theta\theta^{-1/2} = \mathbf{I}$$

so that if we compute

$$\mathbf{D} = \mathbf{V}\theta^{-1/2} \qquad\qquad (6.1.4)$$

then

$$\mathbf{D}'\mathbf{R}_{22}\mathbf{D} = \mathbf{I}$$

as required.

It is worthwhile to notice that in forming the matrix product required for canonical analysis an intermediate result is the complete multiple correlation analysis of each variable of each set regressed in the space of all the variables of the other set. The $p_2 \times p_1$ matrix

$$\mathbf{B}_2 = \mathbf{R}_{22}^{-1}\mathbf{R}_{21}$$

contains as column vectors the standardized regression weights for the regression of each element of $\mathbf{z}_1$ on all the elements of $\mathbf{z}_2$. Likewise, the $p_1 \times p_2$ matrix

$$\mathbf{B}_1 = \mathbf{R}_{11}^{-1}\mathbf{R}_{12}$$

contains as column vectors the weights for the regression of each element of $\mathbf{z}_2$ on all the elements of $\mathbf{z}_1$. If we let $\mathbf{R}_{1\cdot2}$ be the p_1-element vector of multiple correlations of each test in $\mathbf{z}_1$ with all the tests of $\mathbf{z}_2$, then

$$\mathbf{R}_{1\cdot2} = [(\mathbf{B}_2'\mathbf{R}_{21})\mathrm{diag}]^{1/2}$$

and correspondingly

$$\mathbf{R}_{2\cdot1} = [(\mathbf{B}_1'\mathbf{R}_{12})\mathrm{diag}]^{1/2}$$

where "diag" means that the vector to the left of the $=$ sign is created from the diagonal elements of the matrix to the right of the $=$ sign. The structure matrix containing the correlations of the p_1 elements of $\mathbf{z}_1$ with the regression functions defined on them for predicting the p_2 elements of $\mathbf{z}_2$ is assembled by dividing all the elements of the jth column of $\mathbf{R}_{12}$ by the quantity $R_{2\cdot1_j}$. Similarly, the pattern matrix containing the correlations of the p_2 elements of $\mathbf{z}_2$ with the regression functions defined on them for predicting the p_1 elements of $\mathbf{z}_1$ is given by dividing the jth column of $\mathbf{R}_{21}$ by the quantity $R_{1\cdot2_j}$.

When we have computed the weights $\mathbf{d}_j$ for the jth canonical factor of $\mathbf{z}_2$

the corresponding weights for the jth canonical factor of z_1 may be obtained from the relationship

$$c_j = \frac{(R_{11}{}^{-1}R_{12}d_j)}{\sqrt{\lambda_j}}$$ (6.1.5)

For the justification of this formula see T. W. Anderson, 1958, page 303.

Finally, in this formulation of the canonical correlation problem, the jth eigenvalue λ_j is the square of the jth canonical correlation coefficient R_{c_j}.

Bartlett (1941, 1947) has derived a procedure for testing the significance of canonical correlations. He defines a test statistic lambda

$$\Lambda = \prod_{i=1}^{p_2}(1 - \lambda_i)$$

The null hypothesis that z_1 is unrelated to z_2 is tested by a function of Λ that is distributed approximately as chi-square with $p_1 p_2$ degrees of freedom:

$$x^2 = -[n - .5(p_1 + p_2 + 1)] \log_e \Lambda$$ (6.1.6)

where

$$n = N - 1$$

If the null hypothesis can be rejected, the contribution of the first canonical relation to Λ can be removed and the significance of the remaining $p_2 - 1$ canonical relations as a set can be tested by

$$\Lambda' = \prod_{i=2}^{p_2}(1 - \lambda_i)$$

and

$$x^2 = -[n - .5(p_1 + p_2 + 1)] \log_e \Lambda'$$

with $(p_1 - 1)(p_2 - 1)$ degrees of freedom. In general, with r canonical relations removed from Λ

$$\Lambda' = \prod_{i=r+1}^{p_2}(1 - \lambda_i)$$

and

$$x^2 = -[n - .5(p_1 + p_2 + 1)] \log_e \Lambda'$$

with $(p_1 - r)(p_2 - r)$ degrees of freedom.

What we have in canonical correlation analysis is a model for representing the relationship between two sets of measures as n correlations between n factors of the first set and n factors of the second set, with all other correlations among the factors held to zero. For small sample studies n will be the number of statistically significant canonical correlations. For large sample studies n

will be the number of significant canonical correlations judged to be theoretically or practically nontrivial. As a rule of thumb, the authors frequently treat canonical correlations of .30 or less as trivial. The selected factors of each set are interpreted in the light of their correlations with the variables of the set as given in the factor pattern. We speak of n as the rank of the selected solution to the canonical problem. The sum of the redundancy measures for the n factors of a set of measures expresses the total redundancy in the rank n canonical model for that set given the availability of the other set of variables. The redundancy measure is important, because a very large canonical correlation coefficient could be the result of a very large zero-order correlation of just one variable of one set with just one variable of the other set, and the remainder of the two sets could be essentially uninvolved in the canonical structure.

The canonical correlation model appears at first to be a complicated way of expressing the relationship between two measurement batteries. In fact, it is the simplest analytic model that can begin to do justice to this difficult problem of scientific generalization. A useful supplement to, but no substitute for, the canonical structure is provided by the multiple correlation analysis of each variable of each set regressed on all the variables of the other set.

6.2 NUMERICAL EXAMPLE OF CANONICAL CORRELATION

The smallest possible numerical example for illustrating canonical correlation is $p_1 = 2$ and $p_2 = 2$. Starting with the partitioned correlation matrix (based upon $N = 100$),

$$
\mathbf{R} = \begin{bmatrix} 1.00 & .40 & .50 & .60 \\ .40 & 1.00 & .30 & .40 \\ \hline .50 & .30 & 1.00 & .20 \\ .60 & .40 & .20 & 1.00 \end{bmatrix} = \begin{bmatrix} \mathbf{R}_{11} & \mathbf{R}_{12} \\ \hline \mathbf{R}_{21} & \mathbf{R}_{22} \end{bmatrix}
$$

the matrix product in the canonical equation (6.1.3) is formed:

$$
[\mathbf{R}_{22}{}^{-1}\mathbf{R}_{21}\mathbf{R}_{11}{}^{-1}\mathbf{R}_{12}] = \begin{bmatrix} 1.041 & -.208 \\ -.208 & 1.041 \end{bmatrix} \cdot \begin{bmatrix} .50 & .30 \\ .60 & .40 \end{bmatrix} \cdot \begin{bmatrix} 1.190 & -.467 \\ -.476 & 1.190 \end{bmatrix}
$$

$$
\cdot \begin{bmatrix} .50 & .60 \\ .30 & .40 \end{bmatrix}
$$

$$
= \begin{bmatrix} .206 & .251 \\ .278 & .341 \end{bmatrix}
$$

call this matrix product M. Then we solve for the roots (λ_1 and λ_2) of M such that substituting either root in the following determinant yields zero:

$$\begin{vmatrix} .206 - \lambda & .251 \\ .278 & .341 - \lambda \end{vmatrix} = 0$$

One way to find the unknown roots is to expand the determinant, obtaining

$$\lambda^2 - .547\lambda + .0003 = 0$$

and then solving for the roots of that quadratic equation:

$$\lambda_1 = .546 \quad \text{and} \quad \lambda_2 = .001$$

Notice that the sum of the roots is equal to the trace of the matrix M.

Given the roots we now know the two possible canonical correlations

$$R_{c_1} = \sqrt{\lambda_1} = .74 \quad \text{and} \quad R_{c_2} = \sqrt{\lambda_2} = .03$$

Applying Bartlett's test we first obtain Wilks' lambda

$$\Lambda = (1.0 - .546)(1.0 - .001) = .454$$

then substituting into (6.1.6)

$$\chi^2 = -[99 - .5(2 + 2 + 1)] \log_e .454$$
$$\chi^2 = 77.0$$

where $ndf = 4$. Therefore we can reject the null hypothesis that Z_1 is unrelated to Z_2, since $P(\chi_4^2 = 77) > .999$. After removing the largest root,

$$\Lambda' = (1.0 - .001) = .999$$

and Bartlett's test yields:

$$\chi^2 = -96.5 \log_e .999$$
$$= .09$$

where $ndf = 1.0$, which indicates that only the first canonical correlation is significantly different from zero.

Now to be able to talk about this first, significantly related pair of canonical variates we need to solve for the unknown vectors $\mathbf{c}$ and $\mathbf{d}$, and then the corresponding structures s_1 and s_2. Equation 6.1.3 can now be written:

$$\begin{bmatrix} -.340 & .251 \\ .278 & -.205 \end{bmatrix} \begin{bmatrix} \mathbf{d}_1 \\ \mathbf{d}_2 \end{bmatrix} = \begin{bmatrix} 0 \\ 0 \end{bmatrix}$$

The unknown vector $\mathbf{d}$ is proportional to the vector formed by finding the cofactors of any row of $(\mathbf{M} - \lambda\mathbf{I})$. Finding the cofactors of the first row. $\mathbf{v} = [-.205 \quad -.278]$. Then

$$[-.205 \quad -.278] \cdot \begin{bmatrix} 1.00 & .20 \\ .20 & 1.00 \end{bmatrix} \begin{bmatrix} -.205 \\ -.278 \end{bmatrix} = \theta$$

where θ is the variance of the canonical variate on the right hand side using $\mathbf{v}$ as the canonical coefficients. The desired vector $\mathbf{d}$ is then found using equation 6.1.4,

$$\mathbf{d} = \mathbf{v}\theta^{-1/2} = \begin{bmatrix} .545 \\ .737 \end{bmatrix}$$

Notice that corresponding elements of $\mathbf{d}$ are proportional to $\mathbf{v}$:

$$\frac{-.205}{.545} = \frac{-.278}{.737}$$

Given vector $\mathbf{d}$ we can find $\mathbf{c}$ following (6.1.5):

$$\mathbf{c} = \begin{bmatrix} 1.190 & -.476 \\ -.476 & 1.190 \end{bmatrix} \cdot \begin{bmatrix} 50 & .60 \\ .30 & .40 \end{bmatrix} \cdot \begin{bmatrix} .545 \\ .737 \end{bmatrix} \cdot \frac{1}{\sqrt{.546}}$$

$$\mathbf{c} = \begin{bmatrix} .856 \\ .278 \end{bmatrix}$$

Now we have two sets of coefficients, $\mathbf{c}$ for the left hand variables and $\mathbf{d}$ for the right hand variables, such that

$$.856z_{11i} + .278z_{12i} = x_i$$

and

$$.545z_{21i} + .737z_{22i} = y_i$$

For i from 1 to N, x and y have unit variance and correlate .74, the canonical correlation.

If a second canonical relationship had proven significant, a second pair of coefficients vectors, $\mathbf{c}_2$ and $\mathbf{d}_2$, associated with that second root could have been found in a similar fashion.

One remaining task is to find the correlations between our new canonical variates and the original variables. As we saw in (6.1.1) and (6.1.2), these are vectors $\mathbf{s}_1$ and $\mathbf{s}_2$ and are found as follows:

$$\mathbf{s}_1 = \mathbf{R}_{11}\mathbf{c} = \begin{bmatrix} 1.00 & .40 \\ .40 & 1.00 \end{bmatrix} \cdot \begin{bmatrix} .856 \\ .278 \end{bmatrix} = \begin{bmatrix} .967 \\ .620 \end{bmatrix}$$

$$\mathbf{s}_2 = \mathbf{R}_{22}\mathbf{d} = \begin{bmatrix} 1.00 & .20 \\ .20 & 1.00 \end{bmatrix} \cdot \begin{bmatrix} .545 \\ .737 \end{bmatrix} = \begin{bmatrix} .692 \\ .846 \end{bmatrix}$$

So, for example, .620 is the correlation between the second, left side variable, z_{12}, and the first, left side canonical variate X.

Given s_1, the proportion of the left side battery extracted by the first canonical variate is:

$$\frac{s_1's_1}{p_1} = [.967 \quad .620] \cdot \begin{bmatrix} 967 \\ .620 \end{bmatrix} \cdot \tfrac{1}{2} = .660$$

Thus the redundancy of battery 1 given battery 2 is $R_{d_x} = (.660)(.546) = .36$. That is, 66 percent of battery 1 is explained by the first canonical factor, and since 55 percent of its variance is "explained" by the first factor of battery 2, 36 percent of battery 1 variance is explained by the first factor of battery 2.

Similarly, .597 is the proportion of battery 2 variance explained by its first canonical factor, and $R_{d_y} = .326$ is the proportion of battery 2 variance explained by the first factor of battery 1. As pointed out in Section 6.1, R_{d_x} is not necessarily equal to R_{d_y}, as is the case in this example.

Another possible way of describing canonical results is in terms of the correlations between the variables of battery 1 and the canonical variates of battery 2, and vice versa. These are r_{z_1y} and r_{z_2x}, where $x = c'z_1$ and $y = d'z_2$. The vector $r_{z_1y} = s_1R_c$ and $r_{z_2x} = s_2R_c$. In this numerical example

$$r_{z_1y} = \begin{bmatrix} .967 \\ .620 \end{bmatrix} \cdot (.739) = \begin{bmatrix} .715 \\ .458 \end{bmatrix}$$

$$r_{z_2x} = \begin{bmatrix} .692 \\ .846 \end{bmatrix} \cdot (.739) = \begin{bmatrix} .511 \\ .625 \end{bmatrix}$$

So, for example, the correlation between the first variable on the left and the first canonical variable on the right is .715. Similarly, $[(.715)^2 + ((.458)^2]/2 = .360$ is the proportion of battery 1 variance explained by the first canonical variate of battery 2, which is (R_{d_x}). This then, gives us another way of obtaining and seeing what is meant by R_{d_x} and R_{d_y}.

6.3 RESEARCH EXAMPLES

The research example of principal components presented in Chapter Four was a study titled "Redundancy in Student Records" (Lohnes and Marshall, 1965). Scores from 13 standardized tests and 8 course marks were correlated for 230 junior high school students. The first principal factor accounted for 68 percent of the variance in the 21 measures, and the argument was advanced that this large g factor based on both tests and marks indicated extensive informational overlap between the two domains of indicators. The authors also employed canonical correlation to display the structure of the redundancy between the domains. They found that the correlation between first canonical

TABLE 6.1   Structure Coefficients for Two Canonical Factors of 13 Tests, and Corresponding Two Canonical Factors of 8 School Marks

Test	Factor I $R_c = .90$	Factor II $R_c = .66$		Factor I $R_c = .90$	Factor II $R_c = .66$
PGAT Verbal	.79	−.06	Seventh grade		
PGAT Reasoning	.83	.16	English	.85	.32
PGAT Number	.71	.46	Eighth grade		
MAT Word			English	.80	.45
Knowledge	.80	.03	Seventh grade		
MAT Reading	.82	−.06	Arithmetic	.95	−.14
MAT Spelling	.89	−.19	Eighth grade		
MAT Language	.92	−.12	Arithmetic	.88	−.24
MAT Study Skills			Seventh grade		
Language	.84	.07	Social Studies	.90	−.13
MAT Arithmetic			Eighth grade		
Computation	.90	.21	Social Studies	.74	.00
MAT Arithmetic			Seventh grade		
Problems	.84	.35	Science	.80	−.03
MAT Social			Eighth grade		
Studies	.75	−.05	Science	.73	.08
MAT Study Skills					
Social Studies	.80	.36			
MAT Science	.73	.19			
Factor			Factor		
Redundancy	.54	.02	Redundancy	.57	.02
Total Redundancy		.59	Total Redundancy		.61

variates of the two domains was a whopping .90. Table 6.1 shows these first canonical variates to be g-type factors of their respective batteries. The redundancy of the tests, given the marks, that is contained in this g factor of the tests is .54. Similarly, the redundancy of the marks given the tests that is contained in the g factor of the marks is .57. When the authors published this analysis they did not have access to the Stewart and Love redundancy measure, and they tended to overgeneralize from their canonical correlation of .90.

The second canonical factors were related by a coefficient of .66. Table 6.1 shows them to be bipolar factors of their domains that do not have any obvious construct validities as a result of their correspondence. English marks residuals from g correlated with arithmetic tests residuals from g? This is not as nonsensical as it seems, in the light of Project TALENT results showing arithmetic tests loading on an English skills factor, rather than on a

Mathematics factor (Lohnes, 1966, Chapter III). The main point is that the redundancy values for these factors are .02, indicating that practically no overlap of the batteries is packaged in these factors. Each of these factors extracts only .05 of the variance from its battery, and this multiplied by $R_c^2 = .44$ yields the .02 values. The fact is that the total redundancy of the tests given the marks is only .59, and the total redundancy of the marks given the tests is only .61, so that the g-type canonical variates account for almost all of the information overlap between the domains. This overlap is less than the authors thought it was in the days when they were without the redundancy statistic. Nevertheless, the extent and structure of the informational overlap between the domains as now revealed may still justify the authors' contention that school tests and marks records should be reorganized in terms of a g score and some orthogonal residual factors. Or, perhaps a g score and nothing else.

For our second research example of canonical correlation we turn again to the study of the internal structure of the 1960 Project TALENT high school measurement battery (Lohnes, 1966). In Chapter Five we presented a factor theory for the 38 motives domain scales in terms of eleven orthogonal factors of individual differences plus the control factors of sex and grade. Lohnes used the same strategy to factor the 60 scales of the abilities domain battery, and reported eleven orthogonal individual differences factors for that domain also. These MAP abilities are listed in Table 6.2, which places

TABLE 6.2 Abilities Domain Factors

Mnemonic	Factor Name	Variance Extracted (in percent)
VKN	Verbal Knowledges	18.7
GRD	Grade	7.8
ENG	English Language	6.6
SEX	Sex	5.7
VIS	Visual Reasoning	5.3
MAT	Mathematics	4.1
PSA	Perceptual Speed and Accuracy	3.6
SCR	Screening	3.3
H-F	Hunting-Fishing	2.2
MEM	Memory	2.1
COL	Color, Foods	1.9
ETI	Etiquette	1.6
GAM	Games	1.5

(13 factors extract 64.6% of variance)

the control factors of Sex and Grade relative to the eleven factors of individual differences. In a different sample of 3100 subjects randomly drawn from TALENT's ninth and twelfth grade data files (both sexes) the structure of relationships between the domains was explored by means of canonical correlations of the two sets of factors. Since the factor theories for the domains apply within the sex-grade groups, the appropriate correlation matrix for canonical analysis was thought to be based on the pooled-within or error matrix of the Manova model for sex and grade effects. This notion of getting from Manova an estimate of the correlation matrix that is assumed to be common to several populations will be clarified in Chapter Eight. Since factors that are computed as orthogonal on one sample will never be exactly orthogonal on any other sample, it is not surprising that R_{11} and R_{12} in the resulting 22nd order correlation matrix were not precisely identity matrices. When elements less than $|.06|$ are edited out, the R matrix for CANON looks like Table 6.3, which makes it apparent that the between-domains structure

TABLE 6.3 Pooled within Groups Correlation Matrix for Four Cell

	Factors	VKN	ENG	VIS	MAT	PSA	SCR	H-F	MEM	COL	ETI	GAM
	VKN	100										
	ENG		100									
	VIS			100								
	MAT				100							
	PSA					100						
Abilities	SCR						100					
	H-F							100				
	MEM								100			
	COL									100		
	ETI										100	
	GAM											100
	CON											
	BUS											
	OUT											
	SCH											
	CUL											
Motives	SCI											
	ACT											
	LEA											
	IMP											
	SOC											
	INT											

is weak. Computation showed that only three of the motives factors had sizable multiple correlations with all eleven of the abilities, namely .53 for Scholasticism, .42 for Science Interests, and .35 for Conformity Needs. Looked at the other way, the only abilities factors that had sizable multiple correlations with all eleven motives factors were Verbal Knowledges (.54), Mathematics (.47), and English Language (.36). This set of six variables, three motives and three abilities, seems to provide a core of educationally relevant factors of adolescent personality.

Table 6.4 reports the four largest canonical relations. The first relation is obviously between academic achievement and academic orientation. The second relation pairs a feminine-masculine abilities variate with a feminine-masculine motives variate. The other two are quite obscure. The main finding is that the relations between the domains are quite weak, and this is supported by total redundancy measures of ten percent for the abilities and eleven percent for the motives. The basic research hypothesis of the separateness of

Design, 3100 Subjects (Upper Triangular, Empty Cell Indicates $|r| \leq .05$)

CON	BUS	OUT	SCH	CUL	SCI	ACT	LEA	IMP	SOC	INT
13	−16	−08	31	24	27	−20			06	07
13	07		15			−22		−17	06	
		−14	14		09		−06	−12		
			38	−07	30			10	06	−17
17	06		07				06	06	11	09
18		09	−09	−15						
	−06	22	−07	−08			−07			
		06								
			07							
11	06			−06	−07				12	−07
		−12	−09	−12				−14	15	
100										
	100									
		100								
			100		19		15		−19	−13
				100						
					100					
						100				
							100			
								100		
									100	−06
										100

TABLE 6.4 Canonical Correlations, F ratios, and Pattern Coefficients for the Interdomain Structures (Coefficients $\leq.20$ Suppressed)

	$R_c = .66$ $F_\infty^{120} = 35$	$R_c = .45$ $F_\infty^{100} = 22$	$R_c = .39$ $F_\infty^{80} = 18$	$R_c = .34$ $F_\infty^{60} = 15$
Abilities				
VKN	.77			−.36
ENG	.30	−.57		
VIS			−.60	−.24
MAT	.55	.46		.56
PSA				.52
SCR				.25
H-F		.24	−.48	
MEM				
COL				
ETI		−.32		.37
GAM		−.45		
Motives				
CON	.24	−.47	−.27	.38
BUS			.33	.40
OUT		.30	−.56	
SCH	.76			.26
CUL	.26		.46	−.59
SCI	.55	.29	−.29	
ACT	−.34	.37		.29
LEA			.23	.29
IMP		.36	.36	.29
SOC		−.55		
INT				

maximum performance and typical performance measurement domains appears to be supported by these results.

A third research example of canonical correlation concerns the relationships between interests measured at grade 9 and interests measured at grade 12. Data necessary for this investigation are available for 1466 Project TALENT males (Cooley, 1967). Although there were no major changes from grade 9 to grade 12 in either the means or the variances for these 17 interest scales, there was a slight tendency for boys to become more interested in business management. If the amount of variance in interests changed at all, it increased slightly between grade 9 and grade 12. The correlations between corresponding grade 9 and grade 12 interests were all about the same order of magnitude, in the .40's or .50's.

Canonical correlations between grade 9 and grade 12 interests were computed in order to identify pairs of factors, one from each of the two sets, that were more highly correlated than any pair of the original interest scales. The loadings as reported for a pair of canonical variates are the correlations between the 17 interest scales and the corresponding canonical variates.

Table 6.5 shows that the first factor (i.e., the largest canonical relationship) extracted from the grade 9 interest scales correlated .75 with a similar grade 12 factor. Inspection of the loadings reveals that this fairly stable interest dimension measures whether or not the boy would prefer professional or at least "white-collar" work to nonprofessional or skilled labor of the "blue-collar" type.

The second pair of canonical variates for the males showed high positive loadings for masculine scales such as farming, hunting-fishing, and the biological-medical sciences, with the highest negative loadings for art, music, sales, and office work interests. A grade 9 masculine-femine factor loaded in this way correlated .64 with a grade 12 factor similarly loaded. The third pair of canonical variates (with a canonical correlation of .63) tended to represent introversion at the positive end and extroversion at the negative end.

The fourth canonical, which was larger than any of the 17 single retest correlations, had a coefficient of .58. On the positive side were physical science, computational, and mechanical-technical, with assorted nonscience scales on the negative side.

These canonical correlation results have revealed several dimensions of vocational interest which are more temporally stable than are the individual interest scales. The most stable factor was whether or not a professional career was preferred. The more stable interest factors may be getting closer to basic motivations and away from the changing adolescent self-concept. These results would prove useful in the development of new interest scales.

Although canonical correlations have been with us since Hotelling (1935) told us how to find "the most predictable criterion," the psychological literature is not exactly full of examples of their application. One reason is the availability of other methods for studying the relationships among two sets of variables. One can examine the interset intercorrelations between all possible pairs of the two sets $(\mathbf{R}_{12})$ and simply discuss the bivariate relationships between pairs. Another procedure would be to factor analyze the two sets combined in one supermatrix and examine the factors to determine those which have high loadings for both sets of variables. A third procedure is to factor analyze each of the two sets separately and relate the resulting factors. A fourth method is to compute the multiple correlations between each of the left-hand variables and the complete right-hand set and then vice versa.

Another reason for the lack of examples of canonical correlations is probably the difficulty of interpreting canonical variates. This has partly

TABLE 6.5 Restest Canonical Correlations and Loadings (N = 1466 Males)

| Canonical Variate | 1 | | 2 | | 3 | | 4 | |
| Canonical R | .75 | | .64 | | .63 | | .58 | |
Scale	Grade 9	Grade 12	Grade 9	Grade 12	Grade 9	Grade 12	Grade 9	Grade 12
P-701 Physical Science	.42	.32	.30	.32	.24	.28	.50	.55
P-702 Biological Science	.53	.55	.40	.42	.00	.10	.09	.08
P-703 Public Service	.49	.50	.00	.02	−.34	−.34	.13	.11
P-704 Literary-Linguistic	.56	.65	.01	.00	−.21	−.09	−.22	−.22
P-705 Social Service	.28	.35	.03	.03	−.39	−.37	−.12	−.14
P-706 Artistic	.33	.36	−.20	−.16	.17	.24	−.35	−.32
P-707 Musical	.41	.50	−.25	−.21	.12	.23	−.18	−.19
P-708 Sports	.14	.15	.32	.32	−.49	−.49	.09	.09
P-709 Hunting and Fishing	−.20	−.20	.51	.51	−.04	−.05	−.05	−.07
P-710 Business Management	.25	.28	−.18	−.18	−.52	−.49	.16	.18
P-711 Sales	.17	.19	−.17	−.22	−.53	−.55	−.01	−.01
P-712 Computation	.24	.20	−.02	−.07	−.36	−.43	.32	.40
P-713 Office Work	.06	.04	−.16	−.20	−.39	−.43	.02	.06
P-714 Mechanical-Technical	−.35	−.53	−.03	−.07	.09	.13	.32	.33
P-715 Skilled Trades	−.49	−.59	−.04	−.06	−.19	−.15	−.06	−.07
P-716 Farming	−.45	−.46	.49	.54	−.08	−.07	−.32	−.30
P-717 Labor	−.47	−.51	−.05	−.01	−.29	−.27	−.09	−.04

TABLE 6.6 Interest-Activities Scales

Interest Inventory
- 701 Physical Science, Engineering, and Math
- 702 Biological Science and Medicine
- 703 Public Service
- 704 Literary-Linguistic
- 705 Social Service
- 706 Artistic
- 707 Musical
- 708 Sports
- 709 Outdoor Recreation
- 710 Business-Management
- 711 Sales
- 712 Computation
- 713 Office Work
- 714 Mechanical-Technical
- 715 Skilled Trades
- 716 Farming
- 717 Labor

Activities Inventory
- 601 *Sociability*
  I'd rather be with a group of friends than at home by myself.
- 602 *Social Sensitivity*
  I never hurt another person's feelings if I can avoid it.
- 603 *Impulsiveness*
  I like to do things on the spur of the moment.
- 604 *Vigor*
  I am a fast walker.
- 605 *Calmness*
  I can usually keep my wits about me even in difficult situations.
- 606 *Tidiness*
  I have a definite place for all of my things.
- 607 *Culture*
  I feel that good manners are very necessary for everyone.
- 608 *Leadership*
  I have held a lot of elected offices.
- 609 *Self-Confidence*
  I am usually at ease.
- 610 *Mature Personality*
  When I say I'll do something I get it done.

TABLE 6.7   Largest Product-Moment
Correlations ($N = 1088$)

Occupational Interest	Personal Activities	r
708. Sports	604. Vigor	.35
704. Lit.-Ling.	607. Culture	.31
705. Soc. Serv.	607. Culture	.29
701. Phys. Sci.	610. Mature	.29
710. Business	601. Sociable	.29
707. Musical	607. Culture	.29
708. Sports	601. Sociable	.28
706. Artistic	607. Culture	.28
704. Lit.-Lang.	608. Leader	.28
703. Pub. Service	608. Leader	.27
705. Soc. Serv.	602. Soc. Sens.	.26
702. Bio. Sci.	607. Culture	.25

been because available programs have provided only the coefficients for computing the canonical variates, and the structure, that is, the correlations between the original variables and the derived canonical variates, have not been available.

The purpose of this fourth research example is to illustrate these various methods of studying two sets of variables. Project TALENT data on occupational preferences and personal activities preferences are used here and are referred to as the left-hand and right-hand set respectively.

In the 1960 national testing program of Project TALENT a 205-item-occupational interest inventory was administered resulting in 17 interest scales, and a 150-item activities inventory resulting in 10 scales of self-description for "temperament" variables. The students rated each occupation and activity item on a five-point scale. The resulting 27 variables are listed in Table 6.6. The interest inventory was primarily composed of items of related occupational titles, and illustrative items are presented for the activities inventory. The numbers refer to test number in the Project TALENT Battery.

The 27 by 27 correlation matrix upon which these analyses are based can be found in Flanagan, et al. (1964) pages 8-6 and 8-7. The sample upon which the following analyses were based consists of 1088 grade 12 males. The 12 largest zero-order correlations are presented in Table 6.7. The types of relationships seen there seem very reasonable. Those preferring sports occupations prefer vigorous and sociable personal activities. Those preferring cultural activities tend to prefer occupations in one or more of the following areas: literary-linguistic, social service, music, art, and biological science. However, none of the correlations is particularly high, the largest one being only .35. The question is, are there dimensions within the interest domain and the activities domain which are more highly correlated than these zero-order relationships?

A principal components analysis of the combined 27 interest and activities variables resulted in seven factors with latent roots greater than 1. Table 6.8 presents the resulting factor loadings for those seven factors, the usual general factor and six bipolar factors. The second, bipolar factor is rather interesting

TABLE 6.8   Interest-Activities Factor Loadings

Variable	Factor						
	1	2	3	4	5	6	7
601 Sociability	.51	−.36	.29	−.25	.23	.15	.00
602 Social Sensitivity	.59	−.45	.22	−.06	.03	−.19	−.02
603 Impulsiveness	.13	−.20	.09	.01	.45	.02	−.73
604 Vigor	.52	−.36	.41	.06	.14	.14	.01
605 Calmness	.55	−.44	.27	−.06	−.22	−.09	.10
606 Tidiness	.50	−.44	.21	−.15	−.28	−.16	.18
607 Culture	.61	−.45	.08	.01	−.03	−.27	.05
608 Leadership	.53	−.42	.06	.01	.16	.00	−.18
609 Self-Confidence	.45	−.36	.15	.01	−.02	.08	−.02
610 Mature Personality	.60	−.45	.29	−.01	−.22	−.04	−.06
701 Physical Science	.53	.16	−.10	.37	−.51	.36	−.21
702 Biological Science	.60	.14	−.20	.41	−.20	.21	−.10
703 Public Service	.62	.12	−.37	−.02	.24	.24	−.01
704 Literature-Linguistics	.71	.17	−.46	.22	.10	−.12	.02
705 Social Service	.68	.26	−.19	.00	.18	−.14	.18
706 Artistic	.55	.24	−.28	.38	−.07	−.29	−.11
707 Musical	.48	.10	−.32	.32	.13	−.51	.07
708 Sports	.48	.21	.23	.16	.29	.33	.30
709 Outdoor Recreation	.17	.34	.44	.39	.24	.24	.28
710 Business-Management	.67	.31	−.21	−.30	.16	.21	−.05
711 Sales	.55	.38	−.19	−.43	.14	.10	.03
712 Computation	.55	.31	−.17	−.43	−.32	.19	.02
713 Office Work	.46	.51	−.05	−.46	−.16	−.14	.04
714 Mechanical-Technical	.16	.60	.48	.11	−.34	.06	−.27
715 Skilled Trades	.14	.75	.44	−.08	−.03	−.25	−.14
716 Farming	.14	.52	.51	.26	.15	−.10	.14
717 Labor	.12	.71	.40	−.19	.10	−.25	−.12
Latent Roots	6.78	4.28	2.37	1.67	1.36	1.22	1.02

since it shows that boys who preferred manual labor occupations tended not to know how to respond to these activities items in a manner consistent with middle class values.

A Varimax rotation (Table 6.9) of these loadings cleaned up the factor structure. Unfortunately very little was learned regarding our question about the relationships among the two sets of variables, since it created a first factor based primarily on the activities items, and the next five factors

TABLE 6.9   Interest-Activities Varimax Loadings

Factor 1	Mature Personality	.81	*Factor 4*	Sales	.81
	Calmness	.78		Business-Management	.80
	Social Sensitivity	.77		Computation	.75
	Tidiness	.75		Office Work	.68
	Culture	.73		Public Service	.60
	Vigor	.68		Social Service	.51
	Sociability	.65			
	Leadership	.59			
	Self-Confidence	.56	*Factor 5*	Physical Science	.89
				Biological Science	.64
Factor 2	Skilled Trades	.89		Mechanical-Technical	.44
	Labor	.82			
	Mechanical-Technical	.78			
	Farming	.62	*Factor 6*	Outdoor Recreation	.76
				Sports	.70
Factor 3	Musical	.84		Farming	.51
	Literature-Linguistics	.73			
	Artistic	.73			
	Social Service	.54	*Factor 7*	Impulsiveness	.88

were based on the interest scales. The seventh Varimax factor appears to be primarily the impulsiveness scale. With the exception of the second principal component, very little was learned about the relationships between the activities and occupational preferences from this factor analysis.

Turning finally to the results of canonical correlation, there are five significant canonical correlations. This means that there are five independent dimensions of the left-hand set which are significantly related to corresponding dimensions of the right-hand set. The correlations between each of the original variates and the derived canonical variates are listed in Table 6.10. Notice that the canonical correlations range from .51 to .24 and that the largest are considerably greater than any of the zero order correlations.

The first right-hand canonical variate is highly positively correlated with all of the activities scales except impulsiveness which had a near-zero correla-

TABLE 6.10   Correlations Between Original Variables and Derived Canonical Variates

	Left-Hand		Right-Hand	
1st Canonical	Lit.-Ling.	.68	Culture	.87
.51	Soc. Service	.66	Leadership	.77
	Pub. Service	.59	Soc. Sensi.	.71
	Bio. Science	.55	Mature Pers.	.62
	Musical	.54	Sociability	.58
	Busi.-Magmt.	.53	Vigor	.57
			Tidiness	.57
	Skilled Trades	−.37	Calmness	.56
	Labor	−.37	Self-Confid.	.56
2nd Canonical	Sports	.67	Sociability	.66
.45	Out. Recrea.	.38	Vigor	.49
	Musical	−.36		
	Artistic	−.34		
3rd Canonical	Phys. Science	.67	Mature Pers.	.59
.36	Mech.-Tech.	.41	Vigor	.47
	Farming	.38	Calmness	.45
	Computation	.32		
4th Canonical	Farming	.44	Vigor	.31
.29	Out. Recrea.	.38		
	Sports	.31		
	Computation	−.47	Tidiness	−.48
	Office Work	−.41	Sociability	−.36
	Busi.-Mgmt.	−.32	Mature Pers.	−.35
5th Canonical	Office Work	−.31	Tidiness	−.32
.24	Pub. Service	.48	Leadership	.49
			Impulsiveness	.48

tion. The left-hand canonical variate was correlated positively with cultural, service, and business interest scales and negatively correlated with the mechanical, skilled worker, and labor interest scales. Since each of the activities scales was composed of a mixture of positively and negatively loaded items, response set probably cannot explain the first canonical. It appears to be related to the fact that the students aspiring to more middle class type of occupations are better able to recognize the types of responses desirable in terms of middle class values for preferred personal activities (i.e. social conformity).

The second canonical factor seems to be a type of extroversion-introversion factor, with the students preferring sociable and vigorous activities tending to prefer sports and outdoor type occupations and tending not to prefer musical and artistic occupations.

The third canonical tells us that students who tend to prefer science-technology type occupations also express a preference for activities which suggest they are very task-oriented, energetic, and not easily rattled. This might be considered a masculinity dimension.

The fourth pair of canonical variates are positively loaded on the vigorous, outdoor scales and negatively loaded on the tidy, computational, office work scales. This seems to be a type of compulsivity factor.

In the fifth canonical we have office work versus public service in the left-hand set and tidiness versus leadership, impulsiveness in the right-hand set. Although the fifth canonical is significant at the .01 level the relationship is too small to be of great interest at this time.

What has been accomplished in this canonical analysis is the determination of five dimensions in the interest domain and five dimensions in the activity domain which appear to be measuring similar underlying traits. One canonical variate extracted from each domain results in a pair of linear functions which are maximally correlated with each other subject to the restriction that they be independent of all other extracted dimensions. The technique located five such pairs which are significantly related.

6.4 EIGENSTRUCTURE OF A NONSYMMETRIC MATRIX

The eigenstructure of a matrix $\mathbf{A}$, as described in Chapter One, consists of a diagonal matrix $\mathbf{L}$ of eigenvalues $(\lambda_j, \ j = 1, 2, \ldots, p)$ and a matrix $\mathbf{V}$ of column eigenvectors that satisfy the equations

$$(\mathbf{A} - \lambda_j \mathbf{I})\mathbf{v}_j = 0$$

and

$$\mathbf{AV} = \mathbf{VL}$$

In Chapter Four, computation of the eigenstructure of a real symmetric matrix by either the standard iterative procedure (Subroutine ITER-S) or by the very powerful methods of Householder, Ortega, and Wilkinson (Subroutine HOW) was discussed. HOW is excellent for factor analysis applications. However, in canonical and discriminant analyses we encounter then problem of the eigenstructure of a non-symmetric matrix of the form

$$(\mathbf{A} - \lambda_j \mathbf{B})\mathbf{v}_j = 0$$

or

$$B^{-1}AV = VL$$

In canonical analysis if we let

$$B = R_{22}$$

and

$$A = R_{21}R_{11}^{-1}R_{12}$$

we note that both A and B are real symmetric but the product $B^{-1}A$ is real nonsymmetric. The method we employ to compute the eigenstructure of $B^{-1}A$ involves successive diagonalization of two symmetric matrices by means of HOW.

First, solve for the eigenstructure of matrix B

$$BU = UK$$

where K is the eigenvalues and $\underline{U}$ is the eigenvectors of B. Then

$$B = UKU' = (UK^{1/2})(K^{1/2}U') = B^{1/2}B^{1/2'}$$

and since

$$B^{-1} = UK^{-1}U' = (UK^{-1/2})(K^{-1/2}U')$$

it follows that

$$B^{-1/2} = UK^{-1/2}$$

Substituting in

$$(A - \lambda_j B)v_j = 0$$

we get

$$(A - \lambda_j UK^{1/2}K^{1/2}U')v_j = 0$$

or

$$(K^{-1/2}U'A - \lambda_j K^{1/2}U')v_j = 0$$

or

$$(K^{-1/2}U'AUK^{-1/2} - \lambda_j I)K^{1/2}U'v_j = 0$$

or

$$(B^{-1/2'}AB^{-1/2} - \lambda_j I)(B^{1/2'}v_j) = 0$$

Solving the eigenstructure of the matrix $B^{-1/2'}AB^{-1/2}$ gives the eigenvalue matrix L containing the eigenvalues of $B^{-1}A$. The resulting eigenvectors $(B^{1/2}V)$ are reduced to the eigenvectors V of $B^{-1}A$ by

$$V = B^{-1/2}(B^{1/2'}V).$$

This algebra has been incorporated by the authors in Subroutine DIRNM ("*di*agonalize *r*eal *n*onsymmetric *m*atrix") which manages the two trips through HOW and the return of L and V to the main program.

CANON

```
C      CANONICAL CORRELATION PROGRAM.  A COOLEY-LOHNES ROUTINE.
C
C      THIS PROGRAM COMPUTES A FULL SET OF CANONICAL CORRELATIONS
C      RELATING M1 VARIABLES ON THE LEFT TO M2 VARIABLES ON THE RIGHT,
C      WHERE M1 AND M2 ARE EACH LESS THAN 51.  TO SAVE TIME, ARRANGE
C      THE DATA SO THAT M2 IS LESS THAN (OR EQUALS) M1. IT DOES NOT MATTER
C      WHICH SET IS CONSIDERED AS THE PREDICTOR SET, IF EITHER.
C
C      INPUT
C
C      1) FIRST TEN CARDS OF DATA DECK DESCRIBE THE PROBLEM IN A TEXT WHICH
C         WILL BE REPRODUCED ON THE OUTPUT. DO NOT USE COLUMN 1 OF THESE
C         CARDS.
C      2) CONTROL CARD (CARD 11)
C         COLS 1-2  M1 = NUMBER OF VARIABLES IN LEFT SET
C         COLS 3-4  M2 = NUMBER OF VARIABLES IN RIGHT SET
C         COLS 5-9  N = NUMBER OF SUBJECTS
C      3) FORMAT CARD (CARD 12) FOR FULL CORRELATION MATRIX.  IF PUNCHED
C         FROM CORREL, FORMAT IS (10X, 7F10.7 / (10X, 7F10.7)).
C      4) R, THE FULL CORRELATION MATRIX, OF ORDER M = M1 + M2,
C         UPPER-TRIANGULAR.
C
C
C      REQUIRED SUBROUTINES ARE MPRINT, MATINV, HOW, AND DIRNM.
C
C
       DIMENSION  TIT(16), R(100,100),  X(50), Y(50), Z(50),
      2    S(50,50),  T(50,50),  V(50,50),  WL(50)
C
 1     WRITE(6,2)
 2     FORMAT (1H1)
       DO 3   J = 1, 10
                      READ(5,4)  (TIT(K),K=1,16)
 3     WRITE(6,4) (TIT(K),K=1,16)
 4     FORMAT (16A5)
       READ(5,5)M1,M2,N
 5     FORMAT (2I2,  I5)
       READ(5,4)  (TIT(K),K=1,16)
       WRITE(6,6)
 6     FORMAT(60H0CANONICAL CORRELATION. A COOLEY-LOHNES MULTIVARIATE ROU
      2TINE)
       M =   M1 + M2
       EM =   M
       EN =   N
       WRITE(6,7)M1
 7     FORMAT (30H0NO. VARIABLES ON LEFT = M1 = I3)
       WRITE(6,8) M2
 8     FORMAT (31H0NO. VARIABLES ON RIGHT = M2 = I3)
       WRITE(6,9) N
 9     FORMAT (20H0NO. SUBJECTS = N = I6)
       DO 12   J = 1, M
 12    READ(5,TIT)  (R(J,K),K=J,M)
       DO 13   J = 1, M
       DO 13   K = J, M
 13    R(K,J) = R(J,K)
       WRITE(6,23)
 23    FORMAT (19H0CORRELATION MATRIX)
       CALL MPRINT (R,M)
C
C
       CALL MATINV (R, M1, DET)
C
       DO 25   J = 1, M2
       J1 = J + M1
       DO 25   K = 1, M1
       S(J,K) = 0.0
```

```
      DO 25    L = 1, M1
  25  S(J,K) = S(J,K) + R(J1,L) * R(L,K)
      DO 26    J = 1, M2
      DO 26    K = 1, M2
      K1 = K + M1
      T(J,K) = 0.0
      DO 26    L = 1, M1
  26  T(J,K) = T(J,K) + S(J,L) * R(L,K1)
C   T NOW CONTAINS   R21 * R11 INVERSE * R12 .
      DO 27    J = 1, M2
      J1 = J + M1
      DO 27    K = 1, M2
      K1 = K + M1
  27  S(J,K) = R(J1,K1)
C   S NOW CONTAINS R22 .
C
      CALL DIRNM (T, M2, S, V, Y, M2)
C
C   Y NOW CONTAINS THE EIGENVALUES AND V NOW CONTAINS THE COLUMN
C   EIGENVECTORS OF    R22 INVERSE * R21 * R11 INVERSE * R12 .
C
      DO 28    J = 1, M2
      DO 28    K = 1, M2
      K1 = K + M1
      S(J,K) = 0.0
      DO 28    L = 1, M2
      L1 = L + M1
  28  S(J,K) = S(J,K) + V(L,J) * R(L1,K1)
      DO 29    J = 1, M2
      DO 29    K = 1, M2
      T(J,K) = 0.0
      DO 29    L = 1, M2
  29  T(J,K) = T(J,K) + S(J,L) * V(L,K)
C   T NOW CONTAINS THE DISPERSION OF THE RIGHT SET COMPONENTS BASED
C   ON UNIT LENGTH VECTORS AS REPORTED BY SUBROUTINE DIRNM.
      DO 30    J = 1, M2
      Z(J)=SQRT(Y(J))
      DO 30    K = 1, M2
  30  V(J,K)=V(J,K)*(1.0/SQRT(T(K,K)))
C   V CONTAINS THE ANDERSON NORMALIZED COLUMN VECTORS FOR THE
C   RIGHT SET, SUCH THAT    V PRIME * R22 * V = I , AN IDENTITY MATRIX.
C
      DO 31    J = 1, M1
      DO 31    K = 1, M2
      K1 = K + M1
      S(J,K) = 0.0
      DO 31    L = 1, M1
  31  S(J,K) = S(J,K) + R(J,L) * R(L,K1)
      DO 32    J = 1, M1
      DO 32    K = 1, M2
      T(J,K) = 0.0
      DO 32    L = 1, M2
  32  T(J,K)  = T(J,K) + S(J,L) * V(L,K)
      DO 33    J = 1, M1
      DO 33    K = 1, M2
  33  T(J,K) = T(J,K) / Z(K)
C   T NOW CONTAINS THE ANDERSON NORMALIZED COLUMN VECTORS FOR THE
C   LEFT SET, SUCH THAT    T PRIME * R11 * T = I , AN IDENTITY MATRIX.
C
C
      WRITE(6,35)
  35  FORMAT(39HOLEFT SET CANONICAL WEIGHTS, COLUMNWISE)
      DO 34    J = 1, M1
  34  WRITE(6,65)J,(T(J,K),K=1,M2)
  65  FORMAT (1X,I5,4X, 10F10.3 / (10X, 10F10.3))
      WRITE(6,36)
```

```
 36    FORMAT (40HORIGHT SET CANONICAL WEIGHTS, COLUMNWISE)
       DO 66    J = 1, M2
 66    WRITE(6,65)J,(V(J,K),K=1,M2)
       CALL MATINV (R, M1, DET)
C   RESTORES R11 .
       DO 37    J = 1, M1
       DO 37    K = 1, M2
       S(J,K) = 0.0
       DO 37    L = 1, M1
 37    S(J,K) = S(J,K) + R(J,L) * T(L,K)
       WRITE(6,39)
 39    FORMAT(80H0FACTOR STRUCTURE FOR LEFT SET. COLUMNS ARE CANONICAL FA
      2CTORS. ROWS ARE TESTS.       )
       DO 38    J = 1, M1
 38    WRITE(6,65)J,(S(J,K),K=1,M2)
       EM1 = M1
       SUM1 = 0.0
       SUM2 = 0.0
       DO 52    J = 1, M2
       WL(J) = 0.0
       DO 52    K = 1, M1
 52    WL(J) = WL(J) + S(K,J) * S(K,J)
       DO 53    J = 1, M2
       WL(J) = WL(J) / EM1
       SUM1 = SUM1 + WL(J)
       X(J) = WL(J) * Y(J)
 53    SUM2 = SUM2 + X(J)
       WRITE(6,54)
 54    FORMAT (40H0FACTOR    VARIANCE EXTRACTED    REDUNDANCY    )
       DO 55    J = 1, M2
 55    WRITE(6,56)J,WL(J),X(J)
 56    FORMAT (I6, 10X, F10.3, 4X, F10.3)
       WRITE(6,57) SUM1
 57    FORMAT(43H0TOTAL VARIANCE EXTRACTED FROM LEFT SET =    F10.3)
       WRITE(6,58) SUM2
 58    FORMAT(51H0TOTAL REDUNDANCY FOR LEFT SET, GIVEN RIGHT SET =    F10.
      23)
       WRITE(6,59)
 59    FORMAT(39H0NOTE THAT ALL VALUES ARE PROPORTIONS.    )
       DO 40    J = 1, M2
       J1 = J + M1
       DO 40    K = 1, M2
       S(J,K) = 0.0
       DO 40    L = 1, M2
       L1 = L + M1
 40    S(J,K) = S(J,K) + R(J1,L1) * V(L,K)
       WRITE(6,41)
 41    FORMAT (80H0 FACTOR STRUCTURE FOR RIGHT SET. COLUMNS ARE FACTORS.
      2ROWS ARE TESTS.               )
       DO 67    J = 1, M2
 67    WRITE(6,65)J,(S(J,K),K=1,M2)
       EM2 = M2
       SUM1 = 0.0
       SUM2 = 0.0
       DO 60    J = 1, M2
       WL(J) = 0.0
       DO 60    K = 1, M2
 60    WL(J) = WL(J) + S(K,J) * S(K,J)
       DO 61    J = 1, M2
       WL(J) = WL(J) / EM2
       SUM1 = SUM1 + WL(J)
       X(J) = WL(J) * Y(J)
 61    SUM2 = SUM2 + X(J)
       WRITE(6,54)
       DO 62    J = 1, M2
 62    WRITE(6,56)J,WL(J),X(J)
```

```
      WRITE(6,63) SUM1
63    FORMAT(44H0TOTAL VARIANCE EXTRACTED FROM RIGHT SET =    F10.3)
      WRITE(6,64) SUM2
64    FORMAT(51H0TOTAL REDUNDANCY FOR RIGHT SET, GIVEN LEFT SET =   F10.
     23)
      WRITE(6,59)
C
C
      J = M2
      WL(J+1) = 1.0
42    WL(J) = WL(J+1) * (1.0 - Y(J))
      J = J - 1
      IF (J)  43, 43,  42
43    WRITE(6,44) WL(1)
44    FORMAT (30HOWILKS LAMBDA FOR TOTAL SET =    F14.7)
      DO 45   J = 1, M2
45    X(J) = -(EN - 1.0 - (EM + 1.0)/2.0) * ALOG(WL(J))
      WRITE(6,46) X(1)
46    FORMAT (24H0CHI SQUARE FOR TOTAL =   F14.7)
      NDT = M1 * M2
      WRITE(6,47) NDT
47    FORMAT(10H0N.D.F. =    I6)
      WRITE(6,48)
48    FORMAT(47H0CHI SQUARE TESTS WITH SUCCESSIVE ROOTS REMOVED)
      WRITE(6,49)
49    FORMAT(76H0ROOTS REMOVED    CANONICAL R   R-SQUARED  CHI-SQUARE
     2N.D.F.  LAMBDA PRIME           )
      J = 0
      WRITE(6,50) J,Z(1),Y(1),X(1),NDT,WL(1)
50    FORMAT (1H0, 10X, I3, 4X, F8.4, 4X, F8.3, 7X, F8.2, 4X, I6, 3X,
     2  F7.4)
      DO 51   J = 2, M2
      J1 = J - 1
      NDT = (M1 - J1) * (M2 - J1)
51    WRITE(6,50) J,Z(J),Y(J),X(J),NDT,WL(J)
      GO TO 1
      END
```

DIRNM

```
      SUBROUTINE DIRNM (A,M,B,X,XL,LVECT)
C     SUBROUTINE DIRNM, DIAGONALIZATION OF A REAL NON-SYMMETRIC MATRIX
C     OF THE FORM B-INVERSE*A.  CODED BY P. R. LOHNES, U. N. H.
C
C     A, M, B, X, AND XL ARE DUMMY NAMES AND MAY BE CHANGED IN THE
C     CALLING STATEMENT.
C     A AND B ARE M BY M INPUT MATRICES.  UPON RETURN VECTOR XL CONTAINS
C     THE EIGENVALUES OF B-1*A, AND MATRIX X CONTAINS THE EIGENVECTORS
C     IN ITS COLUMNS.  SUBROUTINE HOW PACKAGE IS REQUIRED.
C
C     LVECT SPECIFIES THE NUMBER OF EIGENVECTORS TO BE RETURNED.
C
C
      DIMENSION  A(50,50), B(50,50), X(50,50), XL(50), DUM1(50),DUM2(50)
     C , DUM3(50),DUM4(50), E(50)
C
      CALL HOW (M,50,M, B,XL,X, DUM1,DUM2,DUM3,DUM4)
C
      DO 13 I=1,M
      DIAG = SQRT ( ABS (XL(I)))
      DO 13 J=1,M
   13 B(J,I) = X(J,I) * DIAG
   16 DO 1  I = 1, M
    1 XL(I) = 1.0 / SQRT ( ABS (XL(I) ) )
      DO 2  I = 1, M
      DO 2  J = 1, M
    2 B(I,J) = X(I,J) * XL(J)
      DO 3  I = 1, M
      DO 3  J = 1, M
      X(I,J) = 0.0
      DO 3 K = 1, M
    3 X(I,J) = X(I,J) + B(K,I) * A(K,J)
      DO 4  I = 1, M
      DO 4  J = 1, M
      A(I,J) = 0.0
      DO 4  K = 1, M
    4 A(I,J) = A(I,J) + X(I,K) * B(K,J)
C     A NOW CONTAINS BPRIME*A*B OF THE NOTES.
C
C
      CALL HOW (M,50,LVECT, A, XL,X, DUM1,DUM2,DUM3,DUM4)
C
      DO 6  I = 1, M
      DO 6  J = 1, M
      A(I,J) = 0.0
      DO 6  K = 1, M
    6 A(I,J) = A(I,J) + B(I,K) * X(K,J)
      DO 9  I = 1, M
      SUMV = 0.0
      DO 7  J = 1, M
    7 SUMV=SUMV + (A(J,I)**2)
      DEN = SQRT ( SUMV)
      DO 8  J = 1, M
    8 X(J,I)= A(J,I) /DEN
    9 CONTINUE
C
C     COLUMNS OF X(I,J) ARE NOW NORMALIZED.
C
      RETURN
      END
    >
```

```
C       TEST DIRNM (DIAGONALIZE A REAL NONSYMMETRIC MATRIX, A-INVERSE * B)
C       P. R. LOHNES,   PROJECT TALENT,   1966.
C
C       INPUT IS 1) M, ORDER OF THE MATRICES, 2) VARIABLE FORMAT FOR UPPER
C          TRIANGULAR MATRICES, 3) MATRIX A (UPPER TRIANGULAR), 4) MATRIX
C
        DIMENSION  A(50,50), B(50,50), C(50,50), D(50,50), X(50), FMT(16)
C
  1     READ (5, 2)    M
  2     FORMAT (I2)
        READ (5, 4)    (FMT(J),    J = 1, 16)
  4     FORMAT (16A5)
        DO 3   J = 1, M
  3     READ (5, FMT)    (A(J,K),   K = J, M)
        WRITE(6, 5)
  5     FORMAT (15H1INPUT MATRIX A)
        DO 6   J = 1, M
  6     WRITE(6, 7)    (J,  (A(J,K),   K = J, M))
  7     FORMAT (7H0  ROW I3, 7F10.3/ (10X, 7F10.3))
        DO 8   J = 1, M
  8     READ (5, FMT)    (B(J,K),   K = J, M)
        WRITE(6, 9)
  9     FORMAT (15H0INPUT MATRIX B)
        DO 10   J = 1, M
 10     WRITE(6, 7)    (J,  (B(J,K),   K = J, M))
        DO 11   J = 1, M
        DO 11   K = J, M
        A(K,J) = A(J,K)
        B(K,J) = B(J,K)
        C(K,J) = A(J,K)
 11     C(J,K) = A(J,K)
C
        CALL MATINV(C, M, DET)
C
        DO 12   J = 1, M
        DO 12   K = 1, M
        D(J,K) = 0.0
        DO 12   L = 1, M
 12     D(J,K) = D(J,K) + C(J,L) * B(L,K)
C       D NOW CONTAINS A-1 * B.
        WRITE(6, 13)
 13     FORMAT (14H0A-INVERSE * B)
        DO 14   J = 1, M
 14     WRITE(6, 7)    (J,  (D(J,K),   K = 1, M))
C
        CALL DIRNM (B, M, A, C, X, M)
C
        DO 15   J = 1, M
        DO 15   K = 1, M
        B(J,K) = C(J,K)
 15     A(J,K) = C(J,K) * X(K)
C       A NOW CONTAINS V * LAMBDA.
C
        CALL MATINV(C, M, DET)
C
        DO 16   J = 1, M
        DO 16   K = 1, M
        D(J,K) = 0.0
        DO 16   L = 1, M
 16     D(J,K) = D(J,K) + A(J,L) * B(L,K)
        WRITE(6, 17)
 17     FORMAT (23H0V * LAMBDA * V-INVERSE)
        DO 18   J = 1, M
 18     WRITE(6, 7)    (J,  (D(J,K),   K = 1, M))
        DO 19   J = 1, M
        DO 19   K = 1, M
```

TEST DIRNM (*continued*)

```
         A(J,K) = 0.0
         DO 19   L = 1, M
   19    A(J,K) = A(J,K) + C(J,L) * B(L,K)
         WRITE(6, 20)
   20    FORMAT (14HOV * V-INVERSE)
         DO 21   J = 1, M
   21    WRITE(6, 7)     (J, (A(J,K),   K = 1, M))
 C
 C       GIVEN THAT A AND B ARE REAL SYMMETRIC OF THE SAME ORDER, AND THAT
 C       A IS POSITIVE-DEFINITE, AND D = A-1 * B, AND LAMBDA IS THE
 C       DIAGONAL MATRIX OF EIGENVALUES OF D, AND V IS THE MATRIX OF
 C       COLUMN EIGENVECTORS OF D, THEN
 C           D * V = V * LAMBDA
 C           V-1 * D * V = LAMBDA
 C           V * LAMBDA * V-1 = D
 C           V-1 * V = V * V-1 = I
         WRITE(6, 22)
   22    FORMAT (17HOEIGENVALUES OF D)
         DO 23   J = 1, M
   23    WRITE(6, 24)     J, X(J)
   24    FORMAT(2X,I3,5X,F14.7)
         WRITE(6, 25)
   25    FORMAT (37HOEIGENVECTORS OF D, IN ROWS (VPRIME) )
         DO 26   J = 1, M
   26    WRITE(6, 7)     (J, (C(K,J),   K = 1, M))
         WRITE(6, 27)
   27    FORMAT (30HOV-INVERSE, IN ROWS (V-1PRIME))
         DO 28   J = 1, M
   28    WRITE(6, 7)     (J, (B(K,J),   K = 1, M))
         GO TO 1
         END
 >
```

6.7 EXERCISES

1. The concept of structure now pervades most of multivariate analysis as the primary basis for the interpretation of linear composites of original variables. Compare how these "loadings" (correlations between original variates and the derived weighted sum) are produced in

    (a) multiple correlation

    (b) principle components

    (c) canonical correlation

    Answer in matrix notation, defining the matrices with which you begin your explanation.

2. Using the TALENT data, study the canonical relationships between four ability variables (e.g., 12–15) and three motive measures (17–19).

CHAPTER SEVEN

||||||||||||||||||||||||||||||||||||||||||||||||||||||||||||||||||||||

Multiple Partial Correlation

7.1 MATHEMATICS OF MULTIPLE PARTIAL CORRELATION

Sometimes it is desirable to remove the influence of a vector variable z_1 from the elements of a vector variable z_2 and then to analyze the interrelationships among the residuals or unexplained parts of the elements of z_2. The procedure is to compute the multiple regression of each element of z_2 in the space of z_1, then compute the variances and covariances of the regressed variates $\hat{z}_2$ (where the variances are squared multiple correlation coefficients), then compute the variances and covariances of the residuals or unexplained parts $\tilde{z}_2$, and finally to scale the residual variance-covariance matrix into a correlation matrix which represents the intercorrelations among the residuals or unexplained parts $\tilde{z}_2$. This matrix of correlations of residuals we propose to call a multiple partial correlation matrix, identified symbolically as $R_{2 \cdot 1}$.

Another reason for interest in the method of generating a multiple partial correlation matrix is that it is precisely the method that is employed in multivariate analysis of covariance to adjust a criterion or dependent vector variable z_2 for the influence of a set of covariates z_1 prior to performing a multivariate analysis of variance on the adjusted or residual criterion vector variable $\tilde{z}_2$. The only additional aspect of covariance analysis is that two separate multiple partialling jobs are performed, one for the hypothesis or between-groups matrix and one for the error or pooled-within-groups matrix.

Once again, we shall express the model in terms of the partitions of the total correlation matrix for z_1 plus z_2, assuming p_1 elements in z_1 and p_2 elements in z_2, so that the rank of R_{11} is p_1, the rank of R_{22} is p_2, and the order of R_{12} is $p_1 \times p_2$. First, we form the standardized regression weights

$$B = R_{11}{}^{-1}R_{12}$$

The kth column of B contains the weights for regressing the kth element of z_2 on all the elements of z_1, so that

$$\hat{z}_{2_k} = b_k'z_1$$

and

$$\hat{z}_2 = B'z_1$$

Next we form the matrix of variances and covariances of the elements of $\hat{z}_2$ which we call the regression matrix

$$\hat{R}_{22} = R_{21}B = R_{21}R_{11}{}^{-1}R_{12}$$

Then we form the matrix of variances and covariances of the elements of $\tilde{z}_2 = z_2 - \hat{z}_2$, which we call the residual matrix

$$\tilde{R}_{22} = R_{22} - \hat{R}_{22} = R_{22} - R_{21}R_{11}{}^{-1}R_{12}$$

Finally we form the correlations among the elements of $\tilde{z}_2$, which we call the multiple partial correlation matrix, by dividing the j,kth element of $\tilde{R}_{22}$ by the square root of the product of the jth diagonal element and the kth diagonal element of R_{22}

$$R_{2\cdot1} = \{r_{2\cdot1_{jk}}\} = \frac{\tilde{r}_{22_{jk}}}{\sqrt{\tilde{r}_{22_{jj}}\tilde{r}_{22_{kk}}}}$$

In applications of partial correlation, the investigator will frequently have in mind a partition of the z_2 elements of the vector variable into a subset of independent elements and a subset of dependent elements, so that *after* he partials out the influence of the control variables z_1 he will proceed to study the relationships between the residuals of the independent elements and the residuals of the dependent elements by multiple or canonical regression performed on a partitioned $R_{2\cdot1}$. However, it sometimes happens that the investigator wants to partial out the control influences only from the dependent elements and not from the independent elements. Then he starts with an explicit partition of the vector variable into three subsets:

$$z_1 = \text{control variates}$$
$$z_2 = \text{dependent variates}$$
$$z_3 = \text{independent variates}$$

We call the analysis to be performed *part correlation*. The generalized part correlation analysis develops the intercorrelations of residuals of the dependent variates after multiple regression on the control variates, and also the cross-correlations of these residuals of the dependent variates with the unadjusted independent variates.

Part correlation requires the partitioning of the correlation matrix into nine matrices:

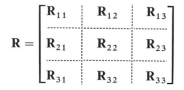

$$\mathbf{R} = \begin{bmatrix} \mathbf{R}_{11} & \mathbf{R}_{12} & \mathbf{R}_{13} \\ \mathbf{R}_{21} & \mathbf{R}_{22} & \mathbf{R}_{23} \\ \mathbf{R}_{31} & \mathbf{R}_{32} & \mathbf{R}_{33} \end{bmatrix}$$

The required intercorrelations of the residuals of the dependent variates are computed exactly as $\mathbf{R}_{2 \cdot 1}$ in the partial correlation model, and may be termed $\mathbf{R}_{2 \cdot 1}$.

The covariances of the residuals of the dependent variates with the unadjusted independent variates are computed by applying the same multiple regression weights that were used to develop $\mathbf{R}_{2 \cdot 1}$, but in this case the weights are applied to $\mathbf{R}_{13}$

$$\tilde{\mathbf{R}}_{23} = \mathbf{R}_{23} - \hat{\mathbf{R}}_{23} = \mathbf{R}_{23} - \mathbf{B}'\mathbf{R}_{13} = \mathbf{R}_{23} - \mathbf{R}_{21}\mathbf{R}_{11}{}^{-1}\mathbf{R}_{13}$$

To scale the j,kth covariance in $\tilde{\mathbf{R}}_{23}$ into a correlation element it is only necessary to divide by the standard deviation of the jth element of $\tilde{\mathbf{z}}_2$, since the standard deviation of the kth element of $\mathbf{z}_3$ has not been changed from unity:

$$\mathbf{R}_{2 \cdot 1, 3} = \{r_{2 \cdot 1, 3}\} = \frac{\tilde{r}_{23_{jk}}}{\sqrt{\tilde{r}_{22_{jj}}}}$$

When the part correlation analysis option of the PARTL program is used the output correlation matrix which is printed and punched is assembled as:

$$\mathbf{R} = \begin{bmatrix} \mathbf{R}_{2 \cdot 1} & \mathbf{R}_{2 \cdot 1, 3} \\ \mathbf{R}_{3, 2 \cdot 1} & \mathbf{R}_{33} \end{bmatrix}$$

It is interesting to note that if $\mathbf{z}_3$ is a vector but z_1 and z_2 are scalar:

$$\tilde{r}_{22} = 1 - r_{21}{}^2$$

and

$$\tilde{r}_{23_j} = r_{23_j} - r_{21}r_{13_j}$$

so that

$$r_{2 \cdot 1, 3_j} = \frac{\tilde{r}_{23_j}}{\sqrt{1 - r_{21}^2}}$$

7.2 NUMERICAL EXAMPLE OF PARTIAL CORRELATION

Four variables from the Project TALENT males can serve as a numerical example of partial correlation. Table 7.1 reports the means, standard deviations, and correlation matrix for (1) Information Part I, (2) Part II, (3) Physical Science Interest, and (4) Socioeconomic Status (SES). The correlation

TABLE 7.1   CORREL Output for PARTL
Numerical Example

TEST	MEAN	S.D.	
1	156.50	35.08	Information Part I
2	79.55	17.84	Information Part II
3	21.30	9.17	Physical Science Interest
4	98.54	9.42	Socieconomic Status

$N = 234$

CORRELATION MATRIX

	1	2	3	4
1	1.00	0.86	0.57	0.48
2	0.86	1.00	0.43	0.49
3	0.57	0.43	1.00	0.36
4	0.48	0.49	0.36	1.00

between (3) and (4) is .36. The research question might be: What is the partial correlation between science interest and SES after removing variance associated with general scholastic achievement as measured by a linear function of Information Part I and Part II? Thus in this example $p_1 = 2$ and $p_2 = 2$.

The partitioned correlation then serves as input to produce the regression matrix $\hat{\mathbf{R}}_{22}$:

$$\hat{\mathbf{R}}_{22} = \mathbf{R}_{21}\mathbf{R}_{11}^{-1}\mathbf{R}_{12} = \begin{bmatrix} .57 & .43 \\ .48 & .49 \end{bmatrix} \cdot \begin{bmatrix} 1.00 & .86 \\ .86 & 1.00 \end{bmatrix}^{-1} \cdot \begin{bmatrix} .57 & .48 \\ .43 & .49 \end{bmatrix}$$

$$\hat{\mathbf{R}}_{22} = \begin{bmatrix} .33 & .26 \\ .26 & .25 \end{bmatrix}$$

The residual matrix, $\tilde{\mathbf{R}}_{22}$, becomes

$$\tilde{\mathbf{R}}_{22} = \mathbf{R}_{22} - \hat{\mathbf{R}}_{22} = \begin{bmatrix} 1.00 & .36 \\ .36 & 1.00 \end{bmatrix} - \begin{bmatrix} .33 & .26 \\ .26 & .25 \end{bmatrix}$$

$$\tilde{\mathbf{R}}_{22} = \begin{bmatrix} .67 & .10 \\ .10 & .75 \end{bmatrix}$$

Then converting the residual matrix, which is a variance-covariance matrix, to a correlation matrix, we obtain $\mathbf{R}_{2 \cdot 1}$:

$$\mathbf{R}_{2 \cdot 1} = \begin{bmatrix} 1.00 & .15 \\ .15 & 1.00 \end{bmatrix}$$

The resulting partial correlation of .15 is the correlation between the residuals of science interest and SES, the residuals being the variance not related to Information Part I and Part II. The interpretation would be that although science interest tends to be slightly related to SES among 12th grade males ($r_{34} = .36$), this relationship is essentially nonexistent among boys of similar scholastic ability ($r_{34 \cdot 12} = .15$).

The other output from the partial correlation program is the table of multiple correlations between each of the dependent variables and the set of control variables. In this example:

	Multiple Correlation	Mult R Squared	F^2_{231} Ratio
Science Interest	.578	.334	57.86
SES	.502	.252	38.83

Thus one-third of the variance in science interest can be predicted from Information Part I and II, while one-fourth of the variance in SES can be similarly accounted for.

7.3 RESEARCH EXAMPLES

As part of a general consideration of how abilities, interests and career plans change during high school (Cooley, 1967), two hypotheses were proposed:

1. There is variance in grade 12 interests not explainable by grade 9 interests which is related to grade 9 abilities (i.e., changes in interest patterns during high school are related to grade 9 abilities)
2. There is variance in grade 12 abilities not explainable by grade 9 abilities which is related to grade 9 interests (i.e., changes in ability patterns during high school are related to grade 9 interests).

These hypotheses may appear to contradict each other. On one hand, it is claimed that interests affect the development of abilities, while at the same time, it is suggested that vocational interests change during high school to become more consistent with abilities. Before turning to the multivariate explorations of these hypotheses, it seems appropriate to explain what is involved by using a simple numerical example.

The four ellipses of Figure 7.1 each represent the total variance of four

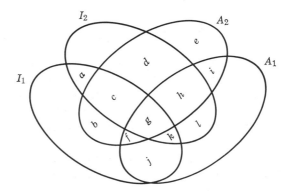

Figure 7.1. Illustration of overlap in variance among interests and abilities.

different variables, grade 9 interest (I_1), grade 12 interest (I_2), grade 9 ability (A_1), and grade 12 ability (A_2). The square of the correlation coefficient indicates the proportion of variance in one variable which is related to another (or proportion of overlapping variance). For example, areas $a + c + g + k$ indicate the extent of the overlap between I_1 and I_2. Areas $a + c$ represent that partial overlapping between I_1 and I_2 that is independent of A_1. The square of the partial correlation between I_1 and I_2 with A_1 removed ($I_1 I_2 \cdot A_1$) tells how extensive that partial overlap actually is.

Table 7.2 presents retest data for a particular interest scale (P-701 Physical Science) and a particular ability scale (R-106 Mathematics Information). The ellipses of Figure 7.1 were drawn to approximate these relationships. Areas $b + c$ and $h + l$ and their corresponding partial correlations ($I_1 A_2 \cdot A_1$ and $I_2 A_1 \cdot I_1$) are of particular interest. The fact that both partials are significantly different from zero allows one to infer that some boys improve their mathematics knowledge during high school because they are interested in science, while others lose interest in science because they are not particularly good at learning mathematics. At least the data are consistent with both of these generalizations.

Having illustrated the problem for a simple two variable case, let us now

TABLE 7.2   Variance Overlap using R-106
(Mathematics Information) and P-701
(Science Interest)

Type of Overlap	Figure 7.1 Areas	Correlation[a]	Proportion Overlap
$I_1 I_2$	a,c,g,k	.53	.28
$A_1 A_2$	f,g,h,i	.72	.52
$I_1 A_1$	f,g,j,k	.34	.12
$I_2 A_2$	c,d,g,h	.41	.17
$I_1 A_2$	b,c,f,g	.43	.18
$I_2 A_1$	g,h,k,l	.32	.10
$I_1 A_2 \cdot A_1$	b,c	.28	.08
$I_2 A_1 \cdot I_1$	h,l	.18	.03

[a] All correlations are significantly different from zero at the
.001 level.

turn to a multivariate test of the hypothesis that a significant portion of the
variation in grade 12 interests that is not related to grade 9 interests is related
to grade 9 abilities, using multiple partial canonical correlation. With this
technique one can remove the variation in grade 9 abilities and grade 12
interests that is associated with grade 9 interests and then see whether any
significant relationships exist between the two sets of residuals. Any resulting
significant canonical relationships will allow inferences regarding the ways in
which interests *change* between grades 9 and 12 to be more consistent with
abilities. Between the 60 TALENT ability tests and the 17 interest scales there
are 17 mathematically possible canonical relationships (the number of
variables in the smaller set). Of these, eight were significantly different from
zero at the .05 level for the male sample. That is, there were at least eight ways
in which boys changed their interests to be more consistent with their
abilities.

The 10 largest variable loadings for each side of the four largest canonical
relationships are summarized in Table 7.3. The largest canonical correlation
was .42. Inspection of the first set of loadings reveals that this first, largest
correlation was between a general ability dimension extracted from 60
TALENT ability tests and a general level of aspiration factor (professional
versus nonprofessional) from the 17 interest scales. Thus, boys who did well
on the grade 9 ability measures but planned nonprofessional careers tended
to change to professional plans by grade 12, while boys low on general ability
but with high expectations tended to modify their aspirations by grade 12.
(Keep in mind that these loadings are for residuals.)

TABLE 7.3   Canonical Loadings and Partial Canonical Correlations Between 60 Grade 9 Abilities and Grade 12 Interests (Control on Grade 9 Interests; $N = 1466$ Males)

Canonical Correlation	10 Highest Ability Loadings		10 Highest Interest Loadings	
	R-250 Reading Compre.	.70	P-704 Literature	.66
	R-102 Vocabulary	.68	P-702 Biological Science	.44
	R-103 Literature	.66	P-707 Musical	.37
	R-142 Bible	.61	P-706 Artistic	.35
.42	R-107 Physical Science	.59	P-701 Physical Science	.32
	R-105 Social Studies	.58	P-703 Public Service	.26
	R-110 Aeronautics & Space	.52	P-715 Skilled Trades	—.14
	R-106 Mathematics	.50	P-717 Labor	—.15
	R-104 Music	.48	P-713 Office Work	—.18
	R-131 Art	.47	P-714 Mechanical-Technical	—.24
	R-115 Sports	.48	P-708 Sports	.44
	R-410 Arithmetic Comp.	.39	P-701 Physical Science	.29
	R-106 Mathematics	.37	P-703 Public Service	.25
	R-312 Math II Introductory	.36	P-712 Computation	.25
.39	R-149 Games (Sedentary)	.31	P-707 Musical	.15
	R-233 Punctuation	.30	P-711 Sales	.14
	R-105 Social Studies	.29	P-702 Biological Science	.11
	R-232 Capitalization	.27	P-709 Hunting & Fishing	—.16
	R-311 Math I Reasoning	.26	P-715 Skilled Trades	—.21
	R-211 Memory for Sentences	.26	P-714 Mechanical-Technical	—.25
	R-111 Electricity & Elect.	.32	P-706 Artistic	.25
	R-270 Mechanical Reasoning	.27	P-702 Biological Science	.10
	R-114 Home Economics	.27	P-710 Business Management	—.10
	R-112 Mechanics	.26	P-703 Public Service	—.16
.33	R-151 Foods	.26	P-701 Physical Science	—.17
	R-440 Object Inspection	.24	P-704 Literary-Linguistic	—.21
	R-231 Spelling	—.20	P-709 Hunting & Fishing	—.22
	R-142 Bible	—.22	P-705 Social Service	—.29
	R-312 Math II Introductory	—.25	P-713 Office Work	—.47
	R-410 Arithmetic Comp.	—.26	P-712 Computation	—.52
	R-270 Mechanical Reasoning	.49	P-701 Physical Science	.49
	R-282 Visualization 3-Dim.	.43	P-716 Farming	.40
	R-145 Hunting	.39	P-714 Mechanical-Technical	.36
	R-281 Visualization 2-Dim.	.32	P-709 Hunting & Fishing	.31
.33	R-146 Fishing	.24	P-703 Public Service	—.17
	R-113 Farming	.23	P-713 Office Work	—.19
	R-312 Math II Introductory	.23	P-710 Business Management	—.20
	R-290 Abstract Reasoning	.20	P-705 Social Service	—.22
	R-111 Electricity & Elect.	.20	P-704 Literary-Linguistic	—.23
	R-136 Journalism	—.22	P-711 Sales	—.30

The second pair of canonical variates utilized sports information and several mathematics scales on the positive side of the ability measures and another aspiration level factor (orthogonal to the previous interest factor) with a different combination of positive loadings on the positive side of the interest factor. Sports Interest dominated the definition of this interest dimension.

The third and fourth canonical coefficients were of equal magnitude (.33). A boy will be high on the third ninth-grade ability canonical if he is higher on knowledges associated with tinkering activities (electricity, mechanical gadgets, cooking) and lower on mathematics abilities than would be expected from his grade 9 interests. If a boy is low on that ability canonical, he will tend to change his interests toward computation and office work and away from interests in art and biology.

The fourth set of canonical loadings emphasized the nonverbal ability measures such as mechanical reasoning and spatial visualization, while the largest positive interest loading was on physical science, engineering, and mathematics (P-701) interest and the largest negative loading on sales interest (P-711). Because these results are in terms of partial correlations, it can be concluded that there are several distinct ways in which interests shift between grade 9 and grade 12, which are in turn related to that variation in grade 9 abilities not related to grade 9 interests.

The second hypothesis relating interest and ability suggests that grade 9 interests (as motives) influence the development of abilities between grades 9 and 12. In terms of multiple partial canonical correlation, this hypothesis was explored by relating grade 9 interests and grade 12 abilities after removing the variation in both sets that is related to grade 9 abilities. With the previous ability-interest hypothesis and analysis, the questions were how and to what extent interests shift to be more consistent with earlier abilities. Now the questions are how and to what extent abilities change in a manner consistent with earlier interests.

The design of the Project TALENT retest study made it necessary to divide the analysis into three parts, each part composed of a different set of ability measures, defined as Batteries A, D, and E. The measures used in each of the three canonical analyses are listed in Tables 7.4, 7.5, and 7.6, together with sample sizes, the significant canonical correlations resulting from each set, and the canonical loadings corresponding to them.

The 38 scales from Information Parts I and II formed two significant (at .01 level) canonical relationships with the 17 interest scales. The largest canonical correlation, as seen in Table 7.4 was .51; the pattern of loadings for the interest side was very similar to the pattern for the largest canonical of Table 9. The ability pattern was slightly different, however, with a bipolar factor, loading Music, Mathematics, and Sports positively and Mechanical and Hunting Information negatively. The fact that there were no negative

TABLE 7.4 Canonical Correlations and Loadings Between Grade 9 Inter
and Grade 12 Abilities with Grade 9 Abilities Removed

(Battery A, $N = 493$ males)   1st Canonical $R = .51$
2nd Canonical $R = .45$

Abilities	Loadings 1st	Loadings 2nd	Interests	Loadings 1st	2nd
R-101 Screening	−.27	.03	P-701 Physical Science	.43	
R-102 Vocabulary	−.09	−.10	P-702 Biological Science	.41	−.
R-103 Literature	.09	.00	P-703 Public Service	.29	−.
R-104 Music	.39	.07	P-704 Literary-Linguistic	.39	−.
R-105 Social Studies	.05	−.09	P-705 Social Service	31	−
R-106 Mathematics	.33	−.53	P-706 Artistic	.25	−.
R-107 Physical Science	.12	−.35	P-707 Musical	.55	−.
R-108 Biological Science	−.05	−.00	P-708 Sports	.26	−.
R-109 Scientific Attitude	.04	.01	P-709 Hunting and Fishing	−.32	−.
R-110 Aeronautics, Space	−.13	−.15	P-710 Business Management	.25	−.
R-111 Electricity, Electr.	−.15	−.45	P-711 Sales	.23	−.
R-112 Mechanics	−.43	.03	P-712 Computation	.36	.
R-113 Farming	−.18	.02	P-713 Office Work	.28	.
R-114 Home Economics	−.11	−.17	P-714 Mechanical-Technical	−.23	.
R-115 Sports	.32	.11	P-715 Skilled Trades	−.20	−.
R-131 Art	.17	.05	P-716 Farming	−.33	−.
R-132 Law	−.01	−.23	P-717 Labor	−26	.
R-133 Health	−.02	−.30			
R-134 Engineering	−.18	−.03			
R-135 Architecture	−.09	−.13			
R-136 Journalism	−.14	−.03			
R-137 Foreign Travel	.04	−.11			
R-138 Military	.10	−.00			
R-139 Acctg., Business	−.04	−.07			
R-140 Practical Knowl.	−.19	.12			
R-141 Clerical	.08	−.15			
R-142 Bible	.21	−.28			
R-143 Colors	.08	−.02			
R-144 Etiquette	−.12	.10			
R-145 Hunting	−.44	−.23			
R-146 Fishing	−.18	−.07			
R-147 Outdoor Activities	−.07	.06			
R-148 Photography	.06	.11			
R-149 Games (Sedentary)	.04	−.22			
R-150 Theater and Ballet	.08	.02			
R-151 Foods	.18	.10			
R-152 Miscellaneous	.17	−.18			
R-162 Vocabulary	.01	−.01			

loadings for the ability variables of the first ability-interest canonical summarized in Table 7.3 means that a general grade 9 ability factor (unrelated to grade 9 interests) was related to shifting interest patterns which, in general, are white-collar versus blue-collar. Ninth-grade boys of high general ability who had expressed blue-collar interests (mechanical-technical, labor, skilled trades) in grade 9 tended to change to more professionally related interests. The converse was true for boys of low general ability.

TABLE 7.5   Canonical Correlations and Loadings Between Grade 9 Interests and Grade 12 Abilities with Grade 9 Abilities Removed

(Battery D, $N = 556$ Males)   Canonical $R = .41$

Abilities	Loadings	Interests	Loadings
R-212 Memory for Words	.11	P-701 Physical Science	.67
R-230 English Total	.29	P-702 Biological Science	.21
R-311 Part I. Arith. Reas.	.33	P-703 Public Service	.19
R-312 Part II. Introd. H.S.		P-704 Literary-Linguistic	.20
Math	.10	P-705 Social Service	−.03
R-333 Part III. Advanced		P-706 Artistic	.16
H.S. Math	.86	P-707 Musical	.08
R-410 Arith. Computation	.71	P-708 Sports	−.06
		P-709 Hunting and Fishing	−.07
		P-710 Business Management	−.21
		P-711 Sales	.04
		P-712 Computation	.32
		P-713 Office Work	−.03
		P-714 Mechanical-Technical	−.05
		P-715 Skilled Trades	−.34
		P-716 Farming	−.19
		P-717 Labor	−.27

In the interest-ability canonical of Table 7.4 on the other hand, was a professional versus nonprofessional grade 9 interest factor (unrelated to grade 9 abilities) correlating .51 with the bipolar ability factor. This result indicates that boys who expressed greater professional interest than would be expected from their grade 9 abilities tended to do better on professionally related grade 12 ability measures than would be expected from grade 9 abilities. Those who were more nonprofessionally interested than expected tended to gain more on the negatively loaded scales (Mechanical Engineering, Hunting and Fishing Information) than other grade 9 boys.

Table 7.5 lists the loadings for the largest and only significant canonical correlation between grade 9 interests and those grade 12 abilities in Battery D

TABLE 7.6 Canonical Correlations and Loadings Between Grade 9 Interests and Grade 12 Abilities with Grade 9 Abilities Removed

(Battery E, $N = 414$ males)   1st Canonical $R = .40$
                                         2nd Canonical $R = .33$

Abilities	Loadings 1st	Loadings 2nd	Interests	Loadings 1st	Loadings 2nd
R-220 Disguised Words	.52	.10	P-701 Physical Science	.23	−.24
R-240 Word Functions			P-702 Biological Science	.34	−.16
in Sentences	.64	−.36	P-703 Public Service	.54	.17
R-250 Reading Compre.	34	−.30	P-704 Literary-Linguistic	.17	.31
R-260 Creativity	.23	−.42	P-705 Social Service	.12	.51
R-270 Mech. Reasoning	−.01	−.85	P-706 Artistic	.27	.50
R-281 Visualization in			P-707 Musical	.17	.29
2 Dimensions	.45	.09	P-708 Sports	.18	.26
R-282 Visualization in			P-709 Hunting and Fishing	−.09	.07
3 Dimensions	.40	−.37	P-710 Business Management	.06	.36
R-290 Abstr. Reasoning	.53	−.26	P-711 Sales	.07	.45
R-420 Table Reading	.04	−.16	P-712 Computation	.26	.40
R-430 Clerical Checking	.45	−.12	P-713 Office Work	.08	.44
R-440 Object Inspect.	.07	−.27	P-714 Mechanical-Technical	−.40	.26
			P-715 Skilled Trades	−.28	.63
			P-716 Farming	−.38	.38
			P-717 Labor	−.37	.62

that were not in Battery A (with corresponding grade 9 abilities partialed out). Once again the white-collar versus blue-collar interest factor appeared, but most heavily loaded on the physical science scale. This factor correlated .41 with a general ability factor dominated by advanced mathematics and arithmetic computation. Thus, changes in ability between grade 12 and grade 9 are related to grade 9 interests. The ninth-grade boy who wanted to become a scientist produced greater gains in the mathematics areas during high school than would be expected from his grade 9 abilities.

Two canonical relationships were significant at the .05 level for the 11 ability measures in Battery E which were not in A or D (Table 7.6). Again, there was the professional versus nonprofessional interest factor, this time emphasizing two of the nonscience professions more than they were emphasized in Table 7.5. The associated ability factor ($R = .40$) loaded positively on everything but Mechanical Reasoning, Table Reading, and Object Inspection, the latter having near-zero loadings. The interpretation here is similar to that for the largest canonical in Tables 7.4 and 7.5.

The second canonical of Table 7.6 is quite interesting. Here the two science scales (P-701 and P-702) were loaded negatively with most of the other interest scales loaded high positive. The ability loadings were zero or negative, with the highest negatives being Mechanical Reasoning, Creativity, and Visualization in Three Dimensions. Thus, these types of abilities tend not to be developed in high school by boys not interested in science, at least not to the degree expected from grade 9 abilities.

The two sets of canonical results show that grade 9 abilities are related to changes in interest during high school (Table 7.3), while at the same time grade 9 interests are related to changes in abilities during high school. These are not contradictory findings but demonstrate the dual nature of the interest measures. The interest scales appear to be tapping both basic motivational dimensions, which direct the development of abilities, and self-concept dimensions, which are modified as the boy achieves greater insights into his own abilities and those needed by different career areas.

The extent of a competent researcher's enthusiasm for partialling can vary from outright hostility to almost Messianic fervor. The enthusiast argues that partialling z_1 from z_2 and z_3 before z_3 is regressed on z_2 most clearly demonstrates the dependence of z_3 on z_2 when z_1 is held constant. See J. C. Nunnally (1967, pp. 151–155) for an informative advocacy of partialling. He says, "partial correlation is the correlation expected between two variables when a third variable is held constant," and that this method tells whether z_2 "actually adds something to what could be explained" by the regression of z_3 on z_1 alone. The hostile researcher argues that z_1 is never constant in nature, and that while regression after partialling may tell whether something is added to the predictability of z_3 by bringing z_2 into play, the method does not accurately represent how much is added. He claims that the actual improvement in prediction of z_3 when z_2 is added to z_1 as an additional predictor is *not* estimated by $R_{23 \cdot 1}$ and that the partial correlation will frequently lead to substantial overestimation of the additional value z_2 has for prediction of z_3 when you already possess z_1 as a predictor. Since both positions are technically correct, it is essential to recognize that different questions are involved. We borrow data from C. T. Nephew (1969) to illustrate the different questions and their answers.

Nephew collected data from two sources on 119 school districts in northeast and northcentral states. From Project TALENT he got MAP factor means for the district's high schools as a domain of output measures, and from local sources he compiled district means on eleven fiscal practices. For a control variable he got the mean socioeconomic status for each district's high school population from Project TALENT. (Only districts in which Project TALENT had tested in all the high schools were included in the sample.) His purpose was to show how the fiscal practices of school districts influenced learning outcomes after adjustment for the regression of outcomes on socioeconomic environment (SEE). Thus, he chose to do part correlation and base his conclusions on the multiple regressions of MAP-measure residuals from SEE on

TABLE 7.7 Nephew Data Correlations Among Socioeconomic Status (S.E.E.), Five TALENT MAP Factor Criteria, and 11 Fiscal Practices Predictor Variables for 119 School Districts, with Correlations with MAP Factor Residuals from Regression on S.E.E. Below Original Correlations

	VKN	SCH	MAT	SCI	VIS	FP1	FP2	FP3	FP4	FP5	FP6	FP7	FP8	FP9	FP10	FP11
S.E.E.																
VKN	73	55	41	14		−02	36	42	27	44	31	25	39	−07	33	19
VKN Resid		54	46	23	−17	−04	41	45	21	44	33	19	40	−10	25	30
		41	31	07	−25	−04	22	21	02	18	15	02	17	−08	02	23
SCH			42	40	17	−13	31	45	07	28	32	20	44	17	26	56
SCH Resid			49	00	11	−14	18	30	−05	11	21	11	31	21	14	53
MAT				48	17	−05	25	43	27	33	27	25	45	16	23	37
MAT Resid				29	12	−05	06	23	14	10	11	14	29	24	05	32
SCI					02	09	14	08	09	25	26	−12	16	−24	06	02
SCI Resid					−12	11	−01	−10	−02	08	15	−24	00	−23	−08	−06
VIS						−08	09	09	21	−01	−08	−09	04	11	−02	00
VIS Resid						−08	04	03	17	−07	−12	−13	−01	12	−07	−02

the space of the eleven fiscal practices. However, since the fiscal practices are also related to SEE, he could have done full partialling of SEE from predictors and criteria. Alternatively, he could have compared the variances in the MAP criteria predictable from full regressions on SEE plus eleven fiscal practices with the variances accounted for by regressions on SEE alone. This latter strategy will be referred to as the full versus restricted model method.

Table 7.7 contains the original (or zero order) correlations among selected MAP criteria and with the SEE and eleven fiscal predictors, and also the partial correlations among the MAP criteria after removal of SEE and the part correlations of MAP criteria residuals with unadjusted fiscal predictors. Note that SEE is strongly related to Verbal Knowledge (VKN), and moderately to negligibly related to the other variables. Can you guess from these zero order and part correlations how the different prediction models are going to turn out? In general it is relationships involving VKN that are most altered by partial or part operations. Table 7.8 displays explained variances for five

TABLE 7.8 Nephew Data Squared Multiple Correlations, or Proportions of Explained Variance, for Five TALENT MAP Factor Criteria in Five Prediction Models (Subjects are 119 School Districts)

Model	Criterion	VKN	SCH	MAT	SCI	VIS
I	MAP factor regressed on socieconomic status (SEE)	.53	.18	.31	.17	.02
II	MAP factor regressed on 11 fiscal practices	.46	.41	.33	.21	.13
III	MAP factor regressed on 11 fiscal practices + SEE	.63	.46	.44	.30	.16
IV	MAP factor residual from SEE regressed on 11 fiscal pract.	.17	.34	.17	.16	.14
V	MAP factor residual from SEE regressed on residuals of 11 fiscal practices from SEE	.21	.35	.19	.16	.14

prediction models. The first model shows what SEE alone can do as a predictor. The high predictability of VKN is consistent with our common knowledge that general intelligence and social class status are interdependent, since VKN has a very high general intelligence saturation. It is problematic why SEE doesn't do more to account for district means on Scholasticism (SCH) and Science Interests (SCI), but the Visual Reasoning (VIS) finding is consistent with the theory that VIS is mostly an innate aptitude which is not especially prestigious in American society. The Mathematics (MAT) variance from SEE is about what we would expect from its correlation with VKN. Note that these criterion variables are essentially uncorrelated in the high school population (Lohnes, 1966), but that does not prevent the district means from being correlated. Also, the standard deviations of district means for SEE and the MAP variables are about one-third of the corresponding standard deviations of individual differences.

With the exception of VIS, the restricted model based on SEE alone disqualifies the second model, which is a restricted model that looks at regressions on fiscal practices with no consideration of SEE, and justifies Nephew's concern to control for SEE. The third model is the full prediction one that uses both SEE and the fiscal practices to account for criterion variance. It might seem that the difference between variance accounted for by model I and by model III represents the additional variance accounted for by adding eleven fiscal practices to SEE. If we had the large sample size that is desirable in multivariate analyses this argument would be convincing. Unfortunately, twelve predictors are too many for only 119 subjects, and the variances estimated by model III would undoubtedly shrink a great deal on a replication sample. Assuming considerable shrinkage, we see that the fiscal practices are not going to add much to the regression on SEE alone, with the possible exception of the Scholaticism criterion. It is interesting that in the regression function for SCH under model III it is teacher salaries that has far and away the largest estimated beta weight. This is consistent with the nature of SCH, as it involves primarily the extent to which students are taking the thoroughly academic preparatory courses in high school which are courses that require expensive teachers.

Nephew's dissertation results are given as model IV. They are subject to shrinkage exactly as much as the model III results, of course. Note that in every case except VIS the model IV explained variance in residual criterion is larger than the difference between the model I and model III explained variances. The part correlation method leads to a more optimistic conclusion regarding the value of the additional predictors than does the full versus restricted model method. Nevertheless, both methods agree in singling out Scholasticism as the MAP criterion for which the fiscal practices of districts have most additional value, and this is of major importance for school administrators. Model V shows that it makes little difference, at least for these data, whether part or partial correlation is chosen. We agree with Nunnally when he concludes, "In most practical problems, it makes more sense to employ the partial correlation than to employ the semipartial correlation" (1967, p. 155). He rather neatly renames part as semipartial. We really suspect that partialling makes more sense in theoretical research and the full versus restricted model gambit makes more sense in practical research.

7.4 THE PARTL PROGRAM

PARTL

```
C         PARTIAL CORRELATION PROGRAM.   A COOLEY-LOHNES ROUTINE.
C
C         THIS PROGRAM COMPUTES A PARTIAL CORRELATION ANALYSIS REGRESSING
C   M2 VARIABLES IN THE RIGHT SET UPON M1 VARIABLES IN THE LEFT SET
C   (THUS THE FIRST M1 VARIABLES ARE THE PREDICTOR OR CONTROL VARIABLES
C   WHICH ARE TO BE PARTIALED OUT OF THE SECOND , FOLLOWING SET OF M2
C   CRITERION VARIABLES). M1 AND M2 ARE EACH LESS THAN 51. M2 IS THE
C   ORDER OF THE PARTIAL CORRELATION MATRIX.   THE MULTIPLE REGRESSION
C   OF EACH OF THE M2 VARIABLES IN THE RIGHT SET UPON ALL M1 VARIABLES
C   IN THE LEFT SET IS COMPUTED AS    R21 * R11 INVERSE * R12  , WHICH
C   IS CALLED THE REGRESSION DISPERSION, OR THE EXPLAINED DISPERSION.
C   THIS IS SUBTRACTED FROM R22 TO GIVE THE RESIDUAL DISPERSION, WHICH
C   IS THEN SCALED TO A PARTIAL CORRELATION MATRIX, AND IS PUNCHED OUT
C   SO THAT IT WILL BE AVAILABLE TO BE READ INTO OTHER PROGRAMS FOR
C   FURTHER ANALYSIS.
C
C
C         THE PROGRAM CONTAINS A PART CORRELATION ANALYSIS OPTION,
C   WHICH ALLOWS A THREE-WAY PARTITION OF THE VECTOR VARIABLE, SO
C   THAT THE LAST OR THIRD SECTION BECOMES AN INDEPENDENT OR EXTERNAL
C   SET OF ELEMENTS WHICH WILL NOT BE REGRESSED ON THE CONTROLS. THE
C   INTERCORRELATIONS AMONG THESE EXTERNAL ELEMENTS WILL NOT BE
C   MODIFIED BUT THEIR CORRELATIONS WITH THE CRITERION ELEMENTS WILL
C   BE MODIFIED.
C
C    INPUT
C
C   1) FIRST TEN CARDS OF DATA DECK DESCRIBE THE PROBLEM IN A TEXT WHICH
C      WILL BE REPRODUCED ON THE OUTPUT. DO NOT USE COLUMN 1 OF THESE
C      CARDS.
C   2) CONTROL CARD (CARD 11)
C         COLS 1-2  M1 = NUMBER OF VARIABLES IN LEFT SET
C         COLS 3-4  M2 = NUMBER OF VARIABLES IN RIGHT SET
C         COLS 5-9  N = NUMBER OF SUBJECTS
C            COL 10  JOPT = 0  FOR PARTIAL CORRELATION,
C                    JOPT = 1  FOR PART CORRELATION.
C         COLS 11-12  M3 = NO. OF VARIABLES IN THIRD SET (IF JOPT = 1).
C   3) FORMAT CARD (CARD 12),  FOR R MATRIX.
C   4) R MATRIX, UPPER-TRIANGULAR.
C
C
C      PUNCHED OUTPUT
C
C   1) RESIDUAL CORRELATION MATRIX, UPPER-TRIANGULAR, TO BE READ BY
C      FORMAT (10X, 7F10.7 / (10X, 7F10.7)).
C
C      REQUIRED SUBROUTINES ARE MPRINT AND MATINV.
C
C
      DIMENSION  TIT(16), R(100,100),  X(50), Y(50), Z(50),
     2   S(50,50),  T(50,50)
C
   1  WRITE (6,2)
   2  FORMAT (1H1)
      DO 3   J = 1, 10
      READ  (5,4) (TIT(K),   K = 1, 16)
   3  WRITE (6,4) (TIT(K),   K = 1, 16)
   4  FORMAT (16A5)
      READ (5,5)       M1, M2, N, JOPT, M3
   5  FORMAT (2I2, I5, I1, I2)
      READ  (5,4) (TIT(K),   K = 1, 16)
      WRITE (6,6)
   6  FORMAT(60H0   PARTIAL CORRELATION. A COOLEY-LOHNES MULTIVARIATE ROU
     2TINE)
      M =  M1 + M2
      EN =  N
```

```
      WRITE (6,7) M1
7     FORMAT (30H0NO. VARIABLES ON LEFT = M1 = I3)
      WRITE (6,8) M2
8     FORMAT (31H0NO. VARIABLES ON RIGHT = M2 = I3)
      WRITE (6,9) N
9     FORMAT (20H0NO. SUBJECTS = N = I6)
C
      IF (JOPT)  11, 11, 10
10    WRITE (6, 14)     M3
14    FORMAT(53H0PART CORRELA OPTION. NO. OF VARIABLES IN THIRD SET =
     C  I3)
      M = M1 + M2 + M3
11    CONTINUE
C
      EM =  M
      DO 12   J = 1, M
12    READ   (5,TIT) (R(J,K),   K = J, M)
      DO 13   J = 1, M
      DO 13   K = J, M
13    R(K,J) = R(J,K)
      WRITE (6,23)
23    FORMAT (19H0CORRELATION MATRIX)
      CALL MPRINT(R,M,100)
C
C
      CALL MATINV(R,M1,DET,100)
C
      DO 25   J = 1, M2
      J1 = J + M1
      DO 25   K = 1, M1
      S(J,K) = 0.0
      DO 25   L = 1, M1
25    S(J,K) = S(J,K) + R(J1,L) * R(L,K)
      WRITE(6,39)
39    FORMAT(19H0REGRESSION WEIGHTS)
      DO 40 J = 1, M1
40    WRITE(6,41) J, (S(K,J), K = 1, M2)
41    FORMAT(4H ROWI3,7F10.4/(7X,7F10.4))
      DO 26   J = 1, M2
      DO 26   K = 1, M2
      K1 = K + M1
      T(J,K) = 0.0
      DO 26   L = 1, M1
26    T(J,K) = T(J,K) + S(J,L) * R(L,K1)
C  T NOW CONTAINS    R21 * R11 INVERSE * R12 .
      WRITE(6,42)
42    FORMAT(18H0REGRESSION MATRIX)
      CALL MPRINT(T,M2,50)
C
      IF (JOPT)  18, 18, 17
17    DO 15   J = 1, M2
      J1 = J + M1
      DO 15   K = 1, M3
      K1 = K + M1 + M2
      R(J1,K1) = 0.0
      DO 15   L = 1, M1
15    R(J1,K1) = R(J1,K1) + S(J,L) * R(L,K1)
      DO 16   J = 1, M2
      J1 = J + M1
      DO 16   K = 1, M3
      K1 = K + M1 + M2
16    R(K1,J1) = R(K1,J1) - R(J1,K1)
C     R32 NOW CONTAINS THE ADJUSTED VARIANCES OF PREDICTORS WITH
C     CRITERIA RESIDUALS.
C
18    CONTINUE
```

```
C
      DO 27   J = 1, M2
      J1 = J + M1
      DO 27    K = 1, M2
      K1 = K + M1
 27   S(J,K) = R(J1,K1)
C    S NOW CONTAINS R22 .
C
C
      EM1 = M1
      DO 28    J = 1, M2
      X(J) = T(J,J)
      Y(J) = SQRT  (X(J))
 28   Z(J) = (X(J) * (EN - EM1 - 1.0))/((1.0 - X(J)) * EM1)
      WRITE (6,29)
 29   FORMAT(50HORIGHT-SET VAR.  MULT R    MULT R SQUARE  F-RATIO    )
      DO 30   J = 1, M2
 30   WRITE (6,31) J,  Y(J),  X(J),   Z(J)
 31   FORMAT (10X, I3, 5X, F5.3, 5X, F5.3, 8X, F9.2)
      N1 = N - M1 - 1
      WRITE (6,32) M1,   N1
 32   FORMAT (15HOFOR F, NDF1 = I3, 12H AND NDF2 = I6)
      DO 33   J = 1, M2
      DO 33   K = 1, M2
 33   S(J,K) = S(J,K) - T(J,K)
C   S NOW CONTAINS THE RESIDUAL DISPERSION FOR THE RIGHT SET, GIVEN
C   ITS MULTIPLE REGRESSIONS ON THE LEFT SET.
      WRITE(6,43)
 43   FORMAT(16HORESIDUAL MATRIX)
      CALL MPRINT(S,M2,50)
      DO 34   J = 1, M2
 34   X(J) = SQRT  (S(J,J))
C
      IF (JOPT)  20, 20, 19
 19   DO 21   J = 1, M2
      J1 = J + M1
      DO 21   K = 1, M2
      K1 = K + M1
 21   R(J1,K1) = S(J,K) / (X(J) * X(K))
C    R22 NOW CONTAINS THE ADJUSTED INTERCORRELATIONS OF CRITERIA RESIDU
C
      DO 22   J = 1, M2
      J1 = J + M1
      DO 22   K = 1, M3
      K1 = K + M1 + M2
      R(K1,J1) = R(K1,J1) / X(J)
 22   R(J1,K1) = R(K1,J1)
C    R23 AND R32 NOW CONTAIN THE ADJUSTED CORRELATIONS OF PREDICTORS
C    WITH CRITERIA RESIDUALS.
C
C    NEXT, R22, R23, R32, AND R3TARE MOVED UP AND TO THE LEFT IN R FOR
C    PRINTING AND PUNCHING.
C
      M23 = M2 + M3
      DO 24   J = 1, M23
      J1 = J + M1
      DO 24   K = 1, M23
      K1 = K + M1
 24   R(J,K) = R(J1,K1)
      WRITE (6,44)
 44   FORMAT(34HOPART CORRELATION ANALYSIS MATRIX    )
      CALL MPRINT (R, M23, 100)
      DO 45   J = 1, M23
 45   WRITE (7,38)  J, (R(J,K),   K = J, M23)
      GO TO 1
```

PARTL (*continued*)

```
 20    CONTINUE
C
       DO 35   J = 1, M2
       DO 35   K = 1, M2
 35    R(J,K) = S(J,K) /(X(J) * X(K))
       WRITE (6,36)
 36    FORMAT (29H0PARTIAL CORRELATION MATRIX      )
       CALL MPRINT(R,M2,100)
       DO 37   J = 1, M2
 37    WRITE (7,38) J,   (R(J,K) ,    K = J, M2)
C   PUNCHES PARTIAL CORRELATION MATRIX IN UPPER-TRIANGULAR FORM.
 38    FORMAT (4H ROW,I3,3X, 7F10.7 / (10X, 7F10.7))
C
       GO TO 1
       END
>
```

7.5 EXERCISES

1. What is the relationship between college plans and scholastic ability for grade 12 males of similar socioeconomic status?

2. What computer runs would be involved if we wanted to study the relationship between two sets of variables after removing variance associated with a third set?

PART III

Multiple Population
Studies

CHAPTER EIGHT

||||||||||||||||||||||||||||||||||||||||||||||||||||||||||||||||||||||||||||||||||

Multivariate Analysis of Variance

8.1 MATHEMATICS OF MULTIVARIATE ANALYSIS OF VARIANCE

Normal distribution statistics compose two continents, that of correlation analysis and that of analysis of variance. Part II summarized some of the multivariate generalizations of Karl Pearson's magnificent creation from the year 1900, the correlation coefficient. Building on the work of Pearson, who also obtained the chi-square distribution in 1900, and of "Student" who gave the t distribution in 1908, Ronald Fisher in 1923 created in the analysis of variance the most widely used and basically useful approach to studying differences among several populations. We will assume that the reader is familiar with anova (analysis of variance) methods for studies involving a single dependent variable measured on several samples that are suspected of arising from different populations. Figure 8.1 depicts the research situation of the simple anova design. The dependent variable is assumed to be normally distributed with the same variance in each population, and the research issues concern the "realness" of the differences among the population means. In other words, the research issues concern whether some or all of the populations are centered at different locations on the continuum provided by the dependent variable.

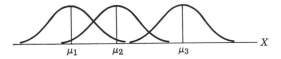

Figure 8.1. The simple anova situation, when the differences among the populations are "real."

In this chapter we are concerned with the simplest multivariate generalization of analysis of variance, which is the study of group differences in location in a multidimensional measurement space. The reader should note that complex anova methods such as factorial designs are multivariate in the independent, or grouping, variables. The distinctive multivariate nature of manova designs is that the dependent variable is a vector variable. This dependent vector variable is assumed to be multivariate normal in distribution with the same dispersion, or variance-covariance matrix, for each population. Equality of dispersions is the manova extension of the assumption of homogeneity of variances in anova designs. Again, the research issues concern the "realness" of the differences among the population centroids, or means vectors. That is, the research issues concern whether some or all of the populations are centered at different locations in the measurement space spanned by the dependent vector variable. Figure 8.2 depicts the research situation of the simple manova design for a bivariate dependent variable.

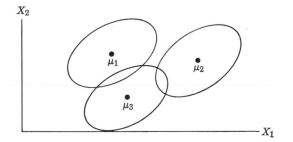

Figure 8.2. The simple manova situation, when the differences among the populations are "real."

The linear model for manova may be written

$$\mathbf{X}_{ki} = \mathbf{m} + (\mathbf{m}_k - \mathbf{m}) + (\mathbf{X}_{ki} - \mathbf{m}_k)$$

where $\mathbf{X}_{ki}$ is the dependent vector variable for the ith subject in the kth sample, $k = 1, 2, \ldots, g$, where g is the number of populations under study; $\mathbf{m}$ is the grand centroid, or vector of total sample means; $\mathbf{m}_k$ is the centroid for sample k. However, we usually are not interested in the grand centroid, and it is conventional to subtract it from both sides of the equation, casting the linear model in terms of individual subjects' deviations from the grand centroid. Letting

$$\mathbf{x}_{ki} = \mathbf{X}_{ki} - \mathbf{m}$$

we may write

$$\mathbf{x}_{ki} = (\mathbf{m}_k - \mathbf{m}) + (\mathbf{X}_{ki} - \mathbf{m}_k)$$

In this partition of the deviation of the ith subject in group k from the grand centroid the first term represents the *hypothesis effect* (differences in locations of groups) and the second term represents the *error effect* (deviations of subjects from their sample centroids). Note that both the hypothesis term and the error term are vectors, since both are defined as vector subtractions. We remind you that all these vectors are construed as column vectors.

Summing over all subjects in the total sample the squares and cross-products of the elements of the score vector and its two partitions yields the fundamental partition theorem of manova.

$$\sum_{k=1}^{g} \sum_{i=1}^{N_k} \mathbf{x}_{ki}\mathbf{x}_{ki}' = \sum_{k=1}^{g} \sum_{i=1}^{N_k} (\mathbf{m}_k - \mathbf{m})(\mathbf{m}_k - \mathbf{m})' + \sum_{k=1}^{g} \sum_{i=1}^{N_k} (\mathbf{X}_{ki} - \mathbf{m}_k)(\mathbf{X}_{ki} - \mathbf{m}_k)'$$

Looking at each of these terms separately, observe that the first term is the matrix of sums of squares and cross-products of deviations of all subjects from the grand centroid. Call this matrix $\mathbf{T}$, for "Total":

$$\mathbf{T} = \sum_{k=1}^{g} \sum_{i=1}^{N_k} \mathbf{x}_{ki}\mathbf{x}_{ki}'$$

The first partition term is the matrix of weighted squares and cross-products of deviations of group centroids from the grand centroid. We will call this $\mathbf{A}$ for "Among-groups":

$$\mathbf{A} = \sum_{k=1}^{g} \sum_{i=1}^{N_k} (\mathbf{m}_k - \mathbf{m})(\mathbf{m}_k - \mathbf{m})'$$

$$= \sum_{k=1}^{g} N_k(\mathbf{m}_k - \mathbf{m})(\mathbf{m}_k - \mathbf{m})'$$

The second partition term is the matrix of squares and cross-products of deviations of subjects from their group centroids, pooled over all groups. This we call $\mathbf{W}$ for "Within-groups":

$$\mathbf{W} = \sum_{k=1}^{g} \sum_{i=1}^{N_k} (\mathbf{X}_{ki} - \mathbf{m}_k)(\mathbf{X}_{ki} - \mathbf{m}_k)'$$

With these mnemonics, we may write the fundamental partition equation as

$$T = A + W$$

Each of the partitions divided by its degrees of freedom is an independent estimator of the common populations dispersion, Δ, when the null hypothesis holds. The estimator based on the group means vector is:

$$D_A = \left(\frac{1}{g-1}\right)A$$

Letting

$$N = \sum_{k=1}^{g} N_k$$

the estimator based on the pooled within-groups deviations is

$$D_w = \left(\frac{1}{N-g}\right)W$$

The null hypothesis is that $\mu_k = \mu$ for $k = 1, 2, \ldots, g$, and if it holds the best estimator of the common populations centroid, μ, is of course m, the grand centroid:

$$m = \frac{1}{N} \sum_{k=1}^{g} \sum_{i=1}^{N_k} X_{ki}$$

We intend to identify the centroids hypothesis as H_2, so that

$$H_2 : \mu_k = \mu$$

because we desire to reserve H_1 as notation for the hypothesis that the populations have a common dispersion, Δ.

When the null hypothesis is rejected the estimator of the research hypothesis effects, or treatment effects, is the matrix of deviations of group means from the grand means, each column of which is formed as $m_k - m$.

The test of H_2, which may be viewed as a test of the discriminating power of the measurement battery for the grouping criterion, or as a test of the effects of the treatments represented by the groups, was formulated by Wilks in 1932 in terms of the distribution of a ratio of determinants. Wilks' determinant ratio test statistic is usually denoted as Lambda (Λ), and is defined as:

$$\Lambda = \frac{|W|}{|T|}$$

Lambda is a family of three-parameter curves, with parameters based on the number of groups, the number of subjects, and the number of elements in the vector variable. Fairly recently Lambda has been tabled for certain special values of its parameters (Hsu, 1940; Roy, 1939; Pillai, 1956; Foster and Rees, 1957), but the general utility of the statistic continues to depend on the availability of transforms of it that distribute approximately as x^2 and F. Bartlett gave the x^2 approximation in 1938, and we have presented it in Chapter Six, where we have already seen Lamdba as a test statistic for canonical correlation analysis. Rao derived an F approximation in 1952 which is superior to the chi-square method in that it gives very close fits to the Lambda cumulants even for very small numbers of degrees of freedom. Lohnes (1961) reports a Monte Carlo study of Lambda and the two approximations.

Notice that as $|T|$ increases relative to $|W|$ the ratio decreases in size with an accompanying increase in the confidence with which we reject H_2. Lambda is inversely related to its cumulative probability, P, and directly related to the achieved α level. It is interesting that Wilks (1932) originally described Lambda in the context of a multivariate generalization of Fisher's correlation ratio, η^2 (eta-square). Wilks' relation was

$$\Lambda = 1 - \eta^2$$

or

$$\eta^2 = 1 - \Lambda$$

Of course, a multivariate correlation ratio is subject to the same difficulties of interpretation that plague its univariate version, but we do see a measure-statistic view of Λ to supplement its test-statistic application. Baggaley and Campbell (1967) employ the correlation ratio interpretation in an effective research operation.

For Rao's F approximation it is necessary to compute a set of functions of the design parameters p (number of tests), g (number of groups), and N (total number of subjects in all groups):

$$s = \sqrt{\frac{p^2(g-1)^2 - 4}{p^2 + (g-1)^2 - 5}}$$

$$n_1 = p(g-1)$$

$$n_2 = s\left[(N-1) - \frac{p + (g-1) + 1}{2}\right] - \frac{p(g-1) - 2}{2}$$

Now let

$$y = \Lambda^{1/s}$$

and

$$F_{n_2}{}^{n_1} = \left(\frac{1-y}{y}\right)\left(\frac{n_2}{n_1}\right) \tag{8.1.1}$$

where n_1 is the number of degrees of freedom associated with the numerator, and n_2 the denominator, of this variance ratio. Table 8.1 summarizes this variance ratio test for special cases in terms of p and q, where $q = g - 1$.

TABLE 8.1   Tests for Wilks' Λ for Special Cases

Parameters p	q	F-ratio	n_1	n_2
any	1	$\left(\dfrac{1-\Lambda}{\Lambda}\right)\left(\dfrac{N-p-1}{p}\right)$	p	$N-p-1$
any	2	$\left(\dfrac{1-\Lambda^{1/2}}{\Lambda^{1/2}}\right)\left(\dfrac{N-p-2}{p}\right)$	$2p$	$2(N-p-2)$
1	any	$\left(\dfrac{1-\Lambda}{\Lambda}\right)\left(\dfrac{N-q-1}{q}\right)$	q	$N-q-1$
2	any	$\left(\dfrac{1-\Lambda^{1/2}}{\Lambda^{1/2}}\right)\left(\dfrac{N-q-2}{q}\right)$	$2q$	$2(N-q-2)$

Rulon and Brooks (1968) show both algebraically and with numerical examples how this test statistic reduces to Hotelling's (1931) T^2 statistic in the case of two groups and p variates, to the univariate anova F in the case of g groups and one variate, and to Student's t in the case of two groups and one variate.

The Lambda test of the null hypothesis of the equality of the mean vectors assumes that the g group dispersion matrices are based on samples of g multivariate normal populations with the same dispersion, Δ. A test criterion for the null hypothesis H_1 of the equality of g group dispersion matrices, extended from a development of Bartlett's (1937), has been presented by Box (1949). Many research workers prefer to ignore the issue of the homogeneity of group dispersions on the grounds that the test of H_2 is probably fairly robust under departures from its assumptions. Also, these multivariate tests are quite powerful, so research on large samples is quite likely to lead to a rejection of H_1 with some consequent embarrassment to a manova theory for the data. However, as Phillip Rulon liked to point out, a finding about group dispersions may have a great deal of value in its own right.

Box defines the test criterion M for $H_1 : \Delta_k = \Delta, k = 1, 2, \ldots, g$.

$$M = (N - g)\log_e |\mathbf{D}_w| - \sum_{k=1}^{g} (N_k - 1)\log_e |\mathbf{D}_k| \qquad (8.1.2)$$

As previously defined, $\mathbf{D}_k$ is the dispersion estimate for the kth sample and $\mathbf{D}_w$ is the pooled-groups estimate based on $\mathbf{W}$. Required functions of design parameters are:

$$A_1 = \left(\sum_{k=1}^{g} \frac{1}{N_k - 1} - \frac{1}{N - g} \right) \frac{2p^2 + 3p - 1}{6(g - 1)(p + 1)}$$

$$A_2 = \left(\sum_{k=1}^{g} \frac{1}{(N_k - 1)^2} - \frac{1}{(N - g)^2} \right) \frac{(p - 1)(p + 2)}{6(g - 1)}$$

If $A_2 - A_1^2$ is positive, then

$$n_1 = \frac{(g - 1)p(p + 1)}{2}$$

$$n_2 = \frac{n_1 + 2}{A_2 - A_1^2}$$

$$b = \frac{n_1}{1 - A_1 - (n_1/n_2)}$$

$$F_{n_2}^{n_1} = \frac{M}{b}$$

If $A_2 - A_1^2$ is negative, use the following:

$$n_1 = \frac{(g - 1)p(p + 1)}{2}$$

$$n_2 = \frac{n_1 + 2}{A_1^2 - A_2}$$

$$b = \frac{n_2}{1 - A_1 + (2/n_2)}$$

$$F_{n_2}^{n_1} = \frac{n_2 M}{n_1(b - M)}$$

Finally, the diagonal elements of $\mathbf{A}$ and $\mathbf{W}$ provide among-groups and within-groups mean squares for manova tests for the separate variates. These are *not independent* tests, however, and should be interpreted only if the manova null hypothesis has been rejected. When the Lambda test has produced a rejection, inspection of the univariate *F*-ratios may suggest which of the elements of the vector variable are contributing most to the discrimination of the groups, or alternatively which variates are most affected by the treatments. Truly independent tests of univariate hypotheses can be computed, but the analysis is very complex and the resulting "stepdown *F*-ratios" contain their own problems of interpretation. For a discussion of independent univariate tests in manova designs see Morrison (1967, Chapter 5) or Bock and Haggard (1968).

Also, it is noteworthy that the **W** (or **D**) matrix can be converted to a correlation matrix which is the best estimator of the common populations correlation matrix. When the manova model applies to data from several samples a total sample **R** matrix based on **T** is *not* meaningful. The Project TALENT factor analysis examples of Chapters Four and Five, drawn from Lohnes (1967), were computed from **R** matrices based on **W** after the linear manova model had accounted for sex and grades effects which were hypothesized to be additive, linear effects in the data.

8.2   NUMERICAL EXAMPLE

The smallest possible numerical example of multivariate analysis of variance is $p = 2$ and $g = 2$. The following example compares the first 14 subjects in the TALENT males with the next 23 subjects (New England vs. Midwest) on Reading Comprehension and Creativity.

The vectors of means and standard deviations are:

Group 1: $N_1 = 14$; $\mathbf{m}_1 = [36.07 \quad 11.00]$; $\mathbf{s}_1 = [9.34 \quad 3.21]$
Group 2: $N_2 = 23$; $\mathbf{m}_2 = [36.96 \quad 11.26]$; $\mathbf{s}_2 = [6.75 \quad 4.04]$
Total: $N = 37$; $\mathbf{m} = [36.62 \quad 11.16]$; $\mathbf{s} = [7.81 \quad 3.75]$

The sums of squares and cross-products for deviations from the grand means yields the matrix **T**;

$$\mathbf{T} = \begin{bmatrix} 2142.7 & 427.3 \\ 427.3 & 493.0 \end{bmatrix}$$

deviations from the group means added together for both groups yields **W**;

$$\mathbf{W} = \begin{bmatrix} 2135.9 & 425.3 \\ 425.3 & 492.4 \end{bmatrix}$$

and the difference $(\mathbf{T} - \mathbf{W})$ is $\mathbf{A}$:

$$\mathbf{A} = \begin{bmatrix} 6.8 & 2.0 \\ 2.0 & 0.6 \end{bmatrix}$$

Matrix $\mathbf{A}$ can be obtained directly from the means as a check. For example, a_{11} is found by:

$$14(36.07 - 36.62)^2 + 23(36.96 - 36.62)^2 = 6.8$$

The variance-covariance matrices formed from $\mathbf{W}_1$, $\mathbf{W}_2$ and $\mathbf{W}$ have the following determinants: $|\mathbf{D}_1| = 626.4$, $|\mathbf{D}_2| = 649.9$ and $|\mathbf{D}| = 711.0$. These are used in Box's test for H_1 as indicated in Equation 8.1.2:

$$M = (37 - 2)\log_e 711.0 - (13 \log_e 626.4 + 22 \log_e 649.9)$$
$$M = 3.622$$

Substituting $g = 2$ and $p = 2$ along with sample sizes into the required functions, the degrees of freedom are found to be

$$n_1 = 3 \qquad n_2 = 26587$$

with a resulting F ratio of 1.125. Since that F ratio is not significant at the .05 level it is reasonable to assume that the dispersions in the two populations represented by these two samples are equal.

Turning to the equality of population centroids, H_2, Wilks' Λ is found to be:

$$\Lambda = \frac{|\mathbf{W}|}{|\mathbf{T}|} = \frac{870837}{873765} = .9967$$

Then in Rao's F approximation, for the particular case for $p = 2$ and $g = 2$,

$$n_1 = 2 \qquad \text{and} \qquad n_2 = 34$$

and finally using equation (8.1.1)

$$F = \left(\frac{1 - .9967}{.9967}\right)\left(\frac{34}{2}\right) = .06$$

which suggests that the difference in population centroids is not significantly different from zero.

The MANOVA program also reports the familiar univariate F ratios for each dependent variable. These are useful for descriptive purposes if it is possible to reject H_2. For variable 1, the numerator of the univariate F ratio is 6.8/1.0 (from matrix $\mathbf{A}$), and the denominator is 2135.9/35 (from $\mathbf{W}$) yielding an F ratio of .11 with $n_1 = 1.0$ and $n_2 = 35$.

8.3   EXAMPLES OF MULTIVARIATE ANALYSIS OF VARIANCE

The construct of vocational maturity has played a central role in research on career development (Super et al. 1957). The first results of the Career Pattern Study were reported as *The Vocational Maturity of Ninth Grade Boys* (Super and Overstreet, 1960). In the Career Development Study (CDS) the basic strategy has been to relate vocational maturity measured on the 111 boys and girls at eighth grade to many criteria of vocational adjustment collected at two year follow-up intervals. The method by which eight Readiness for Vocational Planning (RVP) traits were scaled from the eighth grade interview protocols and the web of predictive validities with criteria from the first three follow-ups are reported in *Emerging Careers* (Gribbons and Lohnes, 1968). Our example of MANOVA is drawn from a report that extends this study to include data from a fourth, or nine-year follow-up, which brings the career patterns up to four years after high school graduation (Gribbons and Lohnes, 1970).

The main purpose of the new study is to demonstrate a method of recovering degrees of freedom for CDS by substituting a univariate scaling of vocational maturity for the RVP syndrome. The new univariate scaling is called Readiness for Career Planning (RCP) and is based on 22 items selected by factor analysis procedures from the original 45 items of the eighth grade interview. The retreat to a single scale for a subset of the 1958 items is justified by the opportunity it creates to team a vocational maturity measure with other

TABLE 8.2   Intercorrelations, Means, and Standard Deviations for 1958 Four Scales

Variable	SES	IQ	RCP	Mean	S.D.
Sex (male = 1, female = 2)	.01	.03	−.18	1.5	.6
Socioeconomic Status (1 = high, 7 = low)		−.35	−.14	4.0	1.6
Otis Beta Form Intelligence			.31	107.9	9.5
Readiness for Career Planning				32.4	10.8
				$N = 110$	

predictors, namely sex, socioeconomic status, and intelligence. Table 8.2 contains the intercorrelations among these variables and their means and standard deviations for total sample. The strongest relationship is a moderate tendency for intelligence to increase with increasing socioeconomic status of family. RCP is not significantly correlated with sex or socioeconomic status and its correlation with intelligence is only .31. These four relatively independent predictors seem to represent a parsimonious yet potentially powerful antecedent measurement space for a longitudinal study with the modest sample size of the CDS.

The eighth-grade career aspirations are classified into a four cell variable reflecting college orientation as well as people versus thing orientation. The discrimination of this criterion in the space of the 1958 four scales is reported in Table 8.3. Sex and RCP are stronger discriminators than socioeconomic

TABLE 8.3   Manova Study of 1958 Four Scales Versus 1958 Four Aspiration Groups ($N = 110$)

| 1958 Four Scales | Four 1958 Aspiration Groups (8th Grade) | | | | Pooled Groups Est. S.D.'s | F^3_{106} | η |
	College Science ($N = 33$) Means	Noncollege Technology ($N = 16$) Means	Noncollege Business ($N = 24$) Means	College Bus-cult ($N = 37$) Means			
Sex	1.3	1.1	1.8	1.7	.5	11.4	.49
SES	3.5	3.8	4.9	3.8	1.6	4.3	.27
IQ	109.8	103.8	105.5	109.6	9.3	2.4	.20
RCP	37.5	24.8	27.0	34.7	9.7	9.5	.39

For equality of dispersions, Manova $F^{30}_{\infty} = 1.6$

For equality of centroids, Manova $F^{12}_{270} = 6.5$; $\eta = .70$

status and intelligence. The same four cell taxonomy is used as criterion for career aspirations expressed in the latest follow-up year, which is 1967. Both socioeconomic status and intelligence are stronger predictors than sex and RCP for the four years out of high school aspirations, as reported in Table 8.4. This exchange of predictive strength among the four scales over a nine year maturation span suggests that phantasy is being replaced by realism in the career development process. The report's two conclusions are (1) that vocational maturity scaled in early adolescence is related to career aspirations on into early adulthood in ways that complement the relations of sex, socioeconomic status, and intelligence to those criteria; and (2) that a univariate

TABLE 8.4   Manova Study of 1958 Four Scales Versus 1967 Four Aspiration Groups ($N = 109$)

1958 Four Scales	Four 1967 Aspiration Groups (HS + 4 Yrs)				Pooled Groups Est. S.D.'s	F^3_{105}	η
	College Science ($N=10$) Means	Noncollege Technology ($N=19$) Means	Noncollege Business ($N=49$) Means	College Bus-cult ($N=31$) Means			
Sex	1.3	1.3	1.8	1.8	.5	6.6	.33
SES	3.7	4.4	4.6	2.8	1.5	9.8	.40
IQ	113.1	101.1	106.3	112.8	8.6	9.2	.38
RCP	32.8	30.2	29.4	37.9	10.2	4.6	.28

For equality of dispersions, Manova $F^{30}_{\infty} = 1.2$

For equality of centroids, Manova $F^{12}_{270} = 6.3$; $\eta = .69$

vocational maturity scale (RCP) can be substituted for the eight RVP scales in CDS research in order to recover needed degrees of freedom.

What about those etas in Tables 8.3 and 8.4? The journal article reporting this research contains the etas but *not* the F-ratios. It argues that "since the 111 subjects are in no sense a sample of some universe, but are instead an interesting group of young people in their own right, it is appropriate to focus attention on descriptive measures of relationship between predictors and criteria, rather than on associated inference statistics. Thus, all Anova results are reported in terms of the correlation ratio η (eta), and Manova results in terms of the Wilks (1932) generalized η. Readers are reminded that η is as close to a correlation coefficient as we can come for relations involving a nominal variable, and that η^2 expresses the proportion of criterion variance explainable by the predictor variance. The Manova η is similar to a multiple correlation."

You recall that in univariate analysis of variance

$$\eta^2 = \frac{\text{Among Sum of Squares}}{\text{Total Sum of Squares}}$$

Wilks simply took the determinants of the Within and Total matrices as appropriate generalizations of these terms, so that

$$\text{Manova } \eta^2 = 1 - \frac{|W|}{|T|} = 1 - \Lambda$$

A useful descriptive statistic? We think so. Our thanks to Andrew Baggaley for calling it to our attention (Baggaley and Campbell, 1967).

What about using sex as an element of the vector variable in these studies? It's hardly continuous normal. The article says that, "Sex was treated as a scaled predictor along with the other predictors for consistency." In defense of the authors, they are not doing any statistical inference, but are seeking an honest and palatable description of trends in their case study data. They only want to assign sex the right rank among the predictors. Probably they have done so.

It would have been nice for us to have given an example of Manova in which the research outcomes depended on the powerful inference tools provided by the model, but this example is more truthful in reflecting our own experience, since we do not do experiments. In fact, we depend on Manova primarily to reduce the data and organize it for multiple group discriminant analysis, a heuristic strategy we turn to next.

Another research example concerns the differences between scientists and science teachers on certain personality dimensions. Lee (1961) has studied certain personality dimensions that are related to the movement into science teaching. His theoretical model postulated that the more person-oriented college science majors would move out of science and that some of them would become science teachers. Twelve relevant dimensions were obtained from three instruments: (1) the Strong vocational interest blank for men, (2) the Allport-Vernon-Lindzey study of values, and (3) the Guilford-Zimmerman temperament survey.

Two groups of subjects were obtained. One group, from the Scientific Careers Study (SCS), consisted of sixty-one ($N_1 = 61$) males who were tested as college seniors. The SCS group continued toward careers in science following graduation. The science teacher group was obtained by contacting "fifth-year" programs at graduate schools of education, selecting those males who majored in one of the sciences but who were then preparing to enter teaching as a career. Lee called this latter group the Liberal Arts Science Teachers (LAST). Testing was completed for sixty-six of them ($N_2 = 66$). Group means and standard deviations are reported in Table 8.5

The sequence of the analyses was as follows:

1. Test of H_1 for the two groups using all twelve dimensions. (No difference in dispersion was found.)
2. Test of H_2 using all twelve dimensions (Results in Table 8.6)

Since (2) was rejected:

3. Test of H_2 using the Strong scales.
4. Test of H_2 using the A-V-L scales.
5. Test of H_2 using the G-Z scales.

Rejection of null hypotheses (3), (4), and (5) (see Table 8.7) made it possible to

TABLE 8.5 Subgroup Means and Standard Deviations for Twelve Dimensions of Personality on Subjects in the Lee Science Teachers Study

Personality Scale	SCS Group, 61 Subjects		LAST Group, 66 Subjects	
	Mean	S.D.	Mean	S.D.
Temperament (G-Z)				
Sociability	16.38	6.27	18.91	5.95
Emotional stability	16.77	6.09	19.47	5.66
Objectivity	19.13	5.30	20.36	4.21
Personal relation	16.10	5.63	18.00	5.02
Values (A-V-L)				
Theoretical	53.65	6.08	51.24	7.37
Aesthetic	40.33	8.95	39.03	8.76
Social	32.84	7.77	38.21	7.89
Interests (SVIB)				
Technical worker	48.18	9.80	41.14	10.25
Welfare worker	36.79	9.30	46.47	7.79
Musical performer	41.44	11.82	42.92	11.36
Business detail	23.84	9.76	22.56	9.45
Business contact	30.15	7.49	30.79	8.05

TABLE 8.6   General Analysis

$$H_1 M = 91.54, \quad F^{78}_{48710} = 1.05, \quad p > .30$$
$$H_2 \Lambda = .614, \quad F^{12}_{114} = 5.98, \quad p < .001$$

TABLE 8.7   Area Analysis: A Comparison of the F Ratios and Probabilities of the Guilford-Zimmerman, Allport-Vernon, and Strong Areas

Area	F	ndf_1	ndf_2	p
Temperament (4 scales)	2.626	4	122	<.05
Values (3 scales)	5.376	3	123	<.01
Interests (5 scales)	11.719	5	121	<.001

examine the scales within each area (interest, values, and temperament) to determine which particular variables were most significant on an individual basis. The corresponding t tests are summarized in Table 8.8.

TABLE 8.8   Individual Variable Analysis: A Summary of the Analyses Performed on the Twelve Individual Variables with the Prediction and Outcome of That Prediction

Area Variable	Prediction	t	p	High Group	Outcome of Prediction
Temperament					
Sociability	SCS < LAST	2.373	$p < .01$	LAST	Correct
Emotional stability	SCS < LAST	2.589	$p < .01$	LAST	Correct
Objectivity	SCS < LAST	1.657	$.025 < p < .05$	LAST	Correct
Personal relation	SCS < LAST	2.013	$.01 < p < .025$	LAST	Correct
Values					
Theoretical	SCS > LAST	1.174	Not significant	SCS	No difference
Aesthetic	SCS < LAST	.825	Not significant	SCS	No difference
Social	SCS < LAST	3.865	$p < .0005$	LAST	Correct
Interests					
Technical worker	SCS > LAST	3.946	$p < .0005$	SCS	Correct
Welfare worker	SCS < LAST	6.379	$p < .0005$	LAST	Correct
Musical performer	SCS > LAST	.720	Not significant	LAST	No difference
Business detail	SCS < LAST	.748	Not significant	SCS	No difference
Business contact	SCS < LAST	.454	Not significant	LAST	No difference

In summary, hypotheses (1) and (2) tested the general model, and (3), (4), and (5) made sure each area was detecting differences and made possible an examination of their relative differences. The univariate t tests made possible the examination of individual variables.[1] The results are shown in Table 8.5.

The results of Lee's analyses supported the contention that "there are identifiable personality attributes associated with persons choosing science teaching as a career. The science teacher group scored higher on all variables concerned with interpersonal relations. The scientist group scored higher on variables which were 'non-person' oriented, such as the Technical Workers scale."

[1] It should be emphasized that the series of hypotheses tested here were all derived from the investigator's theoretical model regarding personality differences between scientists and science teachers. After hypothesis (1) established overall differences on the variables related to the model, the remaining hypotheses explored the exact nature of those differences to make sure they were consistent with the model. These subsequent tests might be considered more descriptive than inferential.

8.4 THE MANOVA PROGRAM

MANOVA

```
C       MULTIVARIATE ANALYSIS OF VARIANCE.    A COOLEY-LOHNES ROUTINE.
C
C       THIS PROGRAM COMPUTES MANOVA TESTS OF H1 (EQUALITY OF DISPERSION
C    AND H2 (EQUALITY OF CENTROIDS),   UNIVARIATE F-RATIOS FOR MEANS,
C    SELECTED SAMPLE STATISTICS,   AND THE W (POOLED WITHIN-GROUP SSCP)
C    AND T(TOTAL SAMPLE SSCP) MATRICES REQUIRED FOR THE DISCRIMINANT
C    ANALYSIS PROGRAM.   THESE MATRICES ARE PUNCHED IN UPPER-TRIANGULAR
C    FORM.   THE PROGRAM WILL PROCESS UP TO 50 VARIABLES AND ANY NUMBER
C    OF GROUPS.
C
C       INPUT
C
C    1) FIRST TEN CARDS OF THE DATA DECK DESCRIBE THE PROBLEM IN A TEXT
C        WHICH WILL BE REPRODUCED ON THE OUTPUT. DO NOT USE COLUMN 1.
C    2) CONTROL CARD (CARD 11)
C        COLS 1-2   M = NUMBER OF VARIABLES
C        COLS 3-5   KG = NUMBER OF GROUPS
C    3) FORMAT CARD (CARD 12)
C    4) EACH GROUP OF SCORE CARDS IS PRECEDED BY A CARD GIVING
C        NG = NUMBER OF SUBJECTS IN THE GROUP (COLS 1-5).
C        THUS, SUBJECTS MUST BE SORTED INTO GROUPS AND THE GROUPS COUNTED
C        BEFORE MANOVA CAN BE RUN.
C
C    PUNCHED OUTPUT IS ALL TO FORMAT (10X, 5E14.7 /(10X, 5E14.7)), AND I!
C
C    1)   GROUP MEANS, FOLLOWED BY GRAND MEANS.
C    2) T MATRIX (TOTAL SAMPLE DEVIATION SSCP MATRIX)
C    3) W MATRIX (POOLED WITHIN-GROUPS DEVIATION SSCP MATRIX)
C    4)   D INVERSE (INVERSE OF POOLED-SAMPLES DISPERSION ESTIMATE)
C
C       SUBROUTINE MATINV IS REQUIRED.
C
C
        DIMENSION  TIT(16), A(50,50), B(50,50), C(50,50),
      2   T(50), U(50), V(50), W(50), X(50), D(50,50)
C
1       WRITE(6,2)
2       FORMAT (33H1MANOVA. A COOLEY-LOHNES PROGRAM     )
        DO 3    J = 1, 10
        READ(5,4)        (TIT(K),    K = 1, 16)
3       WRITE(6,4)       (TIT(K),    K = 1, 16)
4       FORMAT (16A5)
        READ(5,5)      M,  KG
5       FORMAT (I2, I3)
        EM = M
        EKG = KG
        EK = KG
        WRITE(6,6)      M,  KG
6       FORMAT (13H0ANALYSIS FOR I3, 14H VARIABLES ANDI4,7H GROUPS)
        WRITE(6,9)
9       FORMAT (1H0, 25(5H-----))
        DO 7   J = 1, M
        T(J) = 0.0
        DO 7    K = 1, M
        B(J,K) = 0.0
7       C(J,K) = 0.0
        H1LOGS = 0.0
        GA1S = 0.0
        FA1S = 0.0
        N = 0
        READ (5,4)       (TIT(J),    J = 1, 16)
C
        DO 19   IG = 1, KG
        READ(5,8)     NG
8       FORMAT (I5)
        ENG = NG
```

```
        N = N + NG
        WRITE(6,9)
        WRITE (6,10)        IG, NG
10      FORMAT (6H0GROUPI3, 8H,  NG = I6)
        DO 11   J = 1, M
        U(J) = 0.0
        DO 11   K = 1, M
11      A(J,K) = 0.0
        DO 12   NS = 1, NG
        READ(5,TIT)         (V(J),   J = 1, M)
        DO 12   J = 1, M
        U(J) = U(J) + V(J)
        T(J) = T(J) + V(J)
        DO 12   K = 1, M
        A(J,K) = A(J,K) + V(J) * V(K)
12      C(J,K) = C(J,K) + V(J) * V(K)
        DO 13   J = 1, M
        DO 13   K = 1, M
        A(J,K) = A(J,K) - U(J) * U(K) / ENG
        B(J,K) = B(J,K) + A(J,K)
13      A(J,K) = A(J,K) / (ENG - 1.0)
        DO 14   J = 1, M
        U(J) = U(J) / ENG
14      W(J) = SQRT (A(J,J))
        WRITE (6,15)   IG
15      FORMAT (17H0MEANS FOR GROUP   I4)
        WRITE(6,16)        (   U(J),   J = 1, M)
16      FORMAT (1H0, 10(3X, F7.2))
        WRITE(7,30)        IG,  (U(J),   J = 1, M)
30      FORMAT (4H ROWI3,3X, 5E14.7 / (10X, 5E14.7))
        WRITE (6,17)
17      FORMAT (21H0STANDARD DEVIATIONS    )
C
        WRITE (6,16)  ( W(J), J = 1,M)
        CALL MATINV (A, M, DET)
        WRITE(6,18)   DET
18      FORMAT(26H0DISPERSION DETERMINANT =     E14.4)
        H1LOGS = H1LOGS + ((ENG - 1.0) *ALOG   (DET))
        FA1S = FA1S + (1.0 / (ENG - 1.0))
        GA1S = GA1S + (1.0 / ((ENG - 1.0)**2))
19      WRITE(6,9)
C
C
        EN = N
        DO 20   J = 1, M
        DO 20   K = 1, M
        A(J,K) = C(J,K) - T(J) * T(K) / EN
        D(J,K) = A(J,K)
20      C(J,K) = B(J,K) / (EN - EKG)
        DO 21   J = 1, M
        T(J) = T(J) / EN
21      U(J) = SQRT (C(J,J))
        WRITE(6,22)
22      FORMAT (23H0MEANS FOR TOTAL SAMPLE)
        WRITE (6,16)    (   T(J),   J = 1, M)
        KGT = KG + 1
        WRITE(7,30)        KGT,  (T(J),   J = 1, M)
        WRITE (6,23)
23      FORMAT (35H0POOLED-SAMPLES STANDARD DEVIATIONS    )
        WRITE (6,16) (U(J), J = 1,M)
        WRITE (6,9)
        WRITE(6,38)
38      FORMAT (9H0T MATRIX)
        DO 34   J = 1, M
        WRITE (6,30) J, (A(J,K), K = J,M)
34      WRITE(7,30)   J, (A(J,K),   K = J, M)
```

```
        WRITE(6,9)
        DO 35   J = 1, M
        DO 35   K = 1, M
 35     A(J,K) = A(J,K) - B(J,K)
C    A IS NOW THE A (AMONG-GROUPS SSCP) MATRIX.    B IS NOW THE W (WITHIN
C    GROUPS SSCP) MATRIX.    C IS NOW THE POOLED-GROUPS DISPERSION EST.
C
        WRITE(6,28)
 28     FORMAT (9H0A MATRIX)
        DO 29   J = 1, M
 29     WRITE(6,30)      J,  (A(J,K),    K = J, M)
        WRITE(6,9)
C
        WRITE(6,31)
 31     FORMAT(9H0W MATRIX)
        DO 32   J = 1, M
        WRITE (6,30) J,  (B(J,K),  K = J,M)
 32     WRITE(7,30)  J,(B(J,K),  K = J,M)
        WRITE(6,9)
C
        CALL MATINV (C, M, DET)
        DO 33    J = 1, M
 33     WRITE (7,30)   J, (C(J,K),   K = J,M)
C
        WRITE(6,18)        DET
        H1LOG = (EN - EK) * ALOG  (DET)
        XMM = H1LOG - H1LOGS
        F1 = .5 * (EK - 1.0) * EM * (EM + 1.0)
        A1A = (FA1S - (1.0 / (EN - EK))) * ((2.0 * (EM * EM)) + (3.0 *
      C     EM) - 1.0)
        A1 = A1A / (6.0 * (EK - 1.0) *    (EM + 1.0 ))
        A2 = (GA1S - (1.0 / (EN - EK)**2)) * ((EM-1.0) * (EM + 2.0))
      C      / (6.0 *(EK - 1.0))
        DIF = A2 - A1 * A1
        IF (DIF)    24, 24, 25
 24     F2 = (F1 + 2.0) / (A1 * A1 - A2)
        B1 = F2 / (1.0 - A1 + (2.0 / F2))
        F = (F2 * XMM) / (F1 * (B1 - XMM))
        GO TO 45
 25     F2 = (F1 + 2.0) / DIF
        B1 = F1 / (1.0 - A1 - (F1 / F2))
        F = XMM / B1
 45     NDF1 = F1
        NDF2 = F2
        WRITE (6,26)    XMM,F
 26     FORMAT(47H0FOR TEST OF H1 (EQUALITY OF DISPERSIONS), M = F10.3,
      C   10H  AND F = F10.3)
        WRITE (6,27)       NDF1,   NDF2
 27     FORMAT (15H0FOR F, NDF1 = I3, 12H AND NDF2 = I9)
        WRITE (6,9)
C
        N1 = EKG - 1.0
        N2 = EN - EKG
        WRITE(6,9)
        WRITE(6,40)   N1,N2
 40     FORMAT(34H0UNIVARIATE F-RATIOS, WITH NDF1 = I3,12H AND NDF2 = I6)
        WRITE (6,9)
        WRITE (6,41)
 41     FORMAT(71H0VARIABLE   AMONG MEAN SQ     WITHIN MEAN SQ      F-RATIO
      C     ETA SQUARE)
        DO 42 J = 1,M
        ETASQ = A(J,J) / (A(J,J) + B(J,J))
        AMS = A(J,J) / (EKG - 1.0)
        WMS = B(J,J) / (EN - EKG)
        F = AMS/WMS
 42     WRITE (6,43) J,AMS,WMS,F,ETASQ
```

MANOVA (*continued*)

```
43    FORMAT(3X,I3,5X,F9.2,11X,F9.2,10X,F7.2,8X,F5.4)
      WRITE (6,9)
C
      CALL MATINV (B, M, DETW)
C  DETW IS DETERMINANT OF POOLED-SAMPLES DEVIATION SSCP MATRIX, W.
      CALL MATINV (D, M, DETT)
C  DETT IS DETERMINANT OF TOTAL SAMPLE DEVIATION SSCP MATRIX, T.
      XL = DETW / DETT
      YL = 1.0 - XL
      WRITE(6,46) XL, YL
46    FORMAT (16HOWILKS LAMBDA =   F7.4,47H   GENERALIZED CORRELATION RA
     CTIO, ETA SQUARE = F5.4)
      IF (M - 2)   47, 47,  49
47    IF (KG - 3)   48, 48,  49
48    YL = XL
      F1 = 2.0
      F2 = EN - 3.0
      GO TO 50
49    SL = SQRT  (((EM * EM) * ((EKG - 1.0)**2) - 4.0) / ((EM * EM) +
     2  ((EKG - 1.0)**2) - 5.0))
      YL = XL ** (1.0 / SL)
      PL = (EN - 1.0)          - ((EM + EKG) / 2.0)
      QL =  - ((EM *(EKG - 1.0)) - 2.0) / 4.0
      RL = (EM * (EKG - 1.0)) / 2.0
      F1 = 2.0 * RL
      F2 = (PL * SL) + (2.0 * QL)
50    N1 = F1
      N2 = F2
      F = ((1.0 - YL) / YL) * (F2 / F1)
      WRITE(6,51)       F
51    FORMAT(45HOF-RATIO FOR H2, OVERALL DISCRIMINATION, =    F9.2)
      WRITE(6,52)      N1,  N2
52    FORMAT (8HONDF1 = I3,  12H AND NDF2 = I9)
      WRITE (6,9)
      GO TO 1
      END
>
```

8.5 EXERCISES

1. An investigator who wished to contrast 3 treatments on 3 criterion measures randomly divided 90 students into 3 groups of size 30 each. After treatment he measured the 90 subjects with his 3 scales. **W** and **T** matrices were formed yielding determinants of 4.86 and 6.00, respectively.

   (a) What is the probability of drawing 3 samples at least this different from a single multivariate normal distribution in 3-space?

   (b) If the investigator's theory led him to expect treatment effects should he be encouraged by these results?

   (c) What statistical assumption is he making if he claims centroid differences as a result of this above test of significance?

   (d) What numerical results should the investigator display to facilitate interpretation of these results?

2. The numerical example of Section 8.2 did not detect any regional effect for the two regions compared for the variables included. Can you think of any combinations of regions and variables which might produce significant regional effects?

‖‖‖

CHAPTER NINE

Discriminant Analysis

9.1 MATHEMATICS OF DISCRIMINANT FUNCTIONS

In research studies involving several samples from different populations located at different places in a multivariate measurement space, but assumed to be samples from populations having a common dispersion, it can be very interesting to locate the best reduced-rank model for parsimoniously but effectively describing the measured differences of the groups. In multiple discriminant analysis the samples are projected from their places in the complete measurement space into a suitable sub-space. The results of multiple discriminant analysis are phrased in terms of:

1. The number of discriminant functions retained (the rank of the discriminant model) and the relative importance of each discriminant function.
2. The location of each discriminant function as a reference vector spanning a dimension of the selected subspace of the full space, expressed in terms of structure correlation coefficients.
3. The mappings of the groups into the discriminant space, the means and standard deviations of the groups on the functions.

Once again linear functions of a battery are derived which are suitable to doing a specific job. The discriminant model may be interpreted as a special type of factor analysis that extracts orthogonal factors of the measurement battery for the specific task of displaying and capitalizing upon differences among criterion groups. The model derives the components which best separate the cells or groups of a taxonomy in the measurement space. It makes no difference to the formal logic of the model whether the samples of the several populations are viewed as the dependent, criterion variable and the discriminant functions are viewed as the best prediction functions of the independent, predictor vector variable defining the measurement space, or if the groups are viewed as the independent "treatment" variable and the discriminant functions are seen as the most predictable functions of the dependent vector variable. The taxonomic variable is more likely to be the criterion variable in survey science, whereas it is almost certain to be the independent treatment variable in experimental research. In the authors' researches into career development the multivariate normal trait measurement systems have usually been temporally antecedent to the taxonomic adjustment variables, so that the vector variables have been the predictors and the grouping variables have been the criteria.

The possible rank n of the discriminant subspace to be fitted in the measurement space depends on the relative sizes of g, the number of groups, and p, the number of elements in the vector variable. If the quantity $g - 1$ is less than p, then $g - 1$ is the maximum possible rank of the discriminant space. Consider, for example, that the centroids of two groups have to coexist on a single line, regardless of the number of variates in each centroid. The centroids of three groups have to coexist in a plane; the centroids of four groups have to coexist in a three-space, and so forth. On the other hand, if $g - 1$ is equal to or greater than p, then it is possible to fit as many as p discriminant functions. In the quest for parsimony the researcher will usually decide he can make do with a small number of discriminant functions, however, even if the lesser of $g - 1$ and p is quite large. It is surprising how often the authors find in their researches that the best discriminant plane provides an adequate reduced-rank model for their data, although they sometimes have as many as forty groups under study in spaces of twenty to sixty variates. The best discriminant plane has the attraction that graphic maps of the locations of the groups in it can be presented. This consideration of graphics may sometimes lead to underemphasis on the third discriminant function, but we have seen few examples in which the available fourth and further functions were of any real significance. We are going to describe a test of the statistical significance of the remaining discriminant functions after the selection of n of them, but in our experience the research meaningfulness of the functions peters out before statistical nonsignificance is reached.

The geometric interpretation of discriminant analysis can be seen for the case of two groups and two variates with the assistance of Figure 9.1, in which the two sets of concentric ellipses represent the bivariate swarms for the two groups in idealized form. The two variates, X and Y, are moderately positively correlated. Each ellipse is the locus of points of equal density (or frequency) for a group. For example, the outer ellipse for group A might define the region within which 90 percent of group A lies, and the inner ellipse concentric with it might define the region within which 75 percent of group A lies. These ellipses, which we call *centours*, for *cen*tile con*tours*, are further discussed in Chapter Ten. The two points at which corresponding centours intersect define a straight line, *II*. If a second line, *I*, is constructed perpendicular to line *II*, and if the points in the two-dimensional space are projected onto *I*, the overlap between the two groups will be smaller than for any other possible line. The

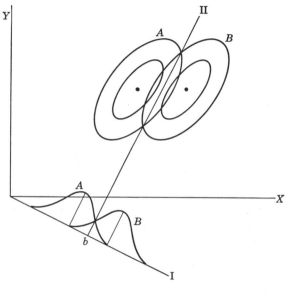

Figure 9.1.

discriminant function therefore transforms the individual test scores to a single discriminant score, and that score is the individual's location along line *I*. The point b where *II* intersects *I* would divide the one-dimensional discriminant space into two regions, one indicating probable membership in group A and the other region for group B. Notice that this diagram depends upon the equality of the two group dispersions. If either the variances of X and Y or the X, Y covariance were different for the two groups the centours for the two groups would not have the same shape and orientation, and the boundary (line *II*) would not be a straight line. The *sizes* of the two populations do not have to be the same, only the dispersions.

As in the case of previous models, this model can best be viewed as a stepwise procedure, so that we look at the extraction of one discriminant function at a time. First, we seek the best discriminant function of the measurement vector **x**, where **x** is a deviation from the grand centroid:

$$\mathbf{x}_{ki} = \mathbf{X}_{ki} - \mathbf{m}$$

(We are employing the notation of Chapter Eight again in this chapter.) We define " best " as that function which will maximize the ratio of the among-groups sum of squares to the within-groups sum of squares, so that among-group differences will be large relative to within-group scatter. Denoting the desired function as

$$y = \mathbf{v}'\mathbf{x}$$

the among-groups sum of squares on the function will be the following quadratic from the matrix of among-groups sums of squares and cross-products, **A**:

$$\mathbf{v}'\mathbf{A}\mathbf{v}$$

Similarly the within-groups sum of squares on the function will be the following quadratic from the matrix of pooled within-groups sums of squares and crossproducts of deviation scores:

$$\mathbf{v}'\mathbf{W}\mathbf{v}$$

The task for the calculus of partial differentiation with respect to the unknown weights in **v** is to maximized the ratio of these two quadratic forms:

$$\lambda = \left.\frac{\mathbf{v}'\mathbf{A}\mathbf{v}}{\mathbf{v}'\mathbf{W}\mathbf{v}}\right|_{maximum}$$

subject to the restriction

$$\mathbf{v}'\mathbf{v} = 1$$

The maximum value λ and the associated vector of weights **v** are shown by the differential calculus to be the largest eigenvalue and its eigenvector of the equation

$$(\mathbf{W}^{-1}\mathbf{A} - \lambda\mathbf{I})\mathbf{v} = 0$$

There are two interesting things about the matrix product $\mathbf{W}^{-1}\mathbf{A}$. The first is that although it is the product of two symmetric matrices it is itself non-symmetric. The implication of this nonsymmetry is that we will not be able to compute the eigenstructure directly with our HOW routine. The second thing is that although $\mathbf{W}$ is of full rank and can therefore be inverted as required, $\mathbf{A}$ is not of full rank when $g - 1$ is less than p. In such cases the rank of $\mathbf{A}$ is exactly $g - 1$, and since the rank of a matrix product is always the rank of the matrix having the lesser rank entering the product, the rank of $\mathbf{W}^{-1}\mathbf{A}$ will be that of $\mathbf{A}$ whenever $\mathbf{A}$ is of reduced rank. Actually, we will compute the eigenstructure of this matrix product through the use of a routine called DIRNM which is not bothered by either the nonsymmetry or the possible reduced rank of the matrix. DIRNM was described in Chapter Six.

The jth discriminant function is a linear component y_{ij} of the deviation of subjects from the grand centroid, such that the grand mean of y_{ij} is zero and the grand variance is θ:

$$\theta_j = \mathbf{v}_j'\left(\frac{1}{N-1}\mathbf{T}\right)\mathbf{v}_j$$

where $j = 1, 2 \ldots, n$.

("Grand" stands for "computed over all subjects and all samples" and $\mathbf{T}$ is the total sums of squares and cross-products of deviations from grand centroid matrix as defined in Chapter Eight. Remember that $\underline{\mathbf{T} = \mathbf{W} + \mathbf{A}}$.) In order to be able to view the discriminant function as a factor of the test battery we have to derive coefficients for a standardized function of a standardized test vector. If we let $\mathbf{D}_{diag}$ be the diagonal matrix formed from the diagonal elements of $(1/(N-1))\mathbf{T}$, then we can approach the desired discriminant factor, $f_{ij} = \mathbf{c}_j'\mathbf{z}_i$, by observing that

$$f_{ij} = \frac{1}{\sqrt{\theta}}\, y_{ij} = \theta^{-1/2}\mathbf{v}_j'\mathbf{x}_i = \theta^{-1/2}\mathbf{v}_j'(\mathbf{X}_i - \mathbf{m})$$

Thus defining $\mathbf{b}_j = \theta_j^{-1/2}\mathbf{v}_j$, the $p \times n$ matrix of coefficients $\mathbf{B}$ is used to transform an $N \times p$ roster of deviation scores to an $N \times n$ roster of standardized discriminant scores. Continuing,

$$\mathbf{x}_i = \mathbf{D}_{diag}^{1/2}\, \mathbf{z}_i$$

$$f_{ij} = \theta^{-1/2}\mathbf{v}_j'\mathbf{D}_{diag}^{1/2}\, \mathbf{z}_i = \mathbf{c}_j'\mathbf{z}_i$$

when

$$\mathbf{c}_j = \mathbf{D}_{diag}^{1/2}\, \mathbf{v}_j\theta^{-1/2} = \mathbf{D}_{diag}^{1/2}\, \mathbf{B}$$

It now follows that

$$\mathbf{c}_j'\mathbf{R}\mathbf{c}_j = 1$$

where $\mathbf{R}$ is a correlation matrix based on $\mathbf{T}$. Also

$$\mathbf{s}_j = \mathbf{Rc}_j$$

is the vector of structure coefficients for the jth discriminant factor. Then $\mathbf{C}$ is a $p \times n$ matrix of coefficients for converting an $N \times p$ roster of standard scores to an $N \times n$ roster of standardized discriminant scores, and $\mathbf{S}$ is the $p \times n$ structure matrix of correlations between the p predictors and the n discriminant functions.

Next, the model derives a second discriminant function which again maximizes λ subject to the requirement that this function be orthogonal to the first, and so on, with each new function required to be orthogonal to all the previous ones. This extraction can go on until a zero eigenvalue of $\mathbf{W}^{-1}\mathbf{A}$ is reached (or until $\mathbf{W}^{-1}\mathbf{A}$ is fully exhausted). The complete eigenstructure problem involved may be stated as

$$(\mathbf{W}^{-1}\mathbf{A})\mathbf{V} = \mathbf{VL}$$

where $\mathbf{L}$ is a diagonal matrix containing the successive values of λ.

Let us assume that n discriminant functions are to be retained from the eigenstructure reported by the DIRNM routine. What we have, then, is the n-square diagonal matrix $\mathbf{L}$ and the $p \times n$ rectangular matrix $\mathbf{V}$. Applying the logic of transformation to factor scores based on a standardized observation vector $\mathbf{z}$, as described above:

$$\theta = \mathbf{V}' \frac{1}{N-1} \mathbf{TV}$$

and

$$\mathbf{f} = \theta^{-1/2} \mathbf{V}' \mathbf{D}_{diag}^{1/2}\, \mathbf{z} = \mathbf{C}'\mathbf{z}$$

where the $p \times n$ matrix of discriminant factor coefficients is

$$\mathbf{C} = \mathbf{D}_{diag}^{1/2} \mathbf{V}\theta^{-1/2}$$

and the matrix of factor structure coefficients (or the factor pattern) is

$$\mathbf{S} = \mathbf{RC}$$

Wilks' Lambda criterion for the discriminating power of the test battery, described as the ratio of determinants $|\mathbf{W}|/|\mathbf{T}|$ in Chapter Eight, may also be computed as a function of the eigenvalues of $\mathbf{W}^{-1}\mathbf{A}$ as follows:

$$\Lambda = \prod_{j=1}^{n} \frac{1}{1 + \lambda_j}$$

A chi-square test for the significance of discrimination afforded by the remaining $n - k$ functions after the acceptance of the first k functions may be computed as

$$\chi^2 = - \left(N - \frac{p + g}{2} - 1\right) \log_e \Lambda'$$

with

$$ndf = (p - k)(g - k - 1)$$

and

$$\Lambda' = \prod_{j=k+1}^{n} \frac{1}{1 + \lambda_j}$$

The proportion of the total discriminating power of the battery contained in the jth discriminant function may be taken to be

$$\frac{\lambda_j}{\sum_{k=1}^{p} \lambda_k}$$

An interesting speculative device is provided by taking

$$R_{c_j} = \sqrt{\frac{\lambda_j}{1 + \lambda_j}}$$

as a canonical correlation coefficient between the discriminant function and the group variable coded as a set of binary dummy variables. This is not as farfetched as it appears at first blush, because in fact it is possible to compute the multiple discriminant analysis as a special case of canonical regression analysis in which a set of binary dummy variables on one side of the canonical equation carry the information about group membership. Bartlett (1938) first introduced the multiple group discriminant analysis this way. The method actually has the computational advantage that the samples do not have to be sorted out and presented to the computer serially, as they do in our MANOVA and DISCRIM programs. We have chosen to present the model in terms of the Fisherian partitioned sums of squares and cross-products because we think that this clarifies the relationship of discriminant analysis to multivariate analysis of variance. The squared canonical correlation coefficient is an eta-square for the jth discriminant function, and may be taken as the proportion of variance in the discriminant function that is in common with the variance in the specific matching linear function of the group membership variable (although we do not compute the latter). The speculative value of this line of reasoning is that the new approach to redundancy analysis described in Chapter Six could be made available in the discriminant analysis situation via this logic. Anyway, it is useful to view the discriminant functions as factors of the measurement battery and to compute all the desirable characteristics of a factoring, including communalities and proportions of variance extracted from the battery. It may even be useful to Varimax or otherwise rotate the chosen discriminant factors rigidly to improve the interpretability of the reference vectors spanning the discriminating sub-space. Under such rotation there will be no loss of discriminating power for the selected reduced-rank model, but of course the relative discrimination afforded by each of the rotated reference vectors will have to be recomputed.

Finally, the centroid of each sample group on the sub-space of the discriminant functions may be computed as

$$\mathbf{m}_{df_k} = \mathbf{B}'(\mathbf{m}_k - \mathbf{m})$$

and the dispersion of each group may be computed on the subspace as

$$\mathbf{D}_{df_k} = \mathbf{B} \mathbf{D}_k \mathbf{B}$$

The discriminant functions are theoretically orthogonal in the populations, and to the extent that the group dispersions do sample a common populations dispersion the nonzero covariances in any $\mathbf{D}_{df_k}$ represent chance and can be ignored.

Notice that we have standardized the discriminant functions and computed their structure coefficients on the basis of the total sample dispersion. Some researchers will prefer to standardize the functions and compute the structure in relation to the pooled-within-groups dispersion, and will therefore substitute the $\mathbf{W}$ matrix for the $\mathbf{T}$ matrix in the appropriate equations. Rao and Slater (1949) and Porebski (1966) illustrate this alternative. Fortunately the computer program is easily modified if this is desired.

9.2 NUMERICAL EXAMPLE OF DISCRIMINANT ANALYSIS

Once again the Project TALENT data can serve as input for a small numerical example. The 196 grade 12 males (from among the first 200) whose school size code[1] was a 2, 3, or 4, sample three populations, based on size of graduating class. The sample sizes, means, and standard deviations on Information Part I and Part II are displayed in Table 9.1.

[1] School size = 1.00 was dropped because there were only four of them among the first 200 cases.

TABLE 9.1   MANOVA Results Prior to Discriminant Analysis

Group	Size Code	N	Means 1	Means 2	Standard Deviation 1	Standard Deviation 2
1	2	60	148.33	74.92	30.65	16.19
2	3	78	159.53	81.15	36.51	17.38
3	4	58	164.22	84.02	34.02	18.26
Total		196	157.49	80.09	34.41	17.54

The MANOVA run also produced the following results:

$$|D_1| = .727 \times 10^5 \qquad \text{For test of } H_1:$$
$$|D_2| = .958 \times 10^5 \qquad\qquad M = 5.032$$
$$|D_3| = .120 \times 10^6 \qquad F^6_{649640} = .826$$
$$|D_W| = .97 \times 10^5$$

$$\mathbf{W} = \begin{bmatrix} .22 \times 10^6 & .97 \times 10^5 \\ .97 \times 10^5 & .58 \times 10^5 \end{bmatrix} \qquad \mathbf{A} = \begin{bmatrix} .80 \times 10^4 & .45 \times 10^4 \\ .45 \times 10^4 & .26 \times 10^4 \end{bmatrix}$$

$$\mathbf{T} = \begin{bmatrix} .23 \times 10^6 & .10 \times 10^6 \\ .10 \times 10^6 & .60 \times 10^5 \end{bmatrix} \qquad \Lambda = \frac{|W|}{|T|} = .96$$

For $H_2 : F^4_{384} = 2.15$

Therefore, H_1 cannot be rejected and H_2 is slightly significant ($.10 > \alpha > .05$), so it makes sense to proceed with discriminant analysis. (At least as a numerical example!) Table 9.2 summarizes the results printed out during the discriminant analysis.

In the DISCRM program, the test of H_2 is repeated, this time using the roots of $[\mathbf{W}^{-1}\mathbf{A}]$ in $\Lambda = \prod_{i=1}^{n} (1/1 + \lambda_i)$. This serves as a check on input to DISCRM from MANOVA at a very small cost of computer time. Thus $(1/1 + .045) = .96$, where .045 is the value of the only nonzero root of $\mathbf{W}^{-1}\mathbf{A}$. Next, the roots are tested using Bartlett's chi-square approximation. This sequence of tests allows one to consider the number of independent functions along which group centroids differ. For the entire set of possible functions here,

$$\chi^2 = -\left(196 - \frac{2+3}{2} - 1\right) \log_e .9568$$
$$= -(196 - 3.5)(-.0439)$$
$$\chi^2 = 8.5 \quad \text{with} \quad ndf = (2 - 0)(3 - 0 - 1) = 4$$

TABLE 9.2   Discriminant Analysis Numerical Example

F—RATIO FOR H2, OVERALL DISCRIMINATION,=2.15
NDF1=4 AND NDF2=384
CHI SQUARE TESTS WITH SUCCESSIVE ROOTS REMOVED

ROOTS REMOVED	CANONICAL R	R SQUARED	EIGENVALUE
0	0.208	0.043	0.045
1	0.008	0.000	0.000

CHI SQUARE	NDF	LAMBDA	PERCENT TRACE
8.51	4	0.96	99.87
0.01	1	1.00	0.13

ROW COEFFICIENTS VECTORS
D F   1   0.0043032   0.0494752
D F   2   -0.0557285   0.0978380

FACTOR STRUCTURE FOR DISCRIMINANT FUNCTIONS
TEST
   1   0.888   -0.449
   2   0.992   0.077

COMMUNALITIES FOR 2 DISCRIMINANT FACTORS
   1   0.990   2   0.990

PERCENTAGE OF TRACE OF R ACCOUNTED FOR BY EACH ROOT
   1   88.611   2   10.372

GROUP CENTROIDS IN DISCRIMINANT SPACE

	FUNCTION	
GROUP	1	2
1	-.295	.004
2	.061	-.010
3	.223	.009

This yields a probability similar to the test of $H_2(.10 > \alpha > .05)$. After removing the group information associated with the first function, the second root yields a chi square of .01 for $ndf = 1$, so only the first discriminant function is of possible interest.

That first function represents a canonical correlation of .21, which is the maximum correlation between a linear function of the two predictors and a linear function of the group membership variables represented as a set of binary criteria. That canonical correlation is computed as $R_{c1} = \sqrt{.045/1.045}$ = .208.

The coefficients for producing standardized (unit variance, centered at zero) discriminant scores from deviation vectors are listed as rows. Thus it represents our $\mathbf{B}'$. Taking the first row of $\mathbf{B}'$ as $\mathbf{b}_1$, and multiplying by the standard deviations $(\mathbf{D}_{diag}^{1/2})$ based upon total, we obtain vector $\mathbf{c}_1$, the coefficients for $\mathbf{Z}$.

$$\mathbf{c}_1 = \mathbf{b}_1 \mathbf{D}_{diag}^{1/2} = [.0043 \quad .0495] \cdot \begin{bmatrix} 34.41 & 0 \\ 0 & 17.54 \end{bmatrix} = \begin{bmatrix} .148 \\ .868 \end{bmatrix}$$

Postmultiplying the total correlation matrix by $\mathbf{c}_1$ yields the structure coefficients:

$$\begin{bmatrix} 1.00 & .85 \\ .85 & 1.00 \end{bmatrix} \cdot \begin{bmatrix} .148 \\ .868 \end{bmatrix} = \begin{bmatrix} .89 \\ .99 \end{bmatrix}$$

since $\mathbf{s} = \mathbf{Rc}$. Then premultiplying $\mathbf{s}$ by $\mathbf{c}$, that is $\mathbf{c}'\mathbf{Rc} = \mathbf{c}'\mathbf{s} = 1.00$

$$[.148 \quad .868] \cdot \begin{bmatrix} .888 \\ .992 \end{bmatrix} = 1.00$$

Thus scores on the first discriminant function correlate .89 with Information Part I and .99 with Part II.

The program reports a few other values to facilitate interpretation. In the table reporting the chi-square test of each root is also found the ratio of each root to the trace of $\mathbf{W}^{-1}\mathbf{A}$, expressed as a percentage. Thus the first function accounts for 99.87% of the discriminating information available in $\mathbf{W}^{-1}\mathbf{A}$.

Another interpretative aid is the table of communalities for each variable, representing the sums of squares of rows of the structure matrix. This will only be of interest when $(g - 1) < p$, since otherwise the communalities are 1.00. In the former case, the variance for those variables with low communalities is not accounted for in the full set of discriminant functions thus indicating they contain little information regarding group differences.

The percentage of trace of $\mathbf{R}$ for each function is found by dividing the sum of squares of each *column* of the structure matrix by the trace of $\mathbf{R}$. Thus $[(.888)^2 + (.992)^2]/2 = .886$. These percentages will sum to 1.00 only when $(g - 1) \geq p$.

TABLE 9.3 Group Means on Nine MAP Predictors for 1965 Five-Year Follow-up Career Plans

Career Plans Group	N	FACTOR: VKN	MAT	VIS	BUS	SCH	OUT	CUL	SCI	SOC
1. Medicine & Ph.D. Biology (MED)	279	61	80	56	44	62	57	42	73	45
2. Medicine & Biology; below Ph.D., M.D. (BIO)	438	57	69	59	45	56	62	38	70	50
3. Physical Science & Mathematics; Ph.D. (RES)	221	61	84	60	45	63	60	38	72	42
4. Physical Science & Engineering (ENG)	939	56	74	61	47	57	64	35	69	48
5. Technical Worker (TEC)	1297	50	59	60	46	49	67	35	61	50
6. Laborer; no post-high school training (LBR)	706	46	54	57	47	46	69	34	56	50
7. Office Worker; no post-high school training (CLK)	530	49	53	55	49	48	64	37	58	52
8. Accountants & other trained non-technical (ACT)	1430	53	57	57	49	49	63	38	60	54
9. Business; college only (BUS)	1214	56	65	56	50	54	61	36	65	53
10. Management; post-college training (MGT)	270	60	75	57	52	59	57	36	68	50
11. Sociocultural; college only (WEL)	1183	57	64	57	48	54	60	43	64	51
12. Sociocultural; research degree (WEL)	815	61	72	55	47	59	57	44	66	49
Total	9322									

Finally the group centroids in discriminant space are reported. This allows a convenient and revealing mapping of the groups if only one or two functions are required to describe group differences. If more than two functions are significant, locating the centroids on each function separately, along with the structure for that function, is a useful interpretive device.

9.3   RESEARCH EXAMPLE

This first research example was drawn from the Project TALENT follow-up studies (Cooley and Lohnes, 1968). That monograph should be examined by the reader interested in seeing the discriminant analysis strategy in the entire research context. In a brief example, the predictors and criterion are simply listed and their relationships, as revealed by the resulting discriminant functions, are described, but the reader is left in the dark regarding how the analysis contributed to the research program, or why those particular predictors were selected, or how it was decided to use those groups as the taxonomic criterion. To tell that whole story requires the whole monograph, so we encourage the reader to see the monograph for the story behind the following example and for dozens of other examples of discriminant analysis and how they contributed to a picture of the career development process.

Table 9.3 lists the twelve career plan categories into which the five-year follow-up careers for 9322 young men were classified. They had been tested in 1960 as twelfth graders, and were contacted five years later to determine career status and plans for the future. The 12 category taxonomy is based upon educational decisions made since high school, work engaged in or planned (if still in school), and advanced degrees sought. The 22 MAP factors (see Sections 3.5 and 5.6 for brief descriptions), based upon the 1960 test battery, served as the predictors of group membership. The group means for the nine best MAP predictors are also listed in Table 9.3. These factors were scaled to have a mean of 50 and a standard deviation of 10 for grade 12 males.

Table 9.4 reports the structure for the three most significant discriminant functions and their predictive potency in terms of canonical correlations between the 22 predictors and the 12 group membership variables.

From the structure coefficients of Table 9.4 and the nature of the MAP factors, the three functions appear to be measuring: (1) science-oriented scholasticism; (2) sociocultural ($+$) versus technical ($-$); (3) cultural ($+$) versus business ($-$). These descriptions are in terms of both abilities and motives.

The location of the 12 group centroids on these three functions also helps to describe the nature of these group differences. These centroids are displayed

TABLE 9.4   Factor-Discriminant Correlations and Canonical Correlations for Five-Year Follow-up Career Plans in Twelfth Grade MAP Space

MAP Factors	Canonical Correlation	Discriminant Functions		
		I .69	II .37	III .23
Abilities				
Verbal Knowledges		.62	.20	−.07
Perceptual Speed, Accuracy		.02	.10	−.17
Mathematics		.73	−.49	−.07
Hunting-Fishing		−.10	−.26	−.03
English		.28	.23	.06
Visual Reasoning		−.01	−.43	−.07
Color, Foods		.08	.10	.15
Etiquette		.05	.07	−.10
Memory		.00	.01	.05
Screening		−.33	−.25	−.05
Games		.10	−.05	−.29
Motives				
Business Interests		−.04	.31	−.51
Conformity Needs		.21	.12	−.08
Scholasticism		.78	−.19	−.06
Outdoors, Shop Interests		−.41	−.42	.07
Cultural Interests		.25	.47	.61
Activity Level		−.22	−.10	−.10
Impulsion		−.01	.08	−.06
Science Interests		.54	−.36	−.23
Sociability		−.19	.47	−.43
Leadership		.28	.22	.04
Introspection		−.06	−.03	.18

DF I: Science-oriented Scholasticism
DF II: Technical (−) versus Sociocultural (+)
DF III: Business (−) versus Cultural (+)

TABLE 9.5   Discriminant Function Centroids for 1965 Five-Year Follow-up Career Plan Groups

Plan	Group	DF I	DF II	DF III
1	MED	64	48	54
2	BIO	55	47	50
3	RES	62	41	52
4	ENG	54	44	48
5	TEC	43	46	51
6	LBR	38	48	53
7	CLK	41	52	50
8	ACT	45	53	49
9	BUS	52	52	46
10	MGT	59	51	45
11	WEL	53	54	52
12	PRF	59	54	53

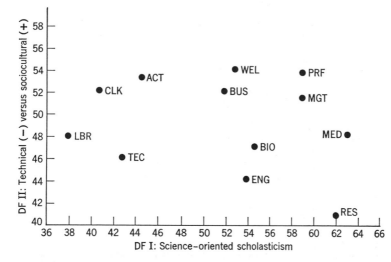

Figure 9.2. Centroids of five-year follow-up career plan groups in discriminant plane.

in Figure 9.2 for the first two functions and are in Table 9.5 for all three functions. Here again the functions have been standardized to a mean of 50 and a standard deviation of 10 in order to facilitate thinking about group differences.

DISCRM

```
C      MULTIPLE GROUP DISCRIMINANT ANALYSIS.  A COOLEY-LOHNES PROGRAM.
C      THIS PROGRAM COMPUTES DISCRIMINANT FUNCTIONS, THEIR CANONICAL
C   CORRELATIONS WITH GROUP MEMBERSHIP DUMMY VARIATES, F-RATIOS FOR
C   THESE, AND CENTROIDS OF GROUPS IN THE STANDARDIZED DISCRIMINANT
C   FUNCTIONS SPACE.  COEFFICIENTS FOR COMPUTING STANDARDIZED
C   DISCRIMINANT FUNCTIONS SCORES FROM DEVIATION TEST SCORES ARE
C   PUNCHED OUT.
C
C      REQUIRED SUBROUTINES ARE DIRNM AND HOW.
C
C      INPUT
C
C   1) FIRST TEN CARDS OF DATA DECK CONTAIN A TEXT DESCRIBING THE
C   JOB, WHICH WILL BE REPRODUCED ON THE OUTPUT. DO NOT USE COLUMN 1.
C   2) CONTROL CARD (CARD 11)
C          COLS 1-2   M = NUMBER OF VARIABLES
C          COLS 3-5   KG = NUMBER OF GROUPS
C          COLS 6-10  N = NUMBER OF SUBJECTS
C          COLS 11-12 KC = NUMBER OF CONTROL VARIABLES PREVIOUSLY
C                          PARTIALED OUT BY COVAR (THIS VALUE WILL BE A
C                          ZERO IF INPUT MATRICES COME FROM MANOVA).
C   3) T MATRIX (TOTAL SAMPLE DEVIATION SSCP, AS PUNCHED BY MANOVA
C      OR COVAR).
C   4) W MATRIX (POOLED WITHIN-GROUPS DEVIATION SSCP MATRIX, AS PUNCHED
C      BY MANOVA OR COVAR).
C   5) GROUP MEANS (AS PUNCHED BY MANOVA OR COVAR).
C   6) GRAND MEANS (AS PUNCHED BY MANOVA).
C
       DIMENSION  A(50,50), B(50,50), C(50,50), T(50), U(50), V(50),
      2 W(50), X(50), Y(50), Z(50),   TIT(16),   D(50,50)
C
   1   WRITE(6,2)
   2   FORMAT(65H1MULTIPLE GROUP DISCRIMINANT ANALYSIS. A COOLEY-LOHNES
      2PROGRAM.        )
       DO 3   J = 1, 10
       READ (5,4)      (TIT(K),   K = 1, 16)
   3   WRITE(6,4)      (TIT(K),   K = 1, 16)
   4   FORMAT (16A5)
       READ(5,9)       M,   KG,   N,   KC
   9   FORMAT (I2, I3, I5, I2)
   5   FORMAT (10X,   5E14.7)
       DO 6   J = 1, M
   6   READ (5,5)      (C(J,K),   K = J, M)
       DO 7   J = 1, M
   7   READ(5,5)       (B(J,K),   K = J, M)
       DO 8   J = 1, M
       DO 8   K = J, M
       C(K,J) = C(J,K)
   8   B(K,J) = B(J,K)
       DO 15   J = 1, M
       DO 15   K = 1, M
  15   A(J,K) = C(J,K) - B(J,K)
C  A NOW CONTAINS THE A MATRIX (AMONG-GROUPS DEVIATION SSCP MATRIX).
C  B CONTAINS THE W MATRIX (WITHIN-GROUPS DEVIATION SSCP MATRIX).
       IF (M - KG) 10, 11, 11
  10   MD = M
       GO TO 12
  11   MD = KG - 1
C
  12   CALL DIRNM (A, M, B, D, T, MD)
C  ROOTS OF W INVERSE * A  ARE IN T AND COLUMN EIGENVECTORS ARE IN D.
C
       EM = M
       EKG = KG
       EN = N
       EKC = KC
```

```
        XL = 1.0
        TRACE = 0.0
        DO 13    J = 1, MD
        U(J) = T(J) / (1.0 + T(J))
        V(J) = SQRT (U(J))
        W(J) = 1.0 / (1.0 + T(J))
        XL = XL * W(J)
13      TRACE = TRACE + T(J)
        DO 14    J = 1, MD
14      Z(J) = 100.0 * (T(J) / TRACE)
        IF (M - 2)    16, 16,   17
16      IF (KG - 3)    18, 18,   17
18      YL = XL
        F1 = 2.0
        F2 = EN - 3.0 - EKC
        GO TO 19
17      SL = SQRT (((EM * EM) * ((EKG - 1.0)**2) - 4.0) / ((EM * EM) +
      2   ((EKG - 1.0)**2) - 5.0))
        YL = XL ** (1.0 / SL)
        PL = (EN - 1.0 - EKC) - ((EM + EKG) / 2.0)
        QL =  - ((EM *(EKG - 1.0)) - 2.0) / 4.0
        RL = (EM * (EKG - 1.0)) / 2.0
        F1 = 2.0 * RL
        F2 = (PL * SL) + (2.0 * QL)
19      N1 = F1
        N2 = F2
        F = ((1.0 - YL) / YL) * (F2 / F1)
        YL = 1.0 - XL
        WRITE(6,201)   XL, YL
201     FORMAT(16HOWILKS LAMBDA = F5.4,46H  GENERALIZED CORRELATION RATIO,
      C ETA SQUARE = F5.4)
        WRITE(6,20)       F
20      FORMAT(45HOF-RATIO FOR H2, OVERALL DISCRIMINATION, =     F9.2)
        WRITE (6,21)     N1,  N2
21      FORMAT (8HONDF1 = I3,   12H AND NDF2 = I6)
        J = MD
        X(J+1) = 1.0
22      X(J) = X(J+1) * W(J)
        J = J - 1
        IF (J)    23, 23,  22
23      DO 24    J = 1, MD
24      Y(J) = - PL *ALOG  (X(J))
        WRITE (6,25)
25      FORMAT(48HOCHI-SQUARE TESTS WITH SUCCESSIVE ROOTS REMOVED    )
        WRITE (6,261)
261     FORMAT(1H0,20X,22H(ETA)     (ETA SQUARE))
        WRITE (6,26)
26      FORMAT(105HOROOTS REMOVED    CANONICAL R     R SQUARED     EIGENVALUE
      C   CHI-SQUARE    N.D.F.    LAMBDA  PERCENT TRACE )
        DO 27    J = 1, MD
        JT = J - 1
        NDF = (M-JT)*(KG - JT - 1.0)
27      WRITE (6,28)     JT,  V(J),  U(J),  T(J),  Y(J),  NDF,  X(J), Z(J)
28      FORMAT(6X, I4, 9X, 2(F6.3,8X), F9.3, 5X,F10.0,4X,I5,2X,F9.2,F8.2)
C
        DO 29    J = 1, MD
        DO 29    K = 1, M
        A(J,K) = 0.0
        DO 29    L = 1, M
29      A(J,K) = A(J,K) + D(L,J) * (C(L,K) / (EN - 1.0))
        DO 30    J = 1, MD
        DO 30    K = 1, MD
        B(J,K) = 0.0
        DO 30    L = 1, M
30      B(J,K) = B(J,K) + A(J,L) * D(L,K)
        DO 31    J = 1, M
```

```
       DO 31   K = 1, MD
  31   D(J,K) = D(J,K) * (1.0 / SQRT (B(K,K)))
       WRITE (6,32)
       WRITE(7,32)
  32   FORMAT (25HOROW COEFFICIENTS VECTORS   )
       DO 33   J = 1, MD
       WRITE(6,49)        J,   (D(K,J) ,   K = 1, M)
  33   WRITE(7,49)        J,   (D(K,J) ,   K = 1, M)
  49   FORMAT (5H D F   I3,2X, 5E14.7 / (10X, 5E14.7))
       DO 34   J = 1, M
  34   Z(J) = SQRT  (C(J,J) / (EN - 1.0))
C   TOTAL SAMPLE STANDARD DEVIATIONS ARE NOW IN Z.
       DO 35   J = 1, M
       DO 35   K = 1, M
  35   C(J,K) = C(J,K) / (EN * Z(J) * Z(K))
C   TOTAL SAMPLE CORRELATION MATRIX IS NOW IN C.
       DO 36   J = 1, M
       DO 36   K = 1, MD
  36   B(J,K) = D(J,K) * Z(J)
       DO 37   J = 1, M
       DO 37   K = 1, MD
       A(J,K) = 0.0
       DO 37   L = 1, M
  37   A(J,K) = A(J,K) + C(J,L) * B(L,K)
       WRITE(6,38)
  38   FORMAT(42HOFACTOR PATTERN FOR DISCRIMINANT FUNCTIONS        )
       DO 39   J = 1, M
  39   WRITE(6,40)        J,  (A(J,K),   K = 1, MD)
  40   FORMAT (5HOTEST I4, .10(3X,F7.3) / (9X, 10(3X,F7.3)))
       DO 41   J = 1, M
       T(J) = 0.0
       DO 41   K = 1, MD
  41   T(J) = T(J) + A(J,K) * A(J,K)
       WRITE (6,42)     MD
  42   FORMAT(19HOCOMMUNALITIES FOR   I5,21H DISCRIMINANT FACTORS   )
       WRITE(6,43)          (J,  T(J),    J = 1, M)
  43   FORMAT (1HO, 10(2X,  I3, F7.3))
       DO 44   J = 1, MD
       T(J) = 0.0
       DO 44   K = 1, M
  44   T(J) = T(J) + A(K,J) * A(K,J)
       WRITE(6,45)
  45   FORMAT(55HOPERCENTAGE OF TRACE OF R ACCOUNTED FOR BY EACH ROOT    )
       DO 46   J = 1, MD
  46   T(J) = 100.0 * (T(J) / EM)
       WRITE (6,43)        (J,  T(J),    J = 1, MD)
C
       KGT = KG + 1
       DO 47   J = 1, KGT
  47   READ(5,5)        (A(J,K),   K = 1, M)
C   READS GROUP MEAN VECTORS AND GRAND MEAN VECTOR INTO COLUMNS OF A.
C   COLUMN KGT CONTAINS THE GRAND MEANS.
       DO 48   J = 1, KG
       DO 51   K = 1, MD
       T(K) = 0.0
       DO 51   L = 1, M
  51   T(K) = T(K) + (A(J,L) - A(KGT,L)) .* D(L,K)
       WRITE (6,50)     J,   MD
  50   FORMAT(20HOCENTROID FOR GROUP   I4,4H IN I4,32H DIMENSIONAL DISCRI
      2MINANT SPACE       )
  48   WRITE(6,43)        (K,  T(K),   K = 1, MD)
       GO TO 1
       END
  >
```

9.5 EXERCISES

1. What matrix equation results when one seeks a normalized column vector **v** such that ratio

$$\frac{\mathbf{v'Av}}{\mathbf{v'Wv}}$$

   is a maximum? (**A** here is the among groups deviation sums of squares and cross-products matrix, **W** is the within.)

2. What is the relationship between the roots (eigenvalues) of $\mathbf{W^{-1}A}$ and:
   (a) Wilks' lambda criterion?
   (b) canonical correlation coefficient?
   (c) the trace of $\mathbf{W^{-1}A}$?

3. How is the among groups deviation sums of squares and cross-products matrix formed in DISCRM program? (This may be expressed in simple matrix notation.)

4. Using the TALENT males of Appendix B, conduct a discriminant analysis with variable 8 (college plans) as a 5-group criterion and a set of the ability measures as predictors. Here the interest is not whether there is a relationship (there had better be!) but the nature of the relationship. Does the rank of the discriminant space tell you something about the scale properties of variable 8?

CHAPTER TEN

||||||||||||||||||||||||||||||||||||||||||||||||||||||||||||||||

Classification Procedures

10.1 MATHEMATICS OF CLASSIFICATION STATISTICS

Classification methods make it possible for the researcher who has data from two or more populations in a common measurement space to demonstrate the validity of the measurements for predicting membership in the populations. The usual procedure will be to estimate the densities of the populations at each point in the space occupied by a member of a replication sample, and assign points to that population having the largest density at each point. If the actual population membership of the points in the replication sample are known, it is then possible to see whether each classification is a *hit* or *miss*. It is possible to compute the classification statistics for the subjects in the research samples from which the population parameters have been estimated, but then the dangers of capitalization on chance have to be considered. Since classification statistics require a great deal of fitting to norming data it is very desirable to make the predictions for an independent replication sample. Only when the research samples are large and carefully randomized will classification of their members be a convincing display of the classification validity of the measurements. The safe way is the replication samples way.

A useful but potentially confusing feature of the classification statistics to be presented is that they allow the use of prior knowledge about the probabilities of group membership as well as information gleaned from the location and dispersion estimates computed on the research samples. Prior knowledge is anything known about the populations from outside this measurement research that permits the formulation of hypothesized probabilities of group membership for a random subject. From census data it may be known that the populations are of specific sizes. Dividing each size by the sum of the sizes then creates a vector of hypothesized probabilities of group membership for a random subject. To the extent that the sizes are disparate this a priori probabilities vector will represent a powerful influence on classifications. If something is known about the subject other than his measurement vector, such as his sex, it may be possible to further improve on the a priori probabilities. Thus, if from census data a different set of population sizes for each sex is available, the appropriate a priori probabilities vector may be selected to match the known sex of the subject who is to be classified.

Classification decisions that involve both a priori probabilities and a posteriori probabilities of group membership are examples of Bayesian logic. In the decision rule the a priori probabilities representing independent knowledge of the relative sizes of the populations (i.e. the unconditional distribution on the taxonomic variable) are multiplied by the a posteriori probabilities representing the relative densities of the swarms for the populations at each point classified, as estimated from sample centroids and dispersions. In other words, the Bayesian feature of classification statistics is that we are able to combine information that is conditional on the measurement vector of the subject (and thus posterior to measurement research) with information that is not conditional on the vector variable (although it may be conditional on some other characteristic of the subject). However, we are not required to provide different prior probabilities by the decision rule, and frequently it may be undesirable to do so. In the latter case, we simply provide $1/g$ as each prior probability.

Consider that even if the posterior information is nil, meaning that the vector variable has no predictive validity for the taxonomic criterion variable, a set of powerful prior probabilities can lead to successful classification of subjects and the illusion of predictive validity. When the object of computing is to display the predictive validity of the measurement system prior probabilities can be misleading.

The classification procedures to be presented all assume a multivariate normal distribution for the vector variable in each of the populations. We use the notion of *swarm* for the plot in the measurement space of points representing all the members of a single population, each point being located by treating the member's vector of measurement scores as coordinates of a single point in p-dimensional space. A multivariate normal swarm is very dense in the region of the population centroid and thins out in all directions away from the centroid. The swarm may be elongated in some directions as a function of covariances among the measurements, and thus the rate of thinning in any direction is a resultant of variances *and* covariances. We speak of a normal swarm as hyper-ellipsoidal, meaning loosely that the projection of the swarm on any plane passing through the centroid is elliptical. Technically, an ellipse is a closed curve and an ellipsoid is a closed surface, so we have to be more technical to get an actual ellipse or ellipsoidal surface. What we do is to conceive of the boundary in the swarm within which P proportion of the subjects will be found when each subject is represented by the deviation of his score vector from the centroid:

$$\mathbf{x}_{ji} = \mathbf{X}_{ji} - \mathbf{m}_j$$

Each such boundary is one out if a set of concentrically nested ellipsoids. Such a boundary is the locus of all points for which a generalized distance function from the centroid is a constant value. The generalized distance function is a quadratic form, involving the inverse of the dispersion, or variance-covariance matrix, D_j:

$$\chi_{ji}{}^2 = \mathbf{x}_{ji}'\mathbf{D}_j{}^{-1}\mathbf{x}_{ji}$$

Notice that if X is univariate the generalized distance function reduces to a squared standard score:

$$\chi_{ji}{}^2 = \frac{(X_{ji} - m_j)^2}{s_j{}^2} = z_{ji}{}^2$$

This may help to explain our use of the chi-square symbol for the distance. To obtain the value of P for a given $\chi_i{}^2$ we can look up or compute the cumulative function of the given $\chi_i{}^2$ with p degrees of freedom. When we have found P we have estimated the proportion of members who reside as close to or closer to the centroid than does individual i.

Actually, we are trying to estimate the probability that individual i is a member of the jth population. We have a set of hypotheses regarding the group membership of the individual subject when we have g populations under study and the ith subject must be classified into one and only one population. The general form for the likelihood of such a hypothesis can be written as:

$$Pr(H_j \,|\, \mathbf{X}_i), i = 1, 2, \ldots, N \qquad \text{and} \qquad j = 1, 2, \ldots, g$$

which reads: The probability of hypothesis j given the score vector $\mathbf{X}_i$. Hypothesis j, which we will abbreviate hereafter as H_j, states that the subject is a member of population j. There will be g such hypotheses to be evaluated for each subject, and the maximum likelihood classification rule will be to assign i to group j if $Pr(H_j \,|\, \mathbf{X}_i)$ is largest.

The relation of the probability of the hypothesis that i belongs to group j to the cumulative probability for the distance function, or $\chi_{ji}{}^2$, may be taken to be simply that:

$$Pr(H_j \,|\, \mathbf{X}_i) = 1 - \mathbf{P}(\chi_{ji}{}^2)$$

By this method the probability of group membership is simply the inverse cumulative function of the generalized distance tabled as chi-square with p degrees of freedom, where p is the number of elements in the vector variable $\mathbf{X}$.

We could, then, collect data on a norming sample from each of g populations; estimate from this data $\mathbf{m}_j$ and $\mathbf{D}_j$ for the populations; using these estimates compute from $\mathbf{X}_i$ a set of g generalized distances; table these as chi-squares and get their inverse cumulants, to be treated as a set of probabilities for membership hypotheses; classify subject i according to the decision rule:

Rule I: H_j if $Pr(H_j \,|\, \mathbf{X}_i) > Pr(H_k \,|\, \mathbf{X}_i)$ for $k = 1, 2, \ldots, g, k \neq j$

Rulon and his associates (1954, 1967) have named this method of evaluating the probability of group membership the *centour* method (for *cen*tile con*tour*). They define a centour score as:

$$100[1 - P(\chi_{ji}{}^2)]$$

That is, the centour for individual i in group j is 100 times the probability of obtaining a larger value of $\chi_{ji}{}^2$, or 100 times the probability of a larger generalized distance from the centroid of group j. The classification rule is to assign the subject to the group for which he has the largest centour, which is our Rule I.

To illustrate, consider a hypothetical bivariate distribution of test scores for a random sample of N graduate students in a university history department. Let the two elements of $\mathbf{X}$ be the Verbal and Quantitative scores on the Graduate Record Examination. The ellipses in Figure 10.1 represent the loci of points for three different generalized distances from the centroid of the sample. For ellipse A the chi-square value is .21 and the centour value is 90. We expect that 90 percent of graduate students in history reside outside ellipse A. Ellipse C is the locus of points outside which 10 percent of the

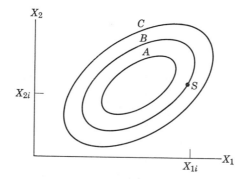

Figure 10.1.

history students are expected to reside ($\chi^2 = 4.60$). Point S is the location of a student whose scores are X_{1i} and X_{2i}. The computed χ^2 for these scores is 2.41 with two degrees of freedom. From a chi-square table we find that $P = .70$, $Pr(H_{\text{history}} | X_i) = .30$, and the centour for S and all other points on ellipse B is 30. It is now possible to give the student an interpretation of his GRE profile in terms of his resemblance to history graduate students. The interpretation is that only 30 percent of history graduate students have GRE score profiles that place them further from the centroid for the group than him. The estimated probability of his membership in the history graduate students population is only .30.

Usually we are interested in situations involving two or more groups, rather than a single group, so that the question of the relative densities of the two populations or several populations in the region of a subject's location arises. Let us add graduate students in physics to the GRE test space of our example. The 90 and 10 centours for the two groups might appear as in Figure 10.2.

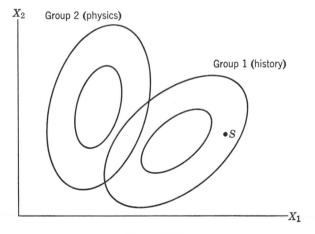

Figure 10.2.

Now we can compute two generalized distances for any subject, based on the centroid and dispersion estimates for history students (χ_{1i}^2) and on the estimates for physics students (χ_{2i}^2). If S is the location for the same subject discussed for Figure 10.1, so that for X_i, $\chi_{1i}^2 = 2.41$ and the centour is 30, suppose that $\chi_{2i}^2 = 6.9$, for which the centour is less than 1. Now we can tell the student not only that he does not look much like a history student, but also that he does not look at all like a physics student, since $Pr(H_{\text{physics}} | X_i) < .01$. This comparison of probabilites of group memberships might be useful to the student as he makes a decision about graduate study. Notice that if we must classify the student into one of the two graduate programs we will place him in history, despite his low probability of .30 for that placement.

Since by the centours classification procedure the probability of membership in each group is inverse monotonic with the chi-square for each group, the decision rule can be stated in terms of the distance functions, and the probabilities (or centours) need not be evaluated:

$$\text{Rule } I_a : H_j \text{ if } \chi_{ji}^2 < \chi_{ki}^2 \text{ for } k = 1, 2, \ldots, g, k \neq j$$

One problem with this rule is that if $|\mathbf{D}_j| > |\mathbf{D}_k|$, group j will tend to be overassigned because a given centour for group j will enclose a larger region of the test space than will the same centour for group k. This rule also makes no provision for incorporating an a priori probabilities vector $\mathbf{q}$ in the calculation. However, if the $\mathbf{D}_j$ were population parameters and all equal, and the $\mathbf{m}_j$ were also parameters, and the prior probabilities were uniform, Rule I or I_a would result in a minimum of misclassifications. Since it will never be the case that $\mathbf{D}_j = \mathbf{D}_k$, and in order to provide for a $\mathbf{q}$ vector, the following rule is needed to minimize misclassifications in the centours approach:

$$\text{Rule II: } H_j \text{ if } \chi_{ji}^2 < \chi_{ki}^2 - \log_e \frac{|\mathbf{D}_j|}{|\mathbf{D}_k|} + 2 \log_e \frac{q_j}{q_k} \text{ for } k = 1, 2, \ldots, g, k \neq j$$

This rule can be elaborated to yield relative densities of the multivariate populations at points $\mathbf{X}_i$ in the measurement space. Recall that the density function for the normal curve is

$$p_i = \frac{1}{\sigma\sqrt{2\pi}} e^{-1/2\chi_i^2}$$

where p_i is the ordinate (height) of the curve at point X_i, and

$$\chi_i^2 = \frac{(X_i - \mu)^2}{\sigma^2}$$

We generalize to the multivariate normal distribution, drop the constant $1/\sqrt{2\pi}$, and substitute sample values for parameters:

$$p_{ji} = q_j |\mathbf{D}_j|^{-1/2} e^{-1/2\chi_{ji}^2}$$

where

$$\chi_{ji}^2 = (\mathbf{X}_i - \mathbf{m}_j)' \mathbf{D}_j^{-1} (\mathbf{X}_i - \mathbf{m}_j)$$

We require very stable estimates of $\mathbf{m}_j$ and $\mathbf{D}_j$ to justify this substitution (i.e., large random samples). The relative density of group j at $\mathbf{X}_i$ is now computed as:

$$Pr_{ji} = Pr(H_j \mid \mathbf{X}_i) = \frac{p_{ji}}{\sum_{k=1}^{g} p_{ki}}$$

Note that

$$\sum_{j=1}^{g} Pr_{ji} = 1.0$$

The decision rule can now be stated:

$$\text{Rule III: } H_j \text{ if } Pr_{ji} > Pr_{ki}, \, k = 1, 2, \ldots, g, \, k \neq j$$

The development of Pr_{ji} and Rule III is credited to Tatsuoka (1957).

One usually arrives at classification questions only after the relation of a vector variable to a taxonomic variable has been established through manova and discriminant analysis stages. By then you are pretty well wedded to a theory for the data that assumes a common populations dispersion, and are not likely to want to revert to statistics that treat differences in sample dispersions as indicative of real differences in population dispersions. Although the null hypothesis of equality of dispersions usually has a prior probability of exactly zero, regardless of how the manova test of H_1 turns out, one proceeds to see what can be accomplished as classification validities by a model that assumes equal dispersions.

Assuming, then, that the distribution for each population is multivariate normal with parameters μ_j and Δ, the density at point $\mathbf{X}_i$ for group j is

$$p_{ji} = \frac{1}{(2\pi)^{1/2p}|\Delta|^{1/2}} \, e^{-1/2(\mathbf{X}_i - \mu_j)'\Delta^{-1}(\mathbf{X}_i - \mu_j)}$$

Given the vector of prior probabilities $\mathbf{q}$ for incorporation in the decision rule, the classification statistic representing the posterior information in the predictive validity of the vector variable is

$$u_{jki} = \log_e \frac{p_{ji}}{p_{ki}} = [\mathbf{X}_i - \tfrac{1}{2}(\mu_j + \mu_k)]'[\Delta^{-1}(\mu_j - \mu_k)]$$

This statistic is derived by T. W. Anderson (1958, Chapter Six, "Classification of Observations"), who points out that

$$u_{kji} = -u_{jki}$$

and that the term

$$\Delta^{-1}(\mu_j - \mu_k)$$

is a coefficients vector for a Fisherian two-group discriminant function.[1] When N_j is reasonably large we may substitute for μ_j and Δ the sample statistics

$$\mathbf{m}_j = \frac{1}{N_j} \sum_{i=1}^{N_j} \mathbf{X}_{ji}$$

and

$$\mathbf{D} = \left[\frac{1}{\sum_{j=1}^{g}(N_j - 1)} \right] \sum_{j=1}^{g} \sum_{i=1}^{N_j} (\mathbf{X}_{ji} - \mathbf{m}_j)(\mathbf{X}_{ji} - \mathbf{m}_j)'$$

Notice that $\mathbf{D}$ is based on $\mathbf{W}$, the pooled within-groups sums of squares and cross-products matrix:

$$\mathbf{D} = \frac{1}{N - g} \mathbf{W}$$

The maximum likelihood classification rule becomes

$$\text{Rule IV: } H_j \text{ if } u_{jki} > \log_e \frac{q_k}{q_j}, \ k = 1, 2 \ \ldots, g, \ k \neq j$$

Rule IV requires large norming samples and computation of an entire g-order matrix of classifications statistics u_{jki} for each subject. Geisser (1964) reviews a class of Bayesian methods for finding posterior odds that a particular observation belongs to one of g multivariate normal populations. The emphasis is on the probabilities of membership, although the classification rule is straightforward. The most relevant model, and the one the authors believe provides the most appropriate classification statistics for a manova-oriented research strategy, is based on small-sample theory, and involves gamma functions. It is Geisser's "Case 8. $\Delta_j = \Delta$ but unknown, μ_j unknown" (1964, p. 73). With $\mathbf{m}_j$ and $\mathbf{D}$ estimates computed as above:

$$p_{ji} = q_j \left[\frac{N_j}{(N_j + 1)} \right]^{1/2p} \frac{\Gamma\frac{1}{2}(N - g + 1)}{\Gamma\frac{1}{2}(N - p - g + 1)}$$

$$\times \left[1 + \frac{N_j(\mathbf{X}_i - \mathbf{m}_i)'\mathbf{D}^{-1}(\mathbf{X}_i - \mathbf{m}_i)}{(N_j + 1)(N - g)} \right]^{-1/2(N-g+1)}$$

[1] Morrison (1967, p. 131) uses the form $u_{jki} = X_i\Delta^{-1}(\mu_j - \mu_k) - 1/2(\mu_j - \mu_k)\Delta^{-1}(\mu_j - \mu_k)$ where the first term is a discriminant score for subject i, and the second term is one-half the distance between the means.

The apparent hairiness of this expression is mitigated by the fact that the ratio of gamma functions is a constant of proportionality and need not be evaluated. From the p_{ji} we compute:

$$Pr(H_j \mid \mathbf{X}_i, q_j) = \frac{p_{ji}}{\sum_{k=1}^{g} p_{ki}} = Pr_{ji} \qquad (10.1.1)$$

The decision rule is again Rule III.

When these decisions have been made for all subjects in the replication sample it is an easy matter to arrange a table in which the rows are Actual Groups, the columns are Assigned Groups, and the cell entries in a row are the frequencies of assignments of each type for actual members of that row group. Given this table, the diagonal cell frequencies are *hits* and the off-diagonal cell frequencies are *misses*. The overall hit rate is the sum of the diagonal frequencies divided by the total sample size.

Another interesting table is created by computing the classification probabilities for each sample centroid. For a given centroid, the vector of probabilities reports on the relative densities of each of the samples at precisely the locus of the center of density for that group. These values give a good picture of the extent of separation of the groups.

10.2 MONTE CARLO STUDY OF CLASSIFICATION STATISTICS

One of the opportunities the electronic digital computer has created for us is the ability to compile huge amounts of experience with the behavior of statistics under controlled sampling conditions, from which we may induce the operating characteristics of the statistics. We frequently use a random number generator as the source of data for such computing, in which case we call the activity a Monte Carlo study. Lohnes and Cooley (1968, Chapter Five) describe the rationale and methods of such studies. One of the authors programmed a Monte Carlo study to compare the efficiency of three classification procedures under a specific set of circumstances, and also to demonstrate the impact of powerful a priori probabilities. The three methods compared are:

1. Tatsuoka's unequal dispersions, large samples procedure
2. Anderson's equal dispersions, large samples procedure
3. Geisser's equal dispersions, small samples procedure.

The program gives the user the opportunity to specify the number of uncorrelated elements in the vector variable, the number of populations, the sample sizes, a constant to separate each population from its nearest neighbors on each element of the variable, and a vector of prior probabilities. For each combination of these setup parameters the program computes 20 sampling experiments and reports the hits and misses table for each experiment and each of the three procedures. Classifications are made on independent replication samples in each experiment. Also reported is the average hit ratio for each of the three procedures over the 20 experiments. We do not list the program because it is long and specialized, but any reader desiring a listing may obtain it from P. R. Lohnes.

TABLE 10.1   First Monte Carlo Classification Study

```
3 VARIATES, 3 GROUPS, .30=CONSTANT SEPARATING GROUPS
STARTING RANDOM FRACTION=.03030000
ACTUAL GROUP SAMPLE SIZES
   20      30      50
PRIOR PROBABILITIES FOR GROUPS
  .15     .20     .65
INDEPENDENT REPLICATION SAMPLES ARE CLASSIFIED.
AVERAGE OF 20 HIT RATIOS FOR TATSUOKA METHOD=.5170
AVERAGE OF 20 HIT RATIOS FOR GEISSER METHOD =.5185
AVERAGE OF 20 HIT RATIOS FOR ANDERSON METHOD=.5215
```

Table 10.1 reports average hit ratios for 20 experiments in which each of three populations was separated from its nearest neighbor(s) by three-tenths of a standard deviation on each of three uncorrelated measurements. Prior probabilities were more extreme than actual sampling ratios but in the right profile. There is only a smidgin of difference in the moderately good performances of the three procedures. Table 10.2 reports a three group, three tests setup in which each group is separated from its nearest neighbor(s) by a whole standard deviation on each test, and the prior probabilities are exactly correct renderings of the sampling ratios. Again the improved performances of the three procedures are quite similar. Many different setups produced results that support the conclusion that the Anderson method is slightly superior in average hit rate, as it should be since all its assumptions are met (if we interpret our sample sizes as "large"), and it was derived to maximize the hit ratio. The Geisser procedure usually comes in a close second, perhaps because our sample sizes are in the gray area between large and small samples. The Tatsuoka procedure is not led far astray by its capitalizations on chance differences in norming samples variances in the conditions of this program.

TABLE 10.2   Second Monte Carlo Classification Study

```
3 VARIATES, 3 GROUPS, 1.00=CONSTANT SEPARATING
  GROUPS STARTING RANDOM FRACTION=.01010000
ACTUAL GROUP SAMPLE SIZES
  50      50      50
PRIOR PROBABILITIES FOR GROUPS
  .33     .33     .33
INDEPENDENT REPLICATION SAMPLES ARE CLASSIFIED.
AVERAGE OF 20 HIT RATIOS FOR TATSUOKA METHOD=.7180
AVERAGE OF 20 HIT RATIOS FOR GEISSER METHOD =.7317
AVERAGE OF 20 HIT RATIOS FOR ANDERSON METHOD=.7317
```

However, if the program allowed for degrees of correlation among the tests (which it does not), the Tatsuoka procedure might suffer more from capitalization on chance differences in covariances.

10.3 NUMERICAL EXAMPLE OF CLASSIFICATION

Perhaps the most useful numerical illustration of classification would be the computation for a single observation. Using the same data as in the discriminant analysis example of Section 9.2, the MANOVA run produced $\mathbf{D}_w^{-1}$ as well as the results displayed in Table 9.1.

$$\mathbf{D}_w^{-1} = \begin{bmatrix} .0031 & -.0052 \\ -.0052 & .0120 \end{bmatrix}$$

The second student (ID $= 2$) in our TALENT males has the following score vector for Information Part I and Part II:

$$\mathbf{X}_2 = [100 \quad 41]$$

His classification χ^2 for group 1 (school size code $= 2$) is computed by premultiplying and postmultiplying $\mathbf{D}_w^{-1}$ by the difference between his score vector and the group 1 centroid.

$$\chi_{12}^2 = [(100 - 148) \quad (41 - 75)] \cdot \begin{bmatrix} .0031 & -.0052 \\ -.0052 & .0120 \end{bmatrix} \cdot \begin{bmatrix} -48 \\ -34 \end{bmatrix}$$

$$\chi_{12}^2 = 4.066$$

Substituting the group 2 centroid, $\chi_{22}^2 = 5.573$, and the group 3 centroid, $\chi_{32}^2 = 6.374$. These chi squares are then used in Geisser's equation to obtain the relative densities of the three groups at the point $\mathbf{X}_2$. Using a priori values of .333 for each of the q_j, p_{12} is obtained as follows:

$$p_{12} = 0.333 \left[\frac{60}{61}\right]\left[1 + \frac{60(4.066)}{61(196-3)}\right]^{-1/2(196-3+1)}$$

$$p_{12} = .328\left[1 + \frac{243.96}{11773}\right]^{-97} = .328(1.021)^{-97}$$

$$\log p_{12} = \log .328 + (-97)\log 1.021$$

$$p_{12} = .0448$$

Repeating this step for groups 2 and 3, we obtain: $p_{22} = .0215$ and $p_{32} = .0148$. The p_{j2} are then summed over j and equation (10.1.1) yields probabilities for the three groups.

$$Pr(H_1 \mid \mathbf{X}_2, q_1) = \frac{.0448}{.0811} = .55$$

$$Pr(H_2 \mid \mathbf{X}_2, q_2) = \frac{.0215}{.0811} = .27$$

$$Pr(H_3 \mid \mathbf{X}_2, q_3) = \frac{.0148}{.0811} = .18$$

Thus if group membership were not known for student 2, he would be classified as a member of group 1, since more boys with his profile are members of group 1 than any other group. Repeating these computations for boys whose group membership is known, the hits and misses table can then be constructed. Table 10.3 reports this for the 30 TALENT boys not in the

TABLE 10.3   Classification Hits and Misses

| | | Predicted Membership | | | Row |
		1	2	3	Totals
	1	3	1	3	7
Actual	2	2	2	6	10
Membership	3	9	1	3	13
		14	4	12	30
		Percent hits = 27			

original MANOVA run, deleting those four boys whose school code was 1. There were only eight correct classifications, indicating that equations based upon the first 196 observations are not particularly useful in classifying the last 30 observations.

10.4 RESEARCH EXAMPLES

One multivariate approach to the study of career development is to examine the extent to which career plans following high school can be predicted from information available during high school. The following Project TALENT example illustrates this approach. The taxonomic criterion was based upon career plans and educational decisions following high school and the predictors were TALENT measurements on grade 9 boys. In the following two analyses, the six groups of Table 10.4, based upon follow-up plans, served as

TABLE 10.4   Six-Category Classification Scheme

Group	Description	Norming Sample Size	Replication Sample Size
PS	Physical science, mathematics (college)	654	653
MB	Medical, biological science (college)	410	410
HU	Humanities, teaching, social (college)	546	545
BC	Business, law, government (college)	797	796
TE	Technical worker (noncollege)	310	309
BNC	Business, sales, office (noncollege)	214	213
		2931	2926

the criterion measure. Twenty-three ability measures were used in the first analysis and 27 motive measures in the second. A sample of 2931 males served as the norming sample, and a different sample of 2926 were then classified as a replication sample.

The two best ability discriminant functions accounted for 93 percent of the discriminating variance available in the 23 ability measures. The information needed to examine the nature of these two functions is summarized in Table 10.5. The first function was essentially general scholastic ability, since it was highly correlated with all of the tests from that general area. The relative positions of the six groups centroids are also shown in Table 10.5. Here the

TABLE 10.5   Ability Discriminant Functions and Centroids

Variables Related to Function I	Structure Coefficients	Variables Related to Function II	Structure Coefficients
R-230 English Total	.83	R-270 Mechanical Reasoning	.63
R-250 Reading Comprehension	.82	R-282 Visualization in Three	
R-105 Social Studies Information	.81	Dimensions	.51
R-340 Mathematics Total	.81	R-112 Mechanical Information	.44
R-106 Mathematics Information	.76	R-107 Physical Science Information	.34
R-102 Vocabulary	.74	R-106 Mathematics Information	.32
R-103 Literature Information	.71	R-230 Mathematics Total	.30
R-107 Physical Science		R-115 Sports Information	−.27
Information	.79		
R-104 Music Information	.64	R-103 Literature Information	−.18

CENTROIDS

Function I

```
          TE   BNC                                    BC HU      MB  PS

    .     .    .    .       .    .    .    .    .    .    .    .    .    .

    3     5    7    9      11   13   15   17   19
```

Within-Group Standard Deviation = 7

Function II

```
              BC HU    MB BNC TE      PS

        .     .    .    .    .    .    .    .    .

        0     1    2    3    4
```

Within-Group Standard Deviation = 3.6

large separation between the four college groups and two noncollege groups is clearly evident.

The second function was primarily based upon information contained in the mechanical and mathematical areas. It is interesting to note that this function has significant negative correlations with Sports and Literature Information. The higher the student's score on this function, the greater the probability that he was planning a career in the science-technology area. Note that the medical-biological group was not particularly high on this function.

Turning to the 27 motive measures, we also find two major discriminant functions. In spite of the fact that they were based upon quite different types of information, these group separations were quite similar to those resulting from the ability information. In Table 10.6, the first function primarily separated the four college groups from the noncollege boys, while the second

TABLE 10.6 Motive Discriminant Functions and Centroids

Variables Related to Function I	Loading	Variables Related to Function II	Loading
P-701 Physical Science	.62	P-714 Mechanical-Technical	.56
P-702 Biological-Medical Science	.59	P-701 Physical Science	.49
R-610 Mature Personality	.49		
R-609 Self-Confidence	.33		
		P-705 Social Service	—.37
P-715 Skilled Trades	—.49	R-601 Sociability	—.35
P-717 Labor	—.47	P-703 Public Service	—.35
P-714 Mechanical-Technical	—.28	P-704 Literary Linguistic	—.32

CENTROIDS

Function I

TE BNC   BC HU PS   MB

```
  .    .    .    .    .    .    .    .    .    .    .    .    .
  1    2    3    4    5    6    7    8    9   10   11   12   13
```
Within Group Standard Deviation = 5

Function II

MB BNC
HU BC   TE   PS

```
   .    .    .    .    .    .    .    .    .    .
   0    1    2    3    4    5    6    7
```
Within Group Standard Deviation = 6

function separated the physical science and technical worker groups from the other four. Looking also at the variables which were primarily defining these two functions, we see they made considerable sense in terms of the types of separations produced.

Although examination of discriminant functions and centroids gives some indication of the nature of group differences, they do not tell how accurate the predictions are for individual students. As was pointed out above, only one-half of the boys were used in the actual discriminant analyses. The other half were set aside to be used as a cross-classification sample to see how well group membership could be predicted for a subsample not included in the determination of the prediction equations. Six probabilities (one for each of the six groups) were determined for each of the 2926 boys in this sample. Group

TABLE 10.7   Predictions Using Ability Measures

		Predicted Group Membership						
		1	2	3	4	5	6	
Actual	1. PS	355	24	33	201	33	7	653
follow-up	2. MB	149	27	40	169	24	1	410
plans	3. HU	132	32	66	268	35	12	545
	4. BC	142	28	67	486	55	18	796
	5. TE	35	1	13	84	164	12	309
	6. BNC	14	0	21	88	75	15	213
		827	112	240	1296	386	65	2926

38 percent hits

TABLE 10.8   Predictions Using Motive Measures

		Predicted Group Membership						
		1	2	3	4	5	6	
Actual	1. PS	387	50	38	137	37	4	653
follow-up	2. MB	110	137	34	116	11	2	410
plans	3. HU	120	63	156	169	28	8	545
	4. BC	146	61	107	417	61	4	796
	5. TE	61	6	25	74	133	10	309
	6. BNC	23	15	26	93	49	7	213
		847	332	386	1006	320	35	2926

42 percent hits

membership was predicted by assigning the boy to the group for which he had the highest probability. In other words, since probabilities tell us the relative group densities at and immediately around a boy's profile, our prediction was that a boy would join that group which most boys like him tended to join.

Tables 10.7 and 10.8 summarize the hits and misses for the ability and motive measures, respectively. Of the 653 boys who had follow-up plans in the physical science area, over one-half (355) were correctly assigned membership in that group by the ability test information. Not all of the groups were so successfully classified by the ability information. Their overall hit rate was 38 percent. The motive measures were a little more successful, classifying 42 percent of the boys correctly.

Predictive validities of assessment batteries are usually developed and applied by means of regression analysis, but the classification validities of assessments can be just as impressive evidence of the practical applications of measurement procedures. Lohnes and McIntire (1967) set out to explore the validities of state and local norms from a particular statewide testing program as those norms might be employed in local school guidance work, and to produce a comparison of the predictive validities of state and local norms. They argued that large-scale testing programs could and should compute and report classification probabilities as a standard procedural service for schools participating in such programs.

1. *Subjects.* The population sampled was 8500 tenth grade students in public schools in New Hampshire in the fall of 1962 who were tested in a University of New Hampshire-sponsored testing program in which all schools in the state participated. The sampling scheme consisted of separating out two random samples of approximately one-tenth of the total population, so that a "norming" and a "replication" sample would be available. All students in each of two high schools in different cities were the basis of "local norms" for those two schools.

2. *Tests.* Six tests from Educational Testing Service were included in the battery, two from the School and College Ability Tests (SCAT Verbal and SCAT Quantitative) and four from the Cooperative English Tests (COOP Reading Vocabulary, COOP Reading Level, COOP Reading Speed, COOP English Expression). Canonical correlation analysis revealed that a linear function of the SCAT tests correlated .9 with a linear function of the COOP tests, and principal components analysis indicated that one verbal ability factor accounted for approximately 75 percent of the total test space variance. There is a high degree of redundancy in the test battery.

3. *Prediction of Two Curriculum Groups Statewide.* Table 10.9 reports means and standard deviations for college preparatory curriculum students and other curricula students in the norming sample. The college prep group was decidedly superior on all the tests. Computed probabilities led to the proportion of correct and incorrect classifications in the norming sample and in the replication sample reported in Table 10.10. It appears that the New Hampshire guidance counselor who predicts college prep versus other curricula placements from these state norms can expect to be right about three times out of four, if the conditions in his high school resemble those in the state as

TABLE 10.9   Means and S.D's for Two Curriculum Groups, 0-Sample

Variable	College Prep. (N = 404)		Noncollege (N = 424)	
	Mean	S.D.	Mean	S.D.
SCAT Verbal	289.2	11.4	274.6	12.1
SCAT Quantitative	302.3	20.0	288.3	19.7
Co-op Reading Vocabulary	156.4	7.2	147.6	8.2
Co-op Reading Level	154.7	7.5	146.0	9.0
Co-op Reading Speed	155.5	8.8	146.0	8.3
Co-op Reading Expression	155.1	8.8	145.5	8.7

TABLE 10.10   Hits and Misses for Two Curriculum Groups 0-Sample Classifications [own norms]

Predicted Groups	Actual Group		
	College	Noncollege	Totals
College	304	112	416
Noncollege	100	312	412
Total	404	424	N = 828

616 hits, or 74%
212 misses, or 26%

1-Sample Predictions [0-Sample Norms]

Predicted Groups	Actual Group		
	College	Noncollege	Totals
College	305	142	447
Noncollege	110	343	453
Total	415	485	N = 900

648 hits, or 72%
252 misses, or 28%

TABLE 10.11   Means and S.D.s for Two Schools, Own Norms

| | School 1 | | | | | | | School 2 | | | | | | |
| | College (N = 239) | | Noncollege (N = 204) | | | College (N = 206) | | Noncollege (N = 156) | |
Variable	Mean	S.D.	Mean	S.D.	Mean	S.D.	Mean	S.D.
SCAT Verbal	293	11.6	276	10.8	283	8.4	274	10.3
SCAT Quantitative	305	23.8	293	12.6	295	10.1	289	12.4
Co-op Reading Vocabulary	159	6.9	149	7.4	154	5.4	149	6.5
Co-op Reading Level	156	7.5	146	7.8	152	6.7	145	8.9
Co-op Reading Speed	159	8.0	149	8.0	152	7.3	146	7.9
Co-op Reading Expression	158	8.4	146	7.9	153	6.6	145	7.8

TABLE 10.12 Hits and Misses for Two City Schools from Own Norms (outside brackets) and from State Norms (inside brackets)

| Predicted Group | School 1 Actual Group | | |
	College	Noncollege	Total
College	182	36	218
	[208]	[67]	[275]
Noncollege	57	168	225
	[31]	[137]	[168]
Total	239	204	$N = 443$

Own Norms: 350 hits, or 79%
93 misses, or 21%
State Norms: 345 hits, or 78%
98 misses, or 22%

| Predicted Group | School 2 Actual Group | | |
	College	Noncollege	Total
College	177	67	244
	[141]	[53]	[194]
Noncollege	29	89	118
	[65]	[103]	[168]
Total	206	156	$N = 362$

Own Norms: 266 hits, or 73%
96 misses, or 27%
State Norms: 244 hits, or 67%
118 misses, or 33%

a whole. Note that in this analysis the norming sample was large enough to permit the apparent hit rate when norming sample members were classified to closely approximate the actual hit rate when replication sample members were classified. There is very little shrinkage in this case.

4. *Prediction from Local Versus State Norms.* All test records from two schools in different cities were collected and local centroids and dispersions for curriculum groups were computed, as reported in Table 10.11. The School 1 college prep students are above the state college prep norms, while the School 2 college prep students are below the state norms. Table 10.12 reports the comparison of hits and misses when probabilities were based on local norms and state norms for the two schools. For School 1 the overall efficiency is about the same for both bases, but state norms produce more hits in the college prep group and more misses in the other curricula group than do local norms. For School 2 the overall efficiency is about five percent better for local norms than for state norms, and state norms rather drastically under-assign to the college prep group. These comparisons were interpreted by the authors as suggesting that in localities where separation of the two major curriculum groups by ability measures is more rigorous than in the state as a whole, state norms may be a very adequate basis for local guidance predictions, while they may be less adequate than local norms in localities where the separation by ability is less rigorous than it is for the state as a whole, assuming that good performance in correct assigning to the college prep group is particularly valued, even at the cost of overassigning this group.

CLASIF

```
C       CLASSIFICATION PROBABILITIES.   A COOLEY-LOHNES PROGRAM.
C
C       THIS PROGRAM COMPUTES CLASSIFICATION PROBABILITIES FOR SUBJECTS,
C    ASSIGNS EACH SUBJECT TO A CRITERION GROUP ON THE BASIS OF HIS
C    HIGHEST PROBABILITY, AND TALLIES THE HITS AND MISSES (IF THE
C    ACTUAL GROUP MEMBERSHIPS OF THE SUBJECTS ARE KNOWN).   THE METHOD IS
C    THAT GIVEN BY S. GEISSER (1964), FOR SEPARATE GROUP CENTROIDS
C    ESTIMATES AND AN ESTIMATED COMMON GROUP DISPERSION MATRIX.
C
C       INPUT
C    1) A TEN CARD TEXT DESCRIBING THE PROBLEM FOR THE PRINTOUT.
C    2) CONTROL CARD (CARD 11),   COLS 1-2   M = NUMBER OF VARIABLES,
C                                 COLS 3-4   NG = NUMBER OF GROUPS,
C                                 COLS 5-9   N = NUMBER OF SUBJECTS,
C                                 COL 10   OUTPUT = 0   TO SUPPRESS PRINT-
C                                 OUT AND PUNCHOUT OF PROBABILITIES FOR
C                                 SUBJECTS, AND TALLY HITS AND MISSES, OR
C                                 OUTPUT = 1   TO PRINT INDIVIDUAL PROBS
C                                 AND SUPPRESS TALLY. (THE ASSUMPTION IS
C                                 THAT WHEN PROBS ARE WANTED ACTUAL
C                                 CRITERION MEMBERSHIP IS NOT AVAILABLE,
C                                 AND VICE VERSA.)
C    3) FORMAT CARD (CARD 12) FOR SCORE VECTORS. NOTE THAT IF OUTPUT = 0,
C       THE PROGRAM ASSUMES THAT ID (READ BEFORE SCORES) IS ACTUAL
C       GROUP MEMBERSHIP CODE, BUT IF OUTPUT = 1, THE PROGRAM ASSUMES
C       THAT ID IS SUBJECT ID NUMBER TO BE ASSOCIATED WITH HIS PROBS.
C    4) GROUP CENTROIDS, FORMAT (10X, 5E14.7).
C    5) INVERSE OF COMMON GROUP DISPERSION, FORMAT (10X, 5E14.7).
C       CENTROIDS AND DISPERSION-INVERSE ARE PUNCHED OUT BY MANOVA.
C    6) PRIOR PROBABILITIES VECTOR (20F4.3).   NOTE THAT IF YOU WANT TO
C       USE DIFFERENT PRIORS FOR DIFFERENT GROUPS OF SUBJECTS, SIMPLY
C       RUN EACH SUCH GROUP AS A SEPARATE JOB. TALLIES CAN BE COMBINED
C       AFTER A SET OF SUCH JOBS.
C    7) NORMING GROUP SAMPLE SIZES, FORMAT (13F6.0).   IF OUTPUT = 0,
C       THIS VECTOR MUST SUM TO N.
C    8) SCORE VECTORS. EACH SUBJECT HAS HIS OWN SET OF CARDS, WITH HIS ID
C       IN THE FIRST FIELD READ.
C
C
       DIMENSION  TIT(16),  G(50,100),  D(100,100),  X(100),  PR(50),
      C EN(50), NC(50,50), Y(100), Z(100), CSQ(50), P(50), R(50),
      C NA(50),  NP(50)
       DIMENSION NCX(50)
C
1      WRITE (6,2)
2      FORMAT(55H1CLASSIFICATION PROBABILITIES.  A COOLEY-LOHNES PROGRAM)
       DO 3   J = 1, 10
       READ  (5,4) (TIT(K),   K = 1, 16)
3      WRITE (6,4) (TIT(K),   K = 1, 16)
4      FORMAT (16A5)
       READ  (5,5) M, NG,  N,  OUTPUT
5      FORMAT(2I2, I5, F1.0)
       READ  (5,4) (TIT(J),   J = 1, 16)
       DO 7   J = 1, NG
7      READ  (5,8) (G(J,K),   K = 1, M)
8      FORMAT (10X, 5E14.7)
       DO 10   J = 1, M
10     READ  (5,8) (D(J,K),   K = J, M)
       DO 9   J = 1, M
       DO 9   K = J, M
9      D(K,J) = D(J,K)
       READ  (5,11) (PR(J),   J = 1, NG)
11     FORMAT (20F4.3)
       READ  (5,20) (EN(J),   J = 1, NG)
20     FORMAT (13F6.0)
       WRITE (6,12) N,  M,  NG
```

```
   12    FORMAT(2H0 I6,10H SUBJECTS,I3,7H TESTS,I3,8H GROUPS. )
         WRITE (6,50)
   50    FORMAT (1H0, 25(5H-----))
C
         DO 13   J = 1, NG
         NA(J) = 0
         NP(J) = 0
         DO 13   K = 1, NG
   13    NC(J,K) = 0
         EM = M
         DF = 0.0
         DO 14   L = 1, NG
         DF = DF +  (EN(L) - 1.0)
   14    R(L) = PR(L)  * ((EN(L)/(EN(L)+1.))**(.5*EM))
C
         DO 26   NS = 1, N
         READ   (5,TIT) ID,   (X(J),   J = 1, M)
         DO 17   L = 1, NG
         DO 15   J = 1, M
   15    Y(J) = X(J) - G(L,J)
         DO 16   J = 1, M
         Z(J) = 0.0
         DO 16   K = 1, M
   16    Z(J) = Z(J) + Y(K) * D(K,J)
         CSQ(L) = 0.0
         DO 17   J = 1, M
   17    CSQ(L) = CSQ(L) + Z(J) * Y(J)
         P2 = 0.0
         DO 18   L = 1, NG
         P(L) = R(L) * ((1.0 + ((EN(L) * CSQ(L)) / ((EN(L) + 1.0) * DF)))
       C    **(-0.5 * (DF + 1.0)))
   18    P2 = P2 + P(L)
         DO 19   L = 1, NG
   19    P(L) = P(L) / P2
C   P(L) NOW CONTAINS GEISSER CLASSIFICATION PROBABILITIES FOR SUBJECT.
C
         IF (OUTPUT)  23, 23,  21
   21    WRITE (6,22) ID,  (P(L),   L = 1, NG)
   22    FORMAT (3H0ID I7,14F5.2 / (10X, 14F5.2))
         WRITE (7,22) ID,  (P(L),   L = 1, NG)
         GO TO 26
C
   23    K = 1
         DO 24   L = 2, NG
         IF (P(L) - P(K))   24,  25, 25
   25    K = L
   24    CONTINUE
         NC(ID,K) = NC(ID,K) + 1
   26    CONTINUE
C
         IF (OUTPUT)   30, 30,  1
   30    DO 27   J = 1, NG
         DO 27   K = 1, NG
         NA(J) = NA(J) + NC(J,K)
   27    NP(J) = NP(J) + NC(K,J)
         NT = 0
         NHITS = 0
         DO 29   J = 1, NG
         NT = NT + NA(J)
   29    NHITS = NHITS + NC(J,J)
         MISSES = NT - NHITS
         ENT = NT
         ENHIT = NHITS
         EMISS = MISSES
         NPHIT = ENHIT / ENT * 100.0
         NPMISS = EMISS / ENT * 100.0
```

```
      WRITE (6,31) NT
31    FORMAT(4H0FOR I7,22H SUBJECTS CLASSIFIED,  )
      WRITE (6,32) NHITS,   NPHIT
32    FORMAT(16H0     N(HITS) =  I6,22H,       PERCENT(HITS) =   I3)
      WRITE (6,33) MISSES,   NPMISS
33    FORMAT(18H0     N(MISSES) =  I6,24H,       PERCENT(MISSES) =   I3)
      WRITE (6,50)
      WRITE (6,34)
34    FORMAT(55H0CLASSIFICATION TABLE, ROWS ARE ACTUAL GROUPS, AND      )
      WRITE (6,28)
28    FORMAT(55H0    COLUMNS ARE PREDICTED GROUPS.                      )
      DO 35   J = 1, NG
35    WRITE (6,36) J,  (NC(J,K),   K = 1, NG)
36    FORMAT(4H0ROW I3,3X, 20I5 / (10X, 20I5))
      WRITE (6,50)
      WRITE (6,37)
37    FORMAT(55H0ROW SUMS FOR TABLE (ACTUAL GROUP SIZES).              )
      WRITE (6,38) (NA(J),   J = 1, NG)
38    FORMAT (10X, 20I5)
      WRITE (6,39)
39    FORMAT(55H0COLUMN SUMS FOR TABLE (PREDICTED GROUP SIZES).        )
      WRITE (6,38) (NP(J),   J = 1, NG)
      WRITE (6,50)
      DO 40   J = 1, NG
      DO 40   K = 1, NG
      ENJ = NA(J)
      ENC = NC(J,K)
40    NC(J,K) = ENC / ENJ * 100.0
      WRITE (6,41)
41    FORMAT(55H0PERCENTAGE OF EACH ROW PREDICTED INTO EACH COLUMN.    )
      DO 42   J = 1, NG
42    WRITE (6,36) J,  (NC(J,K),   K = 1, NG)
      DO 43   J = 1, NG
      ENJ = NA(J)
      ENP = NP(J)
      NCX(J) = PR(J) * 100.0
      NA(J) = ENJ / ENT * 100.0
43    NP(J) = ENP / ENT * 100.0
      WRITE (6,44)
44    FORMAT(55H0ACTUAL GROUP SIZES EXPRESSED AS PERCENTS OF TOTAL     )
      WRITE (6,47)
47    FORMAT(55H0      SAMPLE.                                         )
      WRITE (6,38) (NA(J),   J = 1, NG)
      WRITE (6,50)
      WRITE (6,45)
45    FORMAT(55H0PRIOR PROBABILITIES EMPLOYED IN THIS STUDY.           )
      WRITE (6,38) (NCX(J),   J = 1, NG)
      WRITE (6,50)
      WRITE(6,48) (EN(J),   J = 1, NG)
48    FORMAT(27H0NORMING GROUP SAMPLE SIZES/(10X,20F5.0))
      WRITE(6,50)
      WRITE (6,46)
46    FORMAT(55H0PREDICTED GROUP SIZES EXPRESSED AS PERCENTS OF TOTAL  )
      WRITE (6,47)
      WRITE (6,38) (NP(J),   J = 1, NG)
      WRITE (6,50)
      GO TO 1
      END
```

10.6 EXERCISES

1. How would you change the CLASIF program so that classification chi squares would also be printed for each individual?

2. How well can the TALENT ability, interest and SES information predict the type of college a student plans to attend (variable number 7)? What advantage does CLASIF have over stopping with MANOVA in exploring this question?

‖‖

CHAPTER ELEVEN

Multiple Covariance Analysis

11.1 MATHEMATICS OF MULTIPLE COVARIANCE

Multiple covariance analysis is a model which makes it possible to explore the surplus influences of additional measurements on a taxonomy (or vice versa) when the known influences of a set of related measurements are partialled out. The logic is to regress each of the additional variates (which we will call the *predictors*, p_2 in number) one at a time on the space of the variates that have a known relation to the criterion groups. These already studied variates we will call the *controls*, p_1 in number. The multivariate analysis of variance is then construed in terms of the predictor residuals from multiple regression on the controls. The researcher is studying either the power of the predictor residuals for discriminating among the criterion groups, or the effects on the predictor residuals of the two or more treatments. Although multiple covariance analysis can be generalized to factorial manova designs, we will treat only the simple, one-way-of-grouping manova situation. As we will show in Chapter Twelve, our Fisherian partitioned sums of squares and cross-products notation for manova models is not well suited to complex designs, and we have not considered the alternative general linear regression notation to be well suited for an introductory text.

It is helpful to conceive of multiple covariance as manova on multiple regression residuals, but one would not choose to perform the analysis by first computing a score matrix of residuals and then taking it into MANOVA.

In Chapter Seven we saw how a multiple partial regression matrix could be computed by the equation

$$\tilde{\mathbf{R}}_{22} = \mathbf{R}_{22} - \mathbf{R}_{21}\mathbf{R}_{11}^{-1}\mathbf{R}_{12}$$

or, more abstractly,

$$\tilde{\mathbf{R}}_{22} = \mathbf{R}_{22} - \hat{\mathbf{R}}_{22}$$

Notice that since $\tilde{\mathbf{R}}_{22}$ is a dispersion matrix for residuals it could be denoted $\mathbf{D}_{2.1}$. The same algebra is used to make multiple covariance adjustments to the manova model, but the algebra must be applied twice, once to the total matrix, or total sample sums of squares and cross-products matrix $\mathbf{T}$, and once to the within matrix, or pooled within-groups sums of squares and cross-products matrix $\mathbf{W}$. We compute a super-$\mathbf{T}$ and super-$\mathbf{W}$ matrix by using MANOVA on the $p = p_1 + p_2$ vector variables grouped in separate samples. Then we transfer the super-$\mathbf{T}$ and super-$\mathbf{W}$, each of which is square and symmetric and of order p, to the COVAR program, where they are partitioned as follows:

$$\mathbf{T} = \begin{bmatrix} \mathbf{T}_{11} & \mathbf{T}_{12} \\ \hline \mathbf{T}_{21} & \mathbf{T}_{22} \end{bmatrix}$$

$$\mathbf{W} = \begin{bmatrix} \mathbf{W}_{11} & \mathbf{W}_{12} \\ \hline \mathbf{W}_{21} & \mathbf{W}_{22} \end{bmatrix}$$

Then COVAR computes the total residual matrix

$$\mathbf{T}_{2.1} = \mathbf{T}_{22} - \mathbf{T}_{21}\mathbf{T}_{11}^{-1}\mathbf{T}_{12}$$

and the within residual matrix

$$\mathbf{W}_{2.1} = \mathbf{W}_{22} - \mathbf{W}_{21}\mathbf{W}_{11}^{-1}\mathbf{W}_{12}$$

The Wilks' Lambda criterion is then

$$\Lambda = \frac{|\mathbf{W}_{2.1}|}{|\mathbf{T}_{2.1}|}$$

Rao (1952) shows that his F-ratio approximation to Λ as we gave it in Chapter Eight applies, with the number of degrees of freedom for denominator adjusted by redefining f_2 so that

$$f_2 = s\left[(N - p_1 - 1) - \frac{p_2 + g}{2}\right] - \frac{p_2(g - 1)}{2}$$

Similarly, the univariate F-ratios after covariance adjustments are obtained by using the diagonal elements of $\mathbf{W}_{2 \cdot 1}$ and $\mathbf{T}_{2 \cdot 1}$, but the denominator number of degrees of freedom is adjusted to

$$f_2 = N - g - p_1$$

while f_1 remains $g - 1$. Then the jth univariate F-ratio is

$$F_j = \frac{(t_{2 \cdot 1_{jj}} - w_{2 \cdot 1_{jj}})/f_1}{(w_{2 \cdot 1_{jj}}/f_2)}$$

This test asks the question: Do the additional (predictor) variates add new information about the criterion groups that is not found in the original (control) variates? An analogous test for a continuous criterion is available by utilizing the variance-ratio test for the difference between two multiple correlation coefficients, when one R is based on the p_1 control variates alone and the other R is based on the p_1 control variates and the p_2 predictor variates together. A significant increase in the latter R indicates new information in the p_2 predictor variates for the particular continuous criterion variable. This test for the difference between two multiple R's can be found in McNemar (1962, page 284). Notice that we are not describing covariance as a model for equating unequal groups in an experimental design. Covariance is sometimes misunderstood on this issue, but in fact there is no way to adjust outcomes of an experiment for systematic input differences among the treatment groups. Two excellent discussions of this misunderstanding are provided by Bereiter (1964) and Lord (1967).

Finally, to compute the vector of covariance-adjusted predictor means for group k, namely $\tilde{\mathbf{m}}_{2_k}$, we first extract a matrix of multiple regression coefficients, $\mathbf{B}$, from the computation of $\mathbf{W}_{2 \cdot 1}$:

$$\mathbf{B} = \mathbf{W}_{11}^{-1}\mathbf{W}_{12}$$

Then we apply these coefficients to the deviations of control means for the kth group, namely $\mathbf{m}_{1_k}$, from the grand means of the controls, namely $\mathbf{m}_1$, and subtract the predictable parts of $\mathbf{m}_{2_k}$ from itself:

$$\tilde{\mathbf{m}}_{2_k} = \mathbf{m}_{2_k} - \mathbf{B}'(\mathbf{m}_{1_k} - \mathbf{m}_1)$$

and of course we repeat this operation for all groups, $k = 1, 2, \ldots, g$. The pattern of group differences on the predictor variates after covariance adjustment can be quite different from the pattern before covariance. Particularly interesting is the case where the group differences on a covariance-adjusted variate are much greater than they were before covariance. In such a case covariance is suppressing more of the variance in the predictor that is irrelevant to the criterion than it is suppressing variance relevant to the criterion, so we have an example of the suppression phenomenon in multiple regression.

11.2 NUMERICAL EXAMPLE OF COVAR

Let the 37 twelfth-grade TALENT males planning careers in engineering and the 16 planning business careers represent samples from two populations. Using (1) abstract reasoning and (2) mathematics ability as control variates, one might be interested in knowing whether these two populations differ on (3) sociability and (4) physical science interest, after removing variance associated with the two ability measures. Another way of putting the question: "Is there any information in motive measures (3) and (4) that is not in ability measures (1) and (2) regarding the differences between these two populations, boys planning careers in engineering and those planning careers in business?"

The first step is to run MANOVA on the two groups using the entire four element vector variable. Those results are summarized in Table 11.1. There we see that H_1 could not be rejected while H_2 was significant at the .005 level. Inspecting the means reveals that the main reason for this rejection is the physical science interest difference, the "engineer" mean being more than one standard deviation above the "businessmen's."

The COVAR program then reads in $\mathbf{T}$ and $\mathbf{W}$ from MANOVA and partitions both matrices based upon the number of control variables. First the total residual matrix $\mathbf{T}_{2 \cdot 1}$ is formed:

$$
\mathbf{T}_{2 \cdot 1} = \begin{bmatrix} 423 & 116 \\ 116 & 3947 \end{bmatrix} - \begin{bmatrix} -23.5 & -53.9 \\ 92.1 & 2471 \end{bmatrix} \cdot \begin{bmatrix} 359 & 863 \\ 863 & 5870 \end{bmatrix}^{-1}
$$

$$
\cdot \begin{bmatrix} -23.5 & 92.1 \\ -53.9 & 2471 \end{bmatrix}
$$

$$
= \begin{bmatrix} 423 & 116 \\ 116 & 3947 \end{bmatrix} - \begin{bmatrix} 2 & -4 \\ -4 & 1357 \end{bmatrix} = \begin{bmatrix} 421 & 120 \\ 120 & 2590 \end{bmatrix}
$$

Then the residuals within groups:

$$\mathbf{W}_{2 \cdot 1} = \begin{bmatrix} 420 & 169 \\ 169 & 2859 \end{bmatrix} - \begin{bmatrix} -24.4 & -23.2 \\ 11.1 & 1847 \end{bmatrix} \cdot \begin{bmatrix} 359 & 874 \\ 874 & 5512 \end{bmatrix}^{-1}$$

$$\cdot \begin{bmatrix} -24.4 & 111 \\ -23.2 & 1847 \end{bmatrix}$$

$$= \begin{bmatrix} 420 & 169 \\ 169 & 2859 \end{bmatrix} - \begin{bmatrix} 2 & 9 \\ 9 & 769 \end{bmatrix} = \begin{bmatrix} 418 & 160 \\ 160 & 2090 \end{bmatrix}$$

TABLE 11.1   MANOVA Results Prior to COVAR

		Abstract Reasoning (1)	Math. (2)	Soc. (3)	Sci. Int. (4)
Engineering					
	Means	9.51	30.97	7.70	27.81
	Std. Dev.	2.68	10.94	2.79	7.16
Business					
	Means	9.69	25.31	8.19	17.94
	Std. Dev.	2.57	8.97	3.06	8.23
Univariate F-ratios where $f_1 = 1$ and $f_2 = 51$		.05	3.31	.32	19.43

Test of H_1 (equality of dispersions):

$$M = 12.01 \qquad F = 1.07$$
$$\text{where } f_1 - 10 \text{ and } f_2 = 4015$$

Test of H_2 (equality of centroids):

$$\Lambda = \frac{|\mathbf{W}|}{|\mathbf{T}|} = .70$$

$$F = 5.11 \text{ where } f_1 = 4 \text{ and } f_2 = 48$$

$$p < .005$$

$$\mathbf{T} = \begin{bmatrix} 359. & 863. & -23.5 & 92.1 \\ 863. & 5870. & -53.9 & 2471. \\ -23.5 & -53.9 & 423. & 116. \\ 92.1 & 2471. & 116. & 3947. \end{bmatrix}$$

$$\mathbf{W} = \begin{bmatrix} 359. & 874. & -24.4 & 111. \\ 874. & 5512. & -23.2 & 1847. \\ -24.4 & -23.2 & 420. & 169. \\ 11.1 & 1847. & 169. & 2859. \end{bmatrix}$$

The determinants of $W_{2 \cdot 1}$ and $T_{2 \cdot 1}$ are then used in computing the Wilks' lambda criterion:

$$\Lambda = \frac{|W_{2 \cdot 1}|}{|T_{2 \cdot 1}|} = \frac{.848 \times 10^6}{.108 \times 10^7} = .79$$

Converting this to an F ratio we obtain:

$$F_{48}{}^2 = 6.46$$

Since this F ratio is significant at the .01 level, we conclude that there is information in the two motives measures that is not in the abilities measures regarding the differences between these two groups.

Interpretation of the nature of these group differences is then based upon the adjusted means for the dependent variables. The within-group regression coefficients (**B**) were the result when we had previously multiplied $W_{11}{}^{-1}W_{12}$.

$$B' = \begin{bmatrix} -.09 & .01 \\ -.83 & .47 \end{bmatrix}$$

For group one, the adjusted means are

$$\tilde{m}_{2_1} = \begin{bmatrix} 7.7 \\ 27.8 \end{bmatrix} - \begin{bmatrix} -.09 & .01 \\ -.83 & .47 \end{bmatrix} \begin{bmatrix} (9.51 - 9.57) \\ (30.97 - 29.26) \end{bmatrix}$$

$$\tilde{m}_{2_1} = \begin{bmatrix} 7.7 \\ 27.0 \end{bmatrix}$$

For the business planners, group 2, the adjusted means are:

$$\tilde{m}_{2_1} = \begin{bmatrix} 8.2 \\ 17.9 \end{bmatrix} - \begin{bmatrix} -.09 & .01 \\ -.83 & .47 \end{bmatrix} \begin{bmatrix} (9.69 - 9.57) \\ (25.31 - 29.26) \end{bmatrix}$$

$$= \begin{bmatrix} 8.2 \\ 19.9 \end{bmatrix}$$

11.3 RESEARCH EXAMPLE

An Experimental Study Comparing the Effectiveness of Teaching Deduction in Two Content Areas of Secondary Mathematics

As pointed out earlier, one situation in which the generalized analysis of variance and covariance applies is in experiments involving several experimental and control variables. An experiment by Balomenos (1961) serves to illustrate this application. His general hypothesis was that students who are taught an understanding and appreciation of deduction and deductive systems

TABLE 11.2   Means and Standard Deviations for the Two Groups

	Geometry Content		Algebra Content	
Variable	Mean	S.D.	Mean	S.D.
Experimental				
1 Shaycoft plane geometry, form Bm	52.70	7.73	46.19	9.40
2 Blyth algebra, form Bm	53.73	8.83	56.03	9.77
3 STEP Mathematics, form 2B	54.88	8.10	56.16	8.89
4 Watson-Glaser deduction	17.19	2.66	17.95	2.99
5 Watson-Glaser critical thinking	64.30	8.62	66.57	10.35
Control				
1 CTMM language aptitude	204.30	19.65	207.84	21.34
2 CTMM nonlanguage aptitude	195.63	23.42	194.66	24.11
3 Shaycoft geometry, form Am	40.57	9.90	40.42	10.41
4 Blyth algebra, form Am	50.10	10.00	52.46	11.25
5 STEP Mathematics, form 2A	50.78	11.55	52.04	10.09
6 Watson-Glaser deduction	16.25	2.31	16.58	2.82
7 Watson-Glaser critical thinking	60.97	7.84	62.15	9.50
$\Lambda = .727$, $F^5_{121} = 9.06$, $p = .001$[a]			$N_1 = 67$	$N_1 = 67$

[a] Significance of the difference between the two treatments on the five experimental variables after adjustments for the seven control variables.

TABLE 11.3   Differences on Each of the Experimental Variables

Variable	Λ	F^1_{125}	p
1 Shaycoft plane, geometry form Bm	.7706	36.91	$p < .001$
2 Blyth algebra, form Bm	.9958	.523	—
3 STEP Mathematics, form 2B	.9971	.361	—
4 Watson-Glaser deduction	.9877	1.544	—
5 Watson-Glaser critical thinking	.9936	.799	—

by two different approaches will not differ significantly on five criterion measures after adjustment has been made for differences in aptitude and initial understanding (pretests of the criterion instruments). Seven control variables were used in all. The resulting lambda criterion was highly significant indicating that the two treatments produced over-all differences on the five experimental variables.

The investigation next tested the difference between the two groups on each experimental variable (using all seven controls).[1] Finally, the adjusted means for the two groups on the only significant experimental variable were computed. The coefficients for this adjustment are computed within the covariance program. It involves multiplying $\mathbf{W}_{cc}^{-1}$ by $\mathbf{W}_{cp}$. As can be seen in Table 11.4, the control pretest of the corresponding experimental variate tended to have the highest loading among the adjustment coefficients.

TABLE 11.4   Adjustment Coefficients for the Seven Control Variables

Control Variable	Experimental Variable				
	1	2	3	4	5
1 CTMM language aptitude	.018	−.016	−.017	.011	.021
2 CTMM nonlanguage aptitude	.001	−.012	.045	−.020	−.016
3 Shaycoft geometry, form Am	.412	.067	.016	.036	−.017
4 Blyth algebra, form Am	.149	.517	.020	−.029	.008
5 STEP Mathematics, form 2A	.046	.158	.496	.059	.190
6 Watson-Glaser deduction	.210	−.069	.221	.426	.572
7 Watson-Glaser critical thinking	0.12	.114	.012	.038	.527

Adjusted means for the significant	Content Group	
experimental variables	Geometry	Algebra
Geometry Test	52.95	45.94

---

[1] The program COVAR can be used for one-way analysis of covariance when there is only one experimental or one control variable, as well as for the multivariate case.

11.4 THE COVAR PROGRAM

COVAR

```
C       MANOVA WITH COVARIATES.  A COOLEY-LOHNES PROGRAM.
C
C       THIS PROGRAM COMPUTES THE RESIDUAL W AND T MATRICES FOR THE M2
C       VARIABLES IN THE RIGHT SET AFTER COVARIANCE ADJUSTMENTS FOR THEIR
C       MULTIPLE REGRESSIONS ON THE M1 VARIABLES OF THE LEFT SET. THUS,
C       THE FIRST M1 VARIABLES ARE THE CONTROL VARIABLES, OR COVARIATES,
C       AND THE SECOND M2 VARIABLES ARE THE DEPENDENT VARIABLES. THE PROGRAM
C       READS IN THE W AND T MATRICES OF ORDER  M = M1 + M2  AS PUNCHED BY
C       MANOVA, AND COMPUTES
C           W2.1  =  W22  -  W21 * W11-INVERSE * W12
C       AND
C           T2.1  =  T22  -  T21 * T11-INVERSE * T12        .
C       THESE RESIDUAL MATRICES ARE PUNCHED FOR POSSIBLE INPUT TO DISCRIM-
C       INANT ANALYSIS (IN WHICH CASE THE DISCRIMINANT FUNCTIONS ARE
C       DEFINED UPON RESIDUAL VARIABLES). THE COVARIANCE-ADJUSTED WILKS
C       LAMBDA TEST OF H2 AND THE MULTIVARIATE ANALYSIS OF COVARIANCE
C       F-RATIOS ARE COMPUTED. THE GROUP MEANS AND GRAND MEANS AS
C       PUNCHED BY MANOVA ARE READ IN, AND THE COVARIANCE-ADJUSTED
C       GROUP AND GRAND MEANS ARE COMPUTED AND PUNCHED FOR POSSIBLE ENTRY
C       TO DISCRIMINANT ANALYSIS.
C
C       SUBROUTINES MPRINT AND MATINV ARE REQUIRED.
C
C       INPUT
C
C    1) FIRST TEN CARDS OF THE DATA DECK DESCRIBE THE PROBLEM IN A TEXT
C           WHICH WILL BE REPRODUCED ON THE OUTPUT. DO NOT USE COLUMN 1.
C    2) CONTROL CARD (CARD 11)
C           COLS 1-2   M1 = NUMBER OF CONTROL VARIABLES (LEFT SET)
C           COLS 3-4   M2 = NUMBER OF DEPENDENT VARIABLES (RIGHT SET)
C           COLS 5-7   KG = NUMBER OF GROUPS
C           COLS 8-12  N = NUMBER OF SUBJECTS.
C    3) T MATRIX (AS PUNCHED BY MANOVA)
C    4) W MATRIX (AS PUNCHED BY MANOVA)
C    5) GROUP MEANS (AS PUNCHED BY MANOVA)
C    6) GRAND MEANS (AS PUNCHED BY MANOVA)
C
C       PUNCHED OUTPUT
C
C    1) T2.1  (RESIDUAL T MATRIX)
C    2) W2.1  (RESIDUAL W MATRIX)
C    3)  RESIDUAL GROUP MEANS
C
        DIMENSION   TIT(16), A(50,50), B(50,50), C(50,50), D(50,50),
       2    T(50), U(50)
C
   1    WRITE (6,2)
   2    FORMAT(52H1MANOVA WITH COVARIATES.  A COOLEY-LOHNES PROGRAM.     )
        DO 3   J = 1, 10
        READ  (5,4) (TIT(K),   K = 1, 16)
   3    WRITE (6,4) (TIT(K),   K = 1, 16)
   4    FORMAT (16A5)
        READ  (5,5) M1,  M2,  KG,  N
   5    FORMAT (2I2, I3, I5)
        M = M1 + M2
        EM = M
        EM1 = M1
        EM2 = M2
        KGT = KG + 1
        EKG = KG
        EN = N
        WRITE (6,9) M1
   9    FORMAT (30HONO. VARIABLES ON LEFT = M1 =   I3)
        WRITE (6,10) M2
  10    FORMAT (31HONO. VARIABLES ON RIGHT = M2 =   I3)
        WRITE (6,11) N
```

```
11      FORMAT (20H0NO. SUBJECTS = N =    I6)
        WRITE (6,39)
39      FORMAT(52H0---------------------------------------------------.
        IT = 1
30      DO 6    J = 1, M
6       READ  (5,7) (A(J,K),    K = J, M)
7       FORMAT (10X, 5E14.7)
        DO 8    J = 1, M
        DO 8    K = J, M
8       A(K,J) = A(J,K)
C
        CALL MATINV (A, M1, DET)
        DO 12    J = 1, M2
        J1 = J + M1
        DO 12    K = 1, M1
        D(J,K) = 0.0
        DO 12    L = 1, M1
12      D(J,K) = D(J,K) + A(J1,L) * A(L,K)
        DO 13    J = 1, M2
        DO 13    K = 1, M2
        K1 = K + M1
        C(J,K) = 0.0
        DO 13    L = 1, M1
13      C(J,K) = C(J,K) + D(J,L) * A(L,K1)
        DO 14    J = 1, M2
        J1 = J + M1
        DO 14    K = 1, M2
        K1 = K + M1
14      B(J,K) = A(J1,K1) - C(J,K)
        IF (IT - 1)    15, 15,  18
15      WRITE (7,16)
        WRITE (6,16)
16      FORMAT (25H0T2.1, RESIDUAL T MATRIX    )
        DO 17    J = 1, M2
17      T(J) = B(J,J)
        GO TO 21
18      WRITE (7,19)
        WRITE (6,19)
19      FORMAT (25H0W2.1, RESIDUAL W MATRIX    )
        DO 20    J = 1, M2
20      U(J) = B(J,J)
C
21      CALL MPRINT (B, M2)
        WRITE (6,39)
        DO 22    J = 1, M2
22      WRITE (7,23) J,  (B(J,K),    K = J, M2)
23      FORMAT (4H ROWI3,3X, 5E14.7 / (10X, 5E14.7))
        CALL MATINV (B, M2, DET)
        IF (IT - 1)    24, 24,  25
24      DETT = DET
        GO TO 26
25      DETW = DET
26      IT = IT + 1
        IF (IT - 2)    30, 30,  31
C
31      WRITE (6,39)
        XL = DETW  /  DETT
        WRITE (6,32) XL
32      FORMAT (16H0WILKS LAMBDA =    F7.4)
        IF (M2 - 2)    33, 33,  35
33      IF (KG - 3)    34, 34,   35
34      YL = XL
        F1 = 2.0
        F2 = EN - 3.0 - EM1
        GO TO 36
35      SL = SQRT ((((EM2*EM2)*((EKG -1.0)**2) -4.0)/((EM2*EM2)
```

```
      2   + ((EKG - 1.0)**2) - 5.0))
          YL = XL ** (1.0 / SL)
          PL = (EN - 1.0 - EM1) - ((EM2 + EKG) / 2.0)
          QL =  - ((EM2 * (EKG - 1.0)) - 2.0)/ 4.0
          RL = (EM2 * (EKG - 1.0)) / 2.0
          F1 = 2.0 * RL
          F2 = (PL * SL) + (2.0 * QL)
   36     N1 = F1
          N2 = F2
          F = ((1.0 - YL) / YL) * (F2 / F1)
          WRITE (6,37) F
   37     FORMAT(45H0F-RATIO FOR H2, OVERALL DISCRIMINATION, =     F9.2)
          WRITE (6,38) N1,  N2
   38     FORMAT (8H0NDF1 = I3,  12H AND NDF2 = I6)
          N1 = KG - 1
          N2 = N - KG - M1
          EN1 = N1
          EN2 = N2
          WRITE (6,39)
          WRITE (6,40) N1,  N2
   40     FORMAT(34H0UNIVARIATE F-RATIOS, WITH NDF1 = I3,12H AND NDF2 = I6)
          WRITE (6,41)
   41     FORMAT(55H0    THESE ARE FOR COVARIANCE-ADJUSTED VARIABLES.        )
          WRITE (6,42)
   42     FORMAT(55H0VARIABLE  AMONG MEAN SQ     WITHIN MEAN SQ      F-RATIO)
          WRITE (6,39)
          DO 43   J = 1, M2
          AMS = (T(J) - U(J)) / EN1
          WMS = U(J) / EN2
          F = AMS  / WMS
   43     WRITE (6,44) J,  AMS,  WMS,  F
   44     FORMAT(3X,I3,5X,F9.2,  11X,F9.2,  10X,F7.2)
          WRITE (6,39)
          WRITE (6,39)
C
          DO 45   J = 1, M2
   45     U(J) = SQRT  (U(J) / (EN - EKG))
          WRITE (6,46)
   46     FORMAT (35H0POOLED SAMPLES STANDARD DEVIATIONS      )
          WRITE (6,53)       KGT, (U(J),   J = 1, M2)
          WRITE (6,39)
          DO 47   J = 1, KGT
   47     READ  (5,7) (A(J,K),   K = 1, M)
C   READS GROUP MEANS AND GRAND MEANS INTO A.  FIRST KG ROWS ARE
C   GROUP MEANS VECTORS, AND ROW KGT IS GRAND MEANS VECTOR.
          DO 48   J = 1, M2
          DO 48   K = 1, KG
          B(K,J) = 0.0
          DO 48   L = 1, M1
   48     B(K,J) = B(K,J) + D(J,L) * (A(K,L) - A(KGT,L))
C   FIRST KG ROWS OF B ARE PREDICTED MEAN DEVIATIONS VECTORS
C   FOR M2 RIGHT SET DEPENDENT VARIABLES.
          DO 50   J = 1, M2
          J1 = J + M1
          DO 50   K = 1, KG
   50     A(K,J) = A(K,J1) - B(K,J)
          WRITE (6,51)
   51     FORMAT(55H0COVARIANCE-ADJUSTED GROUP MEANS FOR THE DEPENDENT X.  )
          DO 52   J = 1, KG
          WRITE (6,53) J,   (A(J,K),   K = 1, M2)
   52     WRITE (7,23) J,   (A(J,K),   K = 1, M2)
   53     FORMAT (6H0GROUP I3,1X, 10F10.1 / (10X, 10F10.1))
          WRITE (6,39)
          GO TO 1
          END
```

11.5 EXERCISES

1. Using the Appendix B Project TALENT males, see if there is any information regarding college plans (variable 8) in the socioeconomic status index that is not in reading comprehension and mathematics ability (variables 12 and 16). That is, is the residual variance in SES (that variance not related to reading and math ability) related to college plans? Treating college plans as a nominal scale may allow you to detect some possible qualitative difference among these five plan groups.

||||||||||||||||||||||||||||||||||||||||||||||||||||||||||||||||||||||||||||||||

CHAPTER TWELVE

Factorial Discriminant Analysis

12.1 MATHEMATICS OF FACTORIAL DISCRIMINATION

Fisher's theorem for partitioning the sum of squares in anova into orthogonal, additive components permits the number of hypothesis partitions to be extended as far as $g - 1$, where g is the number of interaction cells; thus it is possible to test several hypotheses simultaneously. In this chapter we consider the easiest, most obvious, and perhaps most useful extension of manova, which is the two-factors-with-interaction balanced design. This is the only complex manova design we shall treat, because a more general treatment would require abandoning the partitioned sums of squares and cross-products notation that we feel to be facilitative in an introductory text. For a more general treatment cast in the notation of the general linear regression model the reader may go to Bock (1966) or Bock and Haggard (1968). Moreover, we title our discussion of the factorial design "factorial discriminant analysis" because we shall emphasize the heuristic value of the best linear functions of the measurement vector variable for describing group differences (or treatment effects, in the case of an experiment), rather than the tests of the statistical hypotheses.

To keep the algebra and the interpretation of outcomes simple we shall insist on a design that is balanced by employing exactly the same sample size for all cells, or interaction groups. We shall denote cell sample size as n, the number of levels of the row basis for grouping, or row effect, as r, and the number of levels of the column basis for grouping, as c. Thus, there are n subjects in each interaction cell; there are $r \times c$ such cells in the lattice design; the total number of subjects is

$$N = r \times c \times n$$

Once again, we let p be the number of elements in the vector variable $\mathbf{X}$. Given this notation, the total sample sums of squares and cross-products matrix becomes

$$\mathbf{T} = \sum_{j=1}^{r} \sum_{k=1}^{c} \sum_{i=1}^{n} (\mathbf{X}_{jki} - \mathbf{m})(\mathbf{X}_{jki} - \mathbf{m})'$$

where $\mathbf{m}$ is again the grand centroid, or vector of total sample means. The pooled within-groups sums of squares and cross-products of deviations of subjects from cell means becomes

$$\mathbf{W} = \sum_{j=1}^{r} \sum_{k=1}^{c} \sum_{i=1}^{n} (\mathbf{X}_{jki} - \mathbf{m}_{jk})(\mathbf{X}_{jki} - \mathbf{m}_{jk})'$$

where $\mathbf{m}_{jk}$ is the centroid for the j,kth cell. This $\mathbf{W}$ matrix will be our *error* matrix for the model. We will employ MANOVA to compute $\mathbf{T}$, $\mathbf{W}$, and the cell and grand centroids, which we will read into the FACDIS program. The MANOVA run will provide the omnibus test of H_2, and also the appropriate Bartlett's test of homogeneity of cell dispersions. FACDIS will compute the row levels centroids and column levels centroids by manipulating the cell centroids. (Be sure you figure out how this can be done.) Then the program computes the three hypothesis matrices required by the following methods.

Letting $\mathbf{A}_r$ be the among rows sums of squares and cross-products matrix for the rows effect hypothesis:

$$\mathbf{A}_r = nc \sum_{j=1}^{r} (\mathbf{m}_j - \mathbf{m})(\mathbf{m}_j - \mathbf{m})'$$

where $\mathbf{m}_j$ is the row centroid for the jth row. Letting $\mathbf{A}_c$ be the sums of squares and cross-products matrix for the columns effect hypothesis:

$$\mathbf{A}_c = nr \sum_{k=1}^{c} (\mathbf{m}_k - \mathbf{m})(\mathbf{m}_k - \mathbf{m})'$$

where $\mathbf{m}_k$ is the column centroid for the kth column. Letting $\mathbf{A}_i$ be the sums of squares and cross-products matrix for the interaction hypothesis:

$$\mathbf{A}_i = \mathbf{T} - (\mathbf{A}_r + \mathbf{A}_c + \mathbf{W})$$

That is the way FACDIS computes $\mathbf{A}_i$. A definition formula for $\mathbf{A}_i$ is

$$\mathbf{A}_i = n \sum_{j=1}^{r} \sum_{k=1}^{c} (\mathbf{m}_{jk} - \mathbf{m}_j - \mathbf{m}_k + \mathbf{m})(\mathbf{m}_{jk} - \mathbf{m}_j - \mathbf{m}_k + \mathbf{m})'$$

The full partition of the sums of squares and cross-products for the factorial model is then

$$T = A_r + A_c + A_i + W$$

and the number of degrees of freedom are

$$
\begin{aligned}
\text{for } \mathbf{T} \quad & ndf = N - 1 \\
\text{for } \mathbf{A}_r \quad & ndf = r - 1 \\
\text{for } \mathbf{A}_c \quad & ndf = c - 1 \\
\text{for } \mathbf{A}_i \quad & ndf = (r - 1)(c - 1) \\
\text{for } \mathbf{W} \quad & ndf = N - rc
\end{aligned}
$$

The manova test criteria are

$$
\begin{aligned}
\text{for hypothesized row effect} \quad & \Lambda_r = |\mathbf{W}|/|\mathbf{A}_r + \mathbf{W}| \\
\text{for hypothesized column effect} \quad & \Lambda_c = |\mathbf{W}|/|\mathbf{A}_c + \mathbf{W}| \\
\text{for hypothesized interaction effect} \quad & \Lambda_i = |\mathbf{W}|/|\mathbf{A}_i + \mathbf{W}|
\end{aligned}
$$

The F-ratio approximations to Lambda are computed as in Chapter Eight, employing the appropriate number of levels value and appropriate number of subjects at each level in the computations of F for each hypothesis.

Univariate F-ratios are also computed, using the diagonal elements of $\mathbf{W}$ and $\mathbf{A}_r$, $\mathbf{A}_c$ or $\mathbf{A}_i$ and the ndf for these matrices. For example, univariate F-ratios for the row hypothesis are formed by:

$$F^{r-1}_{(r-1)(c-1)} = \frac{(a_{jj}/(r-1))}{(w_{jj}/(r-1)(c-1))}$$

Notice that the $\mathbf{W}$ matrix is the appropriate source of estimates of the common populations standard deviations and correlations. These estimates are reported by FACDIS. The "contrasts" reported for each hypothesis are simply the differences between level means and grand means for that hypothesis.

References to factorial manova employing Fisherian partitions are rare. T. W. Anderson (1958, pp. 217–221) develops the model the Fisherian way for two factors without replications (i.e., no interaction hypothesis), but it is difficult to imagine a research into human behaviors for which such a procedure would be sensible. J. L. Saupe (1965) is the most useful reference, and indeed represents the source of our interest in the problem and our notion of how to proceed on it. He has a convincing research example, too.

12.2 NUMERICAL EXAMPLE OF FACTORIAL DISCRIMINANT ANALYSIS

The Project TALENT males and females who planned careers (variable 5) in accounting, high school teaching, or art provide the data for this example of a 2 × 3 factorial design, using the creativity test and office work interest scale as two dependent variables. Six students were randomly selected for each cell. So that this numerical example can also serve a check example for the FACDIS program, the ID numbers for the 36 students used here are listed in Table 12.1. The row, column and cell means are summarized in Table 12.2,

TABLE 12.1   ID Numbers for Students Included in Example

Sex	Accountant (01)	High School Teacher (07)	Artist or Entertainer (18)
Males (1)	103	17	6
	111	19	46
	140	22	80
	143	82	168
	216	194	202
	230	219	212
Females (2)	50	80	104
	89	163	150
	155	164	198
	200	243	252
	208	266	255
	248	269	257

TABLE 12.2   Means for 2 × 3 Design

Sex (row effect)	Accountant	H.S. Teacher	Art	Row Means
Male	10.00	8.83	11.67	10.17
	21.33	11.83	9.00	14.06
Female	9.67	11.67	9.00	10.11
	31.00	15.67	12.33	19.67
Column Means	9.83	10.25	10.33	10.14
	26.17	13.75	10.67	16.86

listing the creativity score on top and the office work interest score on the bottom in each cell.

The MANOVA program is used to form the total and within cells sums of squares and crossproducts matrices as well as the centroids for cells and for total sample. These are read in by FACDIS to form the three hypotheses matrices needed in factorial discriminant analysis, where they have the relation: $T = A_r + A_c + A_i + W$. These five matrices are shown in Table 12.3.

TABLE 12.3   Basic Matrices for FACDIS

$$T = \begin{bmatrix} 490.30 & 224.70 \\ 224.70 & 4908.31 \end{bmatrix} \qquad W = \begin{bmatrix} 442.83 & 281.17 \\ 281.17 & 2934.83 \end{bmatrix}$$

$$A_r = \begin{bmatrix} .03 & -2.81 \\ -2.81 & 283.36 \end{bmatrix} \qquad A_c = \begin{bmatrix} 1.72 & -52.72 \\ -52.72 & 1615.72 \end{bmatrix}$$

$$A_i = \begin{bmatrix} 45.72 & -.94 \\ -.94 & 74.39 \end{bmatrix}$$

For example, the total sum of squares of the creativity test is partitioned as follows:

$$\begin{array}{ccccccccc} 490.30 & = & .03 & + & 1.72 & + & 45.72 & + & 442.83 \\ \text{total} & & \text{among} & & \text{among} & & \text{among} & & \text{within} \\ & & \text{rows} & & \text{columns} & & \text{cells} & & \text{cells} \end{array}$$

The program performs a discriminant analysis on each of the partitions of A. For example, $[W^{-1}A_r]$ is formed for the row effect, and its roots and vectors are evaluated as in discriminant analysis. Since there are only two levels for the row effect, only a single nonzero root is possible ($\lambda_1 = .104$). Thus the test criterion for the row effect is: $\Lambda_r = 1/(1 + .104) = .906$. Converted to an F-ratio, this becomes $F_{66}{}^2 = 1.68$, which is not significant at the .05 level. Therefore sex is apparently unrelated to the dependent measures.

Similarly, the column effect is analyzed by forming $[W^{-1}A_c]$, which has two nonzero roots, $\lambda_1 = .615$ and $\lambda_2 = .00004$. Here the Wilks criterion is $\Lambda_c = .619$, with a variance ratio $F_{64}{}^4 = 4.33$, which is significant ($p < .01$). Now that a significant column effect is established, the other discriminant results can be examined to explore the nature of that column difference. Only the first root is significantly different from zero, so all of the reliable information regarding column differences is in the first function. The structure,

$$s_{c_1} = \begin{bmatrix} -.12 \\ .97 \end{bmatrix}$$

shows that the function is primarily defined by the office work interest scale. Further indication of this is seen in the univariate F-ratios, which serve as additional *descriptors* of where the major differences lie.

$$\text{Creativity} \qquad F_{30}{}^2 = \frac{.861}{14.761} = .06$$

$$\text{Office work} \qquad F_{30}{}^2 = \frac{807.86}{97.83} = 8.26$$

Although the FACDIS program computes and prints the interaction results last, logically one should turn to them first for interpretive purposes. If the interaction is a significant effect, this affects what you might be willing to say about the main effects.

In this example, the roots of $[\mathbf{W}^{-1}\mathbf{A}_i]$ were $\lambda_1 = .113$ and $\lambda_2 = .0247$, yielding $\Lambda_i = (1/1.113)(1/1.0247) = .877$, and $F_{64}{}^4 = 1.08$, which is not significant at the .05 level. For purposes of illustration, suppose it was important to the investigator's theory that interaction did exist, and he figured this result represented type II error. If replication of this study made it possible to establish confidence in the interaction hypothesis, how then is this effect to be interpreted?

NUMERICAL EXAMPLE OF FACDIS

```
FACTORIAL DISCRIMINANT ANALYSIS,   14 MAY 1969.

TYPE THE NUMBER OF GROUPS(CELLS)  (1 DIGIT)
>6

TYPE THE NUMBER OF VARIABLES (2 DIGITS)
>02

TYPE THE NUMBER OF OBSERVATIONS PER CELL (2 DIGITS)
>06

TYPE THE NUMBER OF ROW EFFECT LEVELS (1 DIGIT)
>2

TYPE THE NUMBER OF COLUMN EFFECT LEVELS (1 DIGIT)
>3

TYPE DATASET NAME CONTAINING MEANS FROM MANOVA
DSNAME=>facmns(167wwc)

TYPE DATASET NAME OF T AND W MATRICES FROM MANOVA
DSNAME=>factw(167wwc)

GROUP MEANS
GROUP 1
     10.000      21.333
```

NUMERICAL EXAMPLE OF FACDIS (*continued*)

```
GROUP 2
      8.833        11.833
GROUP 3
     11.667         9.000
GROUP 4
      9.667        31.000
GROUP 5
     11.667        15.667
GROUP 6
      9.000        12.333

TOTAL SUBJECTS =  36

MEANS FOR TOTAL SAMPLE
     10.139        16.861

STANDARD DEVIATIONS FOR TOTAL SAMPLE
      3.743        11.842

POOLED-SAMPLES STANDARD DEVIATIONS
      3.842         9.891

POOLED-SAMPLES CORRELATIONS ESTIMATE
BASED ON NDF =   30.
   SECTION    1
   ROW              1          2
     1            1.00       0.25
     2            0.25       1.00

MEANS FOR 2   ROW EFFECT LEVELS
ROW LEVEL 1
     10.167        14.056
ROW LEVEL 2
     10.111        19.667

A (HYPOTHESIS SSCP) MATRIX FOR ROW EFFECT
   SECTION    1
   ROW              1          2
     1            0.03      -2.81
     2           -2.81     283.36

ROW CONTRASTS
ESTIMATES FOR ROW LEVEL 1
      0.028       -2.806
ESTIMATES FOR ROW LEVEL 2
     -0.028        2.806

F RATIOS FOR UNIVARIATE TESTS
NUMBER OF DEGREES OF FREEDOM ARE  1. AND    30.

AMONG MEAN SQR    WITHIN MEAN SQR    F RATIO    VARIABLE
      0.028            14.761          0.00         1
    283.360            97.828          2.90         2

PERCENTAGE WHICH EACH ROOT IS OF TRACE
    100.000

CANONICAL CORRELATIONS OF TESTS WITH GROUPS
  1 CANON R =   0.307      VARIANCE EXPLAINED =    0.094

VECTORS TO PRODUCE STANDARD DISCRIMINANT SCORES FROM DEVIATION TEST SCORES
      0.059
     -0.085

DISPERSION IN DISCRIMINANT SPACE
   SECTION    1
```

```
ROW               1
  1             1.00
```

CENTROID FOR GROUP
      0.240

CENTROID FOR GROUP
     -0.240

LAMBDA FOR TEST OF H2 =   0.9056632
F1 =   2.0000000
F2 =   66.0000000

FOR TEST OF H2, F =   1.6761265

FACTOR PATTERN FOR DISCRIMINANT FUNCTIONS
   TEST 1   0.0758
   TEST 2   -0.9756

COMMUNALITIES FOR   1 DISCRIMINANT FACTORS
   TEST 1   0.0057
   TEST 2   0.9519

PERCENTAGE OF TRACE OF R ACCOUNTED FOR BY EACH ROOT
      47.880

MEANS FOR 3 COLUMN EFFECT LEVELS
   COLUMN LEVEL 1
       9.833   26.167
   COLUMN LEVEL 2
      10.250   13.750
   COLUMN LEVEL 3
      10.333   10.667

   A (HYPOTHESIS SSCP) MATRIX FOR COLUMN EFFECT
      SECTION   1
      ROW   1   2
        1   1.72   -52.72
        2   -52.72   1615.72

COLUMN CONTRASTS
   ESTIMATES FOR COLUMN LEVEL 1
      -0.306   9.306
   ESTIMATES FOR COLUMN LEVEL 2
       0.111   -3.111
   ESTIMATES FOR COLUMN LEVEL 3
       0.194   -6.194

F RATIOS FOR UNIVARIATE TESTS
NUMBER OF DEGREES OF FREEDOM ARE 2. AND   30.

   AMONG MEAN SQR   WITHIN MEAN SQR   F RATIO   VARIABLE
       0.861   14.761   0.06   1
     807.860   97.828   8.26   2

PERCENTAGE WHICH EACH ROOT IS OF TRACE
      99.999   0.001

CANONICAL CORRELATIONS OF TESTS WITH GROUPS
   1 CANON R =   0.617   VARIANCE EXPLAINED =   0.381
   2 CANON R =   0.002   VARIANCE EXPLAINED =   0.000

VECTORS TO PRODUCE STANDARD DISCRIMINANT SCORES FROM DEVIATION TEST SCORES
      -0.071   0.262
       0.085   0.009
```
```

```
DISPERSION IN DISCRIMINANT SPACE
  SECTION   1
    ROW             1          2
     1           1.00      -0.02
     2          -0.02       1.00

  CENTROID FOR GROUP
        0.810      -0.000

  CENTROID FOR GROUP
       -0.271       0.003

  CENTROID FOR GROUP
       -0.539      -0.002

LAMBDA FOR TEST OF H2 =      0.6193401
F1 =     4.0000000
F2 =    64.0000000

FOR TEST OF H2, F =      4.3308420

FACTOR PATTERN FOR DISCRIMINANT FUNCTIONS
  TEST  1    -0.1191     0.9950
  TEST  2     0.9652     0.2432

  COMMUNALITIES FOR  2 DISCRIMINANT FACTORS
  TEST  1     1.0042
  TEST  2     0.9907

PERCENTAGE OF TRACE OF R ACCOUNTED FOR BY EACH ROOT
     47.286      52.456

  A (HYPOTHESIS SSCP) MATRIX⁻.for INTERACTION EFFECT
  SECTION   1
    ROW             1          2
     1          45.72      -0.94
     2          -0.94      74.39

  INTERACTION CONTRASTS

  ESTIMATES FOR ROW 1, COLUMN 1 CELL
        0.139      -2.028

  ESTIMATES FOR ROW 1, COLUMN 2 CELL
       -1.444       0.889

  ESTIMATES FOR ROW 1, COLUMN 3 CELL
        1.306       1.139

  ESTIMATES FOR ROW 2, COLUMN 1 CELL
       -0.139       2.028

  ESTIMATES FOR ROW 2, COLUMN 2 CELL
        1.444      -0.889

  ESTIMATES FOR ROW 2, COLUMN 3 CELL
       -1.306      -1.139

F RATIOS FOR UNIVARIATE TESTS
NUMBER OF DEGREES OF FREEDOM ARE  2. AND   30.

  AMONG MEAN SQR    WITHIN MEAN SQR   F RATIO    VARIABLE
       22.861            14.761         1.55        1
       37.197            97.828         0.38        2

  PERCENTAGE WHICH EACH ROOT IS OF TRACE
```

NUMERICAL EXAMPLE OF FACDIS (*continued*)

```
      81.991        18.009

CANONICAL CORRELATIONS OF TESTS WITH GROUPS
  1 CANON R =    0.318    VARIANCE EXPLAINED =   0.101
  2 CANON R =    0.155    VARIANCE EXPLAINED =   0.024

VECTORS TO PRODUCE STANDARD DISCRIMINANT SCORES FROM DEVIATION TEST SCORES
       0.261        0.019
      -0.033        0.083

DISPERSION IN DISCRIMINANT SPACE
   SECTION   1
   ROW            1            2
     1          1.00        -0.18
     2         -0.18         1.00

LAMBDA FOR TEST OF H2 =     0.8770809
F1 =     4.0000000
F2 =    64.0000000

FOR TEST OF H2, F =      1.0844154

FACTOR PATTERN FOR DISCRIMINANT FUNCTIONS
   TEST  1     0.9209      0.2139
   TEST  2    -0.2523      0.9975

COMMUNALITIES FOR  2 DISCRIMINANT FACTORS
   TEST  1     0.8938
   TEST  2     1.0587

PERCENTAGE OF TRACE OF R ACCOUNTED FOR BY EACH ROOT
      45.586        52.041
   STOP        00000
M:END OF JOB
>
```

The structure of the first discriminant function,

$$\mathbf{s}_{i_1} = \begin{bmatrix} .92 \\ -.25 \end{bmatrix}$$

indicates that most of the interaction effect is in the creativity measure, which had shown no noticeable main effects for either row or column groupings. Turning back to the means of Table 12.2, the nature of this creativity interaction can be further seen. The pattern of creativity differences among the career groups depends upon which sex you are talking about. For males, the "high school teachers" have lower scores than the other two career plan groups, and for the females the opposite is true. Of course for large values of p the trends in terms of univariate means become difficult to interpret and the investigator would therefore turn to centroids on discriminant functions and the function structures to facilitate interpretation.

The actual printout (IBM 360 time-sharing version) of this numerical example is presented above.

12.3 EXAMPLE OF FACTORIAL DISCRIMINANT ANALYSIS

High school seniors in 1960 reported to Project TALENT on their plans for post-high school education and training. This variable, called "College Plans," has four levels, as follows:

1. Definitely planning a baccalaureate college education
2. Probably will attend a baccalaureate college
3. Planning some postgraduate education or training, but less than a baccalaureate degree
4. No plans for education or training beyond high school.

A random one percent of the TALENT 12th grade data file was selected and sorted by Sex and College Plans. After it was established that there were 39 subjects with complete test vectors in the smallest cell of the eight cell design (2 Sex groups X 4 Plans Groups), random discards were made from the other seven cells to bring all cell memberships to 39, for a total N of 312. Six TALENT tests from the abilities domain were selected as potential predictors of the design variables, namely:

1. Information Part I (IN1)
2. Information Part II (IN2)
3. English Composite (ENG)
4. Reading Comprehension (RDG)
5. Abstract Reasoning (ABS)
6. Mathematics Composite (MAT)

Table 12.4 reports the means for the eight cells and the pooled-within standard deviations. There are substantial correlations among the variables within the groups, as reported in Table 12.5. These tests are *not* the uncorrelated MAP factors of many of our previous examples. The intercorrelations make it clear that the battery is *g*-saturated and that a multivariate analysis is essential.

The first Manova issue is, of course, the significance of the interaction effects. Fortunately, the results are

$$\text{Interaction } \Lambda = .95 \qquad F_{857}^{18} = .83$$

There is no significant interaction to complicate the interpretation of the remaining data analysis.

The Sex means, univariate F-ratios, and discriminant function structure coefficients (representing the correlations of the tests with the function) are given in Table 12.6, which also reports the Manova F, which is highly significant, and the generalized eta-square. The modest "explained variance" of .4

TABLE 12.4   Cell Means and Pooled-Within-Cells Standard Deviations for TALENT Sex X College Plans Groups (39 Subjects Per Cell)

TALENT Variables	Males				Females				Pooled S.D.
	Def. Coll.	Maybe Coll.	Some Trng.	No Trng.	Def. Coll.	Maybe Coll.	Some Trng.	No Trng.	
Information I	178	164	141	136	155	143	121	127	29.6
Information II	90	82	73	69	83	80	68	75	16.2
English Comp.	87	83	81	75	94	91	86	88	10.7
Reading Compr.	37	36	30	27	37	37	31	33	8.4
Abstract Reas.	10	10	9	8	10	10	9	9	2.7
Mathematics	33	27	20	20	29	24	19	21	8.6

reminds us that the Manova F test is very powerful and is not to be enthused over excessively. The univariate F's suggest that Information Part I and English are the major sources of Sex contrast, and we note that the girls are higher on English while the boys are superior on Information Part I. Information Part I contains several scientific and technical knowledge scales. The structure coefficients reveal that the discriminant function is essentially a bipolar contrast of English and Information Part I. Communalities of the tests on this factor are low, and the discriminant factor accounts for only nine percent of the variance in the battery.

College Plans groups are reported on in Table 12.7. The Plans effects are highly significant, although the Manova eta-square is only .3, and all univariate F-ratios are significant. Information Part I and Mathematics appear from the univariate F's to be the leading discriminators, and they have the largest structure coefficients for the first discriminant function and the largest communalities for the three available functions. The first discriminant function accounts for 86 percent of the discriminating power of the battery for the

TABLE 12.5   Correlations Among Six TALENT Ability Variables for 312 Twelfth Grade Subjects (from Sex X Plans Pooled-Within)

Variables	Info II	English	Reading	Abstract	Mathematics
Information I	.80	.58	.64	.46	.69
Information II		.53	.68	.44	.53
English Composite			.50	.38	.60
Reading Comprehension				.49	.54
Abstract Reasoning					.47

TABLE 12.6   Sex Means, F-ratios, and Discriminant Structure Coefficients for TALENT Sex X Plans Study

TALENT Variables	Male Means	Female Means	Univariate F-ratios (1 and 304 df)	Discriminant Structure r_{zf}
Information I	155	136	29.7	−.47
Information II	79	76	1.8	−.16
English Composite	81	90	47.4	.51
Reading Comprehen.	33	35	3.3	.10
Abstract Reasoning	10	10	0.0	−.03
Mathematics Comp.	25	23	3.3	−.21

Manova $F_{305}^{6} = 35.1$, $\eta^2 = .41$

Plans groups, and it is clearly a measure of Spearman's g. It accounts for 49 percent of the total battery variance. The eta-square for this first discriminant function is .26. That the groups are nearly colinear in the measurement space is indicated by the fact that the means on all six tests rank the groups the same way, and in the expected way. It should be noted that while the three discriminant functions available for the Plans groups (of which we judge only the first to be important) *are* orthogonal and will yield uncorrelated discriminant scores, these column effect discriminants need not be and in general will not be orthogonal with the row effect discriminants (in this case the solitary Sex discriminant) or the interaction discriminants, and should not be expected to yield scores uncorrelated with row discriminant scores or interaction discriminant scores. The rule is that we can expect orthogonality of the functions for

TABLE 12.7   College Plans Means, F-Ratios, and Discriminant Structure Coefficients for TALENT Sex X Plans Study

TALENT Variables	Means				Univariate F-ratios (3 and 304 df)	Discriminant Structure r_{zf}
	Def. Coll.	Maybe Coll.	Some Trng.	No Trng.		
Information I	166	153	131	132	26.3	.89
Information II	86	81	70	72	16.6	.73
English Composite	90	87	84	82	9.9	.43
Reading Comprehen.	37	37	31	30	16.1	.64
Abstract Reasoning	10	10	9	9	5.5	.37
Mathematics Comp.	31	26	20	20	30.7	.93

anova $F_{857}^{18} = 6.5$, $\eta^2 = .30$

any one of the effects of the design, but not across effects. That the eigen-structures of the three eigenequations involved in this model are not independent of each other is indicated by the fact that the three Λ's, for Row, Column, and Interaction, are constrained in that they must sum to the Λ for the simple one-way Manova on the $r \times c$ interaction groups.

In summary, in the absence of significant Sex X Plans interactions, it is clear that Sex groups and the Plans groups are discriminable in the space of the six abilities, with the Sex effects slightly stronger than the Plans effects. However, the Sex effects are concentrated in two tests, English on which girls excel and Information Part I on which boys excel. The Plans effects are strong for all tests, but are concentrated in one factor of all the tests, best represented by Information Part I and Mathematics, and interpreted as Spearman's g.

An observation should be made about Wilks' generalized correlation ratio

$$\text{Manova } \eta^2 = 1 - \Lambda = 1 - \frac{|\mathbf{W}|}{|\mathbf{T}|}$$

and the correlation ratios for discriminant functions, which we have previously described as squared canonical correlations of the discriminant functions with best weighted combinations of group membership variates:

$$\text{Discriminant } \eta_j^2 = R_{c_j}^2 = \frac{A_{dfj}}{T_{dfj}} = \frac{\lambda_j}{1 + \lambda_j}$$

where A_{dfj} and T_{dfj} are the Among-groups and Total sums of squares on the jth discriminant function, and λ_j is the jth eigenvalue of $\mathbf{W}^{-1}\mathbf{A}$. The interesting situation exists that when there are only two groups, as in the Sex factor of the present design, the Manova η^2 is equal to the Discriminant η^2. That is, when λ_1 is the only nonzero eigenvalue of $\mathbf{W}^{-1}\mathbf{A}$:

$$\Lambda = \frac{1}{1 + \lambda_1}$$

and

$$R_{c_1}^2 = 1 - \Lambda = \frac{\lambda_1}{1 + \lambda_1}$$

Noting this is another way of verifying that all the power of a p-element vector variable to discriminate two groups is concentrated in the one available discriminant function, at least under the assumptions of this model.

Since we are nearly to the end of our story, it might be well for us to note again from Chapter One that these procedures are elementary tools in a journeyman's kit. Statistics journals and books and computer program libraries are full of other and perhaps better ways to approach the data analysis problems we have considered, as well as of ways to do things we have not dared to consider. For instance, there are three determinantal equations that can be based on the partition of **T**, the Total sums of squares and cross-products matrix, into **A**, the Among-groups, and **W**, the Within-groups, s.s.c.p. matrices:

(1) $|\mathbf{W}^{-1}\mathbf{A} - \lambda_j \mathbf{I}| = 0$
(2) $|\mathbf{T}^{-1}\mathbf{A} - \alpha_j \mathbf{I}| = 0$
(3) $|\mathbf{T}^{-1}\mathbf{W} - \beta_j \mathbf{I}| = 0$

We have used only the first, but the other two have claims for priority in alternative developments of multivariate analysis of variance and discriminant functions. Wilks' Λ criterion comes most naturally from (3) as:

$$\Lambda = \prod_{j=1}^{k} \beta_j$$

where k is the number of nonzero eigenvalues. When Wilks proposed the Λ statistic in 1932 he also proposed as a direct generalization of the correlation ratio the statistic based on (2):

$$U = \prod_{j=1}^{k} \alpha_j$$

Notice that while

$$\Lambda = \frac{|\mathbf{W}|}{|\mathbf{T}|} \qquad U = \frac{|\mathbf{A}|}{|\mathbf{T}|}$$

(Incidentally, Wilks used the notation W for what we call Λ and it is too bad that his original notation did not stick to honor the inventor of this important statistic.) Readers of Wilks who have been wondering why we have chosen $1 - \Lambda$ as the Manova η^2 rather than U should consider that U becomes frightfully small in cases where the discriminating power of the battery is concentrated in fewer than the available number of discriminant functions. A very small non-zero eigenvalue of (1) or (2) leads automatically to a very small U. In the univariate analysis of variance it is the case that

$$\eta^2 = \frac{A}{T} = 1 - \frac{W}{T}$$

but in the multivariate analysis of variance with three or more groups it is not the case that U is equal to $1 - \Lambda$. In general

$$\frac{|\mathbf{A}|}{|\mathbf{T}|} \neq 1 - \frac{|\mathbf{W}|}{|\mathbf{T}|}$$

Wilks did not discuss the problem of choosing between U and $1 - \Lambda$ as a generalization of η^2, but we have had to choose.

Getting back to the three determinantal equations, some relations among

TABLE 12.8   Design Means on Three Post-tests
(1-familiar; 2-technical verbal; 3-technical pictorial)

		Remote Associates Test (RAT)		Row Means
		High	Low	
Response Mode	Reading	67.8	57.9	62.9
		63.7	54.4	59.1
		62.1	47.9	55.0
	CR	65.3	52.8	59.1
		73.0	59.4	66.2
		70.5	61.4	66.0
	Column Means	66.6	55.3	
		68.3	56.9	
		66.3	54.7	

Pooled within cell standard deviations

POST-TEST	
Familiar	12.7
Technical Verbal	14.7
Technical Pictorial	17.5

their eigenvalues that may help you to see why different computer program can work with different equations are:

$$\lambda_j = \frac{\alpha_j}{1 - \alpha_j} = \frac{1 - \beta_j}{\beta_j}$$

$$\alpha_j = \frac{\lambda_j}{1 + \lambda_j} = 1 - \beta_j$$

$$\beta_j = \frac{1}{1 + \lambda_j} = 1 - \alpha_j$$

Thus

$$\Lambda = \prod_{j=1}^{k} \left(\frac{1}{1 + \lambda_j} \right) = \prod_{j=1}^{k} (1 - \alpha_j) = \prod_{j=1}^{k} \beta_j$$

There are more ways to skin a cat than Cooley and Lohnes have dreamed of !

A second research example is a reanalysis of some data from an experiment conducted by Tobias (in press) which was originally analyzed as a 2 × 2 repeated measures analysis of variance. He was interested in the effect of creativity and response mode on achievement in programmed instruction, using three criterion measures which varied in terms of familiar vs. technical content. Creativity for 100 college students was measured with the Remote

Associates Test (RAT). This variable was dichotomized and the 50 subjects in each group were randomly assigned to two response mode treatments, one group which constructed responses (CR) and another group which simply read programmed materials. Tobias expected that the performance of high RAT scores would not be facilitated by the CR program but that the low scores would. That is, he hypothesized a significant interaction between this dimension of individual differences (creativity) and instructional procedure. He also expected the high RAT group to do better on unfamiliar material than the low RAT group, and that CR mode would be more effective than reading with the technical material but not necessarily better with the familiar material. His FACDIS results are summarized in Tables 12.8 and 12.9.

Even with a simple 2 × 2 design and only three dependent variables, the trends do not "pop-out" at you in a display of means such as Table 12.8. It is Table 12.9 that helps to focus attention on the more important trends. The interaction significance test indicates that once again the notion that how you teach depends upon who you teach is not empirically supportable.

The trends involved in the significant row and column effects are seen in the structure and centroids of their respective discriminant results. Consistent with Tobias' predictions, the constructed response mode group centroid is higher than the reading group on a function which is positively defined by the more technical material and negatively loaded on the familiar material, suggesting that making constructed responses to programmed material leads to higher achievement on technical, though not on familiar, subject matter. For the column effect, the high RAT group scores high on a general factor of the post-tests, with the structure indicating a slight tendency in favor of the more familiar material.

TABLE 12.9   Multivariate Tests of Significance

Effect	Λ	n_1	n_2	F	$P(H_0)$
Row					
(Response mode)	.788	3	96	8.60	<.01
Column (RAT)	.810	3	96	7.51	<.01
Interaction	.954	3	96	0.95	>.05

	DISCRIMINANT RESULTS	
Structure	Row effects	Column Effects
Familiar	−.29	.93
Technical Verbal	.49	.83
Pictorial Verbal	.64	.71
Centroids		
Reading −.46	High RAT	.43
CR .46	Low RAT	−.43

FACDIS

```
C        FACTORIAL DISCRIMINANT ANALYSIS.    A COOLEY-LOHNES PROGRAM.
C
C   THIS PROGRAM COMPUTES A TWO-FACTOR MANOVA FROM A BALANCED DESIGN
C   (SAME NUMBER OF REPLICATIONS IN EVERY CELL), WITH PROVISION FOR
C   AN INTERACTION EFFECT.   THE DATA INPUTS TO THIS PROGRAM ARE AS
C   PUNCHED BY THE MANOVA PROGRAM.   THUS A FACTORIAL DISCRIM JOB
C   IS BEGUN BY RUNNING SCORE VECTORS FOR THE CELLS THROUGH A ONE-WAY
C   MANOVA, GETTING THE APPROPRIATE BARTLETT HOMOGENEITY OF DISPERSIONS
C   TEST AND THE WILKS TEST OF THE SEPARATION OF THE CELL CENTROIDS.
C   THEN THE PUNCHED GROUP AND GRAND CENTROIDS, W MATRIX, AND T MATRIX
C   ARE TRANSFERRED TO THIS PROGRAM.   NOTE THAT IN ORDER FOR THE
C   FACTORIAL DESIGN TO BE ORTHOGONAL, MANOVA MUST BE RUN WITH EQUAL
C   CELL SIZES. THIS CAN BE ARRANGED BY MAKING RANDOM DISCARDS OF
C   SUBJECTS FROM CELLS WITH EXCESSIVE SAMPLE MEMBERSHIPS.
C
C       INPUT.
C
C   1) FIRST TEN CARDS DESCRIBE THE JOB FOR THE OUTPUT. DO NOT USE
C        COLUMN 1 ON THESE CARDS.
C   2) CONTROL CARD (CARD 11)
C        COL 1-2   KG = NUMBER OF INTERACTION CELLS
C        COL 4-5   M = NUMBER OF VARIABLES
C        COL  8    KR = NUMBER OF ROW EFFECT LEVELS
C        COL 10    KC = NUMBER OF COLUMN EFFECT LEVELS
C        COL 11-15 NG = NUMBER OF REPLICATIONS PER CELL
C   3) GROUP CENTROIDS FOLLOWED BY GRAND CENTROID
C      NOTE THAT ALL COLUMN CELL CENTROIDS FOR THE FIRST ROW ARE READ,
C      THEN ALL COLUMN CELL CENTROIDS FOR THE SECOND ROW, ETC.
C   4) T MATRIX
C   5) W MATRIX.
C
C       PUNCHED OUTPUT
C
C   1) POOLED-SAMPLES CORRELATION MATRIX, FORMAT (10X,7F10.7 /
C        (10X, 7F10.7)).   MATRIX IS PUNCHED UPPER-TRIANGULAR.
C
C      SUBROUTINES MPRINT, HOW, AND DIRNM ARE REQUIRED.
C
C      SCRATCH TAPES ON LOGICAL UNITS 3, 4, AND 1 ARE EMPLOYED
C
C
       DIMENSION TITLES(16),  SS(30,30),D(30,30),W(30,30),TS(30,30),
      1 R(30,30),Z(30,30),  SCOL(10,30),SROW(10,30), X(30),XM(30),SD(30),
      2 SUMT(30),SX(30),SDT(30),TEMP1(30), TEMP2(30),GM(30)
C
C
  1    WRITE (6,2)
  2    FORMAT (1H1)
       DO 3   J = 1, 10
       READ   (5,4) (TITLES(K),   K = 1, 16)
  3    WRITE (6,4) (TITLES(K),   K = 1, 16)
  4    FORMAT (16A5)
       READ   (5,5) KG,  M,  KR,  KC,  NG
  5    FORMAT (I2, 1X, I2, 2X, I1, 1X, I1, I5)
       EK = KG
       EM = M
       ENG = NG
       WRITE (6,6)
  6    FORMAT(55H0FACTORIAL DISCRIMINANT ANALYSIS.   COOLEY-LOHNES.      )
       WRITE (6,8) KG
  8       FORMAT(17H0NO. OF GROUPS = I3)
       WRITE (6,181) NG
  181  FORMAT (35H0NUMBER OF REPLICATIONS PER CELL = I5)
       WRITE (6,9) M
  9       FORMAT(20H0NO. OF VARIABLES = I3)
       DO 130   I = 1, KR
```

```
        DO 130   J = 1, M
130     SROW(I,J) = 0.0
        DO 131   I = 1, KC
        DO 131   J = 1, M
131     SCOL(I,J) = 0.0
        NROW = 1
        NCOL = 0
        EKR = KR
        EKC = KC
        REWIND 3
        REWIND 1
        WRITE (6,180)
180     FORMAT (1H0, 25(5H-----))
C
C
        DO 80   II = 1, KG
        READ   (5,10) (SX(J),   J = 1, M)
C   READS A GROUP CENTROID INTO SX.
10      FORMAT (10X, 5E14.7 / (10X, 5E14.7))
        WRITE (6,15) II
15      FORMAT (17HOMEANS FOR GROUP I5)
        WRITE (6,16) (SX(J),   J = 1, M)
16      FORMAT (1X, 10(F11.3,1X))
        NCOL = NCOL + 1
        IF (KC - NCOL)    132, 133, 133
132     NCOL = 1
        NROW = NROW + 1
133     DO 19    I = 1, M
        SROW(NROW,I) = SROW(NROW,I) + SX(I)
19      SCOL(NCOL,I) = SCOL(NCOL,I) + SX(I)
C
        WRITE(3)  SX
        WRITE (1)  SX
        WRITE (6,180)
80      CONTINUE
        WRITE (6,180)
C
C
        READ   (5,10) (X(J),   J = 1, M)
C   READS GRAND CENTROID INTO X.
        DO 11   J = 1, M
11      READ   (5,10) (TS(J,K),   K = J, M)
        DO 12   J = 1, M
12      READ   (5,10) (W(J,K),   K = J, M)
        DO 13   J = 1, M
        DO 13   K = J, M
        TS(K,J) = TS(J,K)
13      W(K,J) = W(J,K)
C   TS(I,J) IS NOW T, THE TOTAL DEVIATION-SCORES CROSS-PRODUCTS.
        N = NG * KG
        EN = N
        DO 24   I = 1, M
24      SX(I) = SQRT  (W(I,I))
C   SQUARE ROOTS OF DIAGONALS OF W ARE PRESERVED IN SX(I).
        DO 25   I = 1, M
25        SD(I) =SQRT (TS(I,I)/(EN-1.0))
C   TOTAL SAMPLE STANDARD DEVIATIONS ARE NOW IN SD(I).
        WRITE (6,26)
26      FORMAT (16H POOLED W MATRIX)
        CALL MPRINT (W, M)
        WRITE (6,180)
        WRITE (6,28) N
28        FORMAT(25HOTOTAL NO. OF SUBJECTS = I8)
        WRITE (6,29)
29      FORMAT (23HOMEANS FOR TOTAL SAMPLE)
        WRITE (6,16) (X(I),I=1,M)
```

```
      WRITE (6,30)
 30   FORMAT (37H0STANDARD DEVIATIONS FOR TOTAL SAMPLE)
      WRITE (6,16) (SD(I),I=1,M)
      WRITE (6,180)
      DO 31   I = 1, M
      GM(I) = X(I)
 31   SDT(I) = SD(I)
C
      DO 43   I = 1, M
      DO 43   J = 1, M
 43   D(I,J) = W(I,J) / (EN-EK)
C     D NOW CONTAINS THE POOLED-SAMPLES DISPERSION ESTIMATE.
C
      DO 60   I = 1, M
 60   SD(I) = SQRT (D(I,I))
      WRITE (6,61)
 61   FORMAT(35H0POOLED-SAMPLES STANDARD DEVIATIONS)
      WRITE (6,16) (SD(I),   I = 1, M)
      WRITE (6,180)
      DO 62   I = 1, M
      DO 62   J = 1, M
 62   R(I,J) = D(I,J) / (SD(I) * SD(J))
      WRITE (6,63)
 63   FORMAT(37H0POOLED-SAMPLES CORRELATIONS ESTIMATE)
      DF = EN - EK
      WRITE (6,64) DF
 64   FORMAT (16H0BASED ON NDF = F6.0)
      CALL MPRINT (R,M)
      WRITE (6,180)
      DO 65   I = 1, M
 65   WRITE (7,66) I,   (R(I,J),   J = 1, M)
 66   FORMAT (4H ROWI3,3X, 7F10.7 / (10X, 7F10.7))
C
      REWIND 4
      WRITE (4)  W, TS
      ENEK = EN - EK
      NANA = 1
      GO TO 121
C
C
 135    CONTINUE
      EK1 = EK - 1.0
      WRITE (6,180)
      WRITE (6,73) EK1, ENEK
 73   FORMAT (38H0    F RATIOS FOR UNIVARIATE TESTS        //
     1 33H0NUMBER OF DEGREES OF FREEDOM ARE    F4.0, 2X 3HAND F6.0)
      WRITE (6,74)
 74   FORMAT(57H0AMONG MEAN SQR    WITHIN MEAN SQR     F RATIO     VARIABL
     1E)
      DO 75   I = 1, M
      AMS = D(I,I) / EK1
      WMS = W(I,I) / ENEK
      F = AMS / WMS
 75   WRITE (6,76) AMS,WMS,F,I
 76   FORMAT (1X,F12.3,F19.3,F15.2,I8)
      WRITE (6,180)
      IF (M - KG)    77, 78, 78
 77   MD = M
      GO TO 79
 78   MD = KG - 1
C
 79   CALL DIRNM(D, M, W, SS , SD,MD)
C
C     ROOTS OF W-1 A ARE NOW IN SD AND VECTORS IN COLUMNS OF SS.
C
      TRACE = 0.0
```

```
         DO 82    I = 1, MD
         TEMP1(I) = SD(I) / (1.0 + SD(I))
         TEMP2(I) = SQRT  (TEMP1(I))
  82     TRACE = TRACE + SD(I)
         DO 83    I = 1, MD
  83     XM(I) = 100.0 * (SD(I)    / TRACE)
         WRITE (6,85)
  85     FORMAT(39H0PERCENTAGE WHICH EACH ROOT IS OF TRACE)
         WRITE (6,16) (XM(I),I=1,MD)
         WRITE (6,93)
  93     FORMAT(44H0CANONICAL CORRELATIONS OF TESTS WITH GROUPS)
         DO 94    I = 1, MD
  94     WRITE (6,95) I,  TEMP2(I),  TEMP1(I)
  95     FORMAT(1H0,I3,11H CANON R = , F7.3,5X, 22H VARIANCE EXPLAINED = ,
       C  F7.3)
         WRITE (6,180)
         DO 91    J = 1, MD
         DO 91    K = 1, M
         R(J,K) = 0.0
         DO 91    I = 1, M
  91     R(J,K) = R(J,K) + SS(I,J) * (TS(I,K) / (EN - 1.0))
         DO 92    J = 1, MD
         DO 92    K = 1, MD
         W(J,K) = 0.0
         DO 92    I = 1, M
  92     W(J,K) = W(J,K) + R(J,I) * SS(I,K)
         DO 97    J = 1, M
         DO 97    K = 1, MD
  97     D(J,K) = SS(J,K) * (1.0 / SQRT  (W(K,K)))
         WRITE (6,98)
  98     FORMAT(75H0VECTORS TO PRODUCE STANDARD DISCRIMINANT SCORES FROM DE
        1VIATION TEST SCORES)
         DO 99    J = 1, M
  99     WRITE (6,16) (D(J,K),   K = 1, MD)
         WRITE (6,180)
  C
         DO 88    J = 1, MD
         DO 88    K = 1, M
         R(J,K) = 0.0
         DO 88    I = 1, M
  88     R(J,K) = R(J,K) + D(I,J) * (TS(I,K) / (EN - 1.0))
         DO 89    J = 1, MD
         DO 89    K = 1, MD
         W(J,K) = 0.0
         DO 89    I = 1, M
  89     W(J,K) = W(J,K) + R(J,I) * D(I,K)
         WRITE (6,96)
  96     FORMAT(33H0DISPERSION IN DISCRIMINANT SPACE)
         CALL MPRINT (W,MD)
         WRITE (6,180)
  C
         IF  (NANA - 3)    172,    171,  171
  172    REWIND 3
         DO 115   II = 1, KG
         READ(3) XM
         DO 116   I = 1, MD
         TEMP1(I) = 0.0
         DO 116   J = 1, M
  116    TEMP1(I) = TEMP1(I) + ((XM(J) - GM(J) ) * D(J,I))
         WRITE (6,117) II,   MD
  117    FORMAT (20H0CENTROID FOR GROUP I4,4H IN I4,31H DIMENSIONAL DISCRIM
        1INANT SPACE)
  115    WRITE (6,16) (TEMP1(I),   I = 1, MD)
  C
  171    XLAMB = 1.0
         DO 100   I = 1, MD
```

```
100   XLAMB = XLAMB * (1.0/(1.0 + SD(I)))
101   FORMAT(25H0LAMBDA FOR TEST OF H2 = F14.7)
      IF (M - 2)   102, 102, 103
102   IF (KG - 3)  104, 104, 103
104   Y = XLAMB
      F1 = 2.0
      F2 = EN - 3.0
      GO TO 105
103     S=SQRT ((((EM**2)*((EK-1.0)**2)-4. )/((EM**2)+((EK-1. )**2)-5.
      Y = XLAMB**(1.0/S)
      XM1 = (EN - 1.0) - ((EM + EK)/2.0)
      XL =-((EM * (EK-1.0)) - 2.0)/4.0
      R1 = (EM * (EK-1.0))/2.0
      F1 = 2.0 * R1
      WRITE (6,180)
      WRITE (6,101) XLAMB
      F2 = (XM1 * S) +(2.0*XL)
105   F = ((1.0 - Y) / Y) * (F2 / F1)
      WRITE (6,106) F1
106   FORMAT (6H0F1 = F14.7)
      WRITE (6,107) F2
107   FORMAT (6H0F2 = F14.7)
      WRITE (6,108) F
108   FORMAT (21H0FOR TEST OF H2, F = F14.7)
      WRITE (6,180)
C
      DO 109   I = 1, M
      DO 109   J = 1, M
109   R(I,J) = TS(I,J) / SQRT  (TS(I,I) * TS(J,J))
C TOTAL CORRELATION MATRIX IS NOW IN R(I,J).
      DO 110   J = 1, M
      DO 110   K = 1, MD
110    SS(J,K) = D(J,K) * SDT(J)
      DO 111   J = 1, M
      DO 111   K = 1, MD
      D(J,K) = 0.0
      DO 111   I = 1, M
111   D(J,K) = D(J,K) + R(J,I) * SS(I,K)
      WRITE (6,112)
112   FORMAT(42H0FACTOR PATTERN FOR DISCRIMINANT FUNCTIONS   )
      DO 113   J = 1, M
113   WRITE (6,114) J, (D(J,K),   K = 1, MD)
114   FORMAT(5H0TEST I4, 10(3X,F7.4)/9X,10(3X,F7.4))
      WRITE (6,180)
C     PARTIAL SET OF VECTORS IS NOW IN TS(I,J)
      DO 122   J = 1, M
      TEMP1(J) = 0.0
      DO 122   K = 1, MD
122   TEMP1(J) = TEMP1(J) + D(J,K) * D(J,K)
      WRITE (6,123) MD
123   FORMAT(19H0COMMUNALITIES FOR I5,21H DISCRIMINANT FACTORS)
      DO 124   J = 1, M
124   WRITE (6,114) J, TEMP1(J)
      WRITE (6,180)
      DO 125   J = 1, MD
      TEMP1(J) = 0.0
      DO 125   K = 1, M
125   TEMP1(J) = TEMP1(J) + D(K,J) * D(K,J)
      DO 126   J = 1, MD
126   TEMP1(J) = (TEMP1(J) / EM) * 100.0
      WRITE (6,37)
37    FORMAT(52H0PERCENTAGE OF TRACE OF R ACCOUNTED FOR BY EACH ROOT)
      WRITE (6,16) (TEMP1(J),   J = 1, MD)
      WRITE (6,180)
      WRITE (6,180)
      NANA = NANA + 1
```

```
C
C
 121   IF (NANA - 2)    134,  136,  137
 134   REWIND 4
       READ (4)   W, TS
       DO 138   I = 1, M
       DO 138   J = 1, M
       Z(I,J) = 0.0
 138   D(I,J) = 0.0
       DO 139   I = 1, KR
       DO 139   J = 1, M
 139   SROW(I,J) = B9OW(I,J) / EKC
       DO 140   I = 1, KR
       DO 140   J = 1, M
       DO 140   K = 1, M
 140   D(J,K) = D(J,K) + (SROW(I,J)-GM(J)) * (SROW(I,K)-GM(K))
       DO 141   I = 1, M
       DO 141   J = 1, M
       D(I,J) = D(I,J) * (ENG * EKC)
 141   Z(I,J) = Z(I,J) + D(I,J)
C  D NOW CONTAINS THE HYPOTHESIS S.S.C.P. MATRIX FOR THE ROW EFFECT.
       WRITE (6,142) KR
 142   FORMAT (11H0MEANS FOR I4,19H  ROW EFFECT LEVELS)
       DO 143   I = 1, KR
       WRITE (6,144) I
 144   FORMAT (11H0ROW LEVEL I4,7H  MEANS)
 143   WRITE (6,16) (SROW(I,J),   J = 1, M)
       WRITE (6,180)
       WRITE (6,145)
 145   FORMAT (42H0A (HYPOTHESIS SSCP) MATRIX FOR ROW EFFECT)
       CALL MPRINT (D, M)
       WRITE (6,180)
       KG = KR
       EK = KG
       REWIND 3
       DO 161   I = 1, KR
       DO 162   J = 1, M
 162   XM(J) = SROW(I,J)
 161   WRITE(3)   XM
       WRITE (6,160)
 160   FORMAT (14H0ROW CONTRASTS)
       DO 158   I = 1, KR
       DO 159   J = 1, M
 159   XM(J) = SROW(I,J) - GM(J)
       WRITE (6,163) I
 163   FORMAT (25H0ESTIMATES FOR ROW LEVEL I4)
 158   WRITE (6,16) (XM(J),   J = 1, M)
       GO TO 135
C
C
 136   REWIND 4
       READ (4)   W, TS
       DO 146   I = 1, M
       DO 146   J = 1, M
 146   D(I,J) = 0.0
       DO 147   I = 1, KC
       DO 147   J = 1, M
 147   SCOL(I,J) = SCOL(I,J) / EKR
       DO 148   I = 1, KC
       DO 148   J = 1, M
       DO 148   K = 1, M
 148   D(J,K) = D(J,K) + (SCOL(I,J)-GM(J)) * (SCOL(I,K)-GM(K))
       DO 149   I = 1, M
       DO 149   J = 1, M
       D(I,J) = D(I,J) * (ENG * EKR)
 149   Z(I,J) = Z(I,J) + D(I,J)
```

```
C   D NOW CONTAINS THE HYPOTHESIS S.S.C.P. MATRIX FOR THE COLUMN EFFECT.
        WRITE (6,150) KC
  150   FORMAT(11HOMEANS FOR I4,22H  COLUMN EFFECT LEVELS)
        DO 151   I = 1, KC
        WRITE (6,152) I
  152   FORMAT (14HOCOLUMN LEVEL I4,7H  MEANS)
  151   WRITE (6,16) (SCOL(I,J),    J = 1, M)
        WRITE (6,180)
        WRITE (6,157)
  157   FORMAT (45HOA (HYPOTHESIS SSCP) MATRIX FOR COLUMN EFFECT)
        CALL MPRINT (D, M)
        WRITE (6,180)
        KG = KC
        EK = KC
        REWIND 3
        DO 166   I = 1, KC
        DO 167   J = 1, M
  167   XM(J) = SCOL(I,J)
  166   WRITE(3)  XM
        WRITE (6,164)
  164   FORMAT (17HOCOLUMN CONTRASTS)
        DO 165   I = 1, KC
        DO 168   J = 1, M
  168   XM(J) = SCOL(I,J) - GM(J)
        WRITE (6,169) I
  169   FORMAT (28HOESTIMATES FOR COLUMN LEVEL I4)
  165   WRITE (6,16) (XM(J),    J = 1, M)
        GO TO 135
C
C
  137   IF (NANA - 3 )  156,  155,  156
  155   REWIND 4
        READ (4)  W, TS
        DO 153   I = 1, M
        DO 153   J = 1, M
  153   D(I,J) = TS(I,J) - (Z(I,J) + W(I,J))
C   D NOW CONTAINS THE HYPOTHESIS SSCP MATRIX FOR INTERACTION.
        WRITE (6,154)
  154   FORMAT(50HOA (HYPOTHESIS SSCP) MATRIX FOR INTERACTION EFFECT)
        CALL MPRINT (D, M)
        WRITE (6,180)
        REWIND 1
        WRITE (6,170)
  170   FORMAT (22HOINTERACTION CONTRASTS)
        DO 173   I = 1, KR
        DO 173   J = 1, KC
        READ (1) TEMP1
        DO 174   K = 1, M
  174   XM(K) = TEMP1(K) - SROW(I,K) - SCOL(J,K) + GM(K)
        WRITE (6,175) I, J
  175   FORMAT(19HOESTIMATES FOR ROW I4,9H, COLUMN I4,5H CELL)
  173   WRITE (6,16) (XM(K),    K = 1, M)
        WRITE (6,180)
        KG = (KR - 1) * (KC - 1) + 1
        EK = KG
        GO TO 135
C
  156   GO TO 1
        END
>
```

12.5 EXERCISES

1. How does FACDIS compute row centroids and column centroids?

2. Recompute the example of section 12.2 using a different random selection of six students per cell.

||||||||||||||||||||||||||||||||||||||||||||||||||||||||||||||||||||||||||||||||||||||||||||||||||||

CHAPTER THIRTEEN

Strategy Considerations in Multivariate Research

As scientists we are obliged to keep statistical methods subordinated to our research problems. We must address ourselves to important questions first, and collect appropriate data (and enough of them) second, and only then decide upon analytic procedures. Decisions as to methods should always be provisional and flexible. We should play our hunches regarding methods that might possibly free the data to speak their messages better than the obvious or favorite methods can. The computer's ability to try lots of schemes on for size is there to be used. "Bending the data to fit the analysis can be vital. . . But bending the question to fit the analysis is to be shunned at all costs" (Tukey, 1969, p. 83). The resources of statistics are rich indeed, and do not sponsor obsessions with specific selections of procedures, even if a textbook has to be selective.

This approach to statistics as an aspect of research is perhaps best expressed by John Tukey (1962; 1969), who also licenses our emphasis on exploratory, heuristic operations rather than on hypothesis testing, confirmatory procedures. Both modes of research are needed, but detective work seems to be the occupation of the scientist more of the time, and is more fun to some of us.

The fitting of linear functions called factors to measurement vectors has been a recurrent feature of the procedures documented in this text. Nunnally has voiced the rationale for such operations with special clarity.

> The effort in science is to find a relatively small set of variables which will suffice to "explain" all other variables. A small set of variables "explains" a larger set if some combination of the smaller set correlates highly with each member of the larger set. . . . To achieve such a small set of "explainer" variables is the essence of scientific parsimony (Nunnally, 1967, p. 151).

The factor structure matrix **S** has been stressed repeatedly as the report of the small set of linear combinations (factors) that explains the many observation variables (measurements). Since the factors are columns of **S** and the measurements are rows, the sum of squares of elements in a row is the "communality" or explained variance for that measurement, and the square root of that number is the correlation of the measurement with the linear function of all the factors that best explains it. In some analyses the factors are fitted to be the small set of combinations that best carry the predictive validities of the many measurements for criteria provided by another set of measurements (canonical analysis) or by a set of group memberships (discriminant analysis). Regardless, the rank of the factor model for the measurement data, given by the number of columns in **S** as the number of factors retained, is of the greatest importance as a research decision. Finding the courage to choose a reasonably small rank for an amplitude of considerably correlated and considerably fallible measurement data is part of maturing as a data analyst, at least in our view. Tukey says, "Occam's razor indicates that we should not think behavior to be complicated if we can make it simple by reexpressing, or reformulating, our variables" (1969, p. 86).

This has been a book about research methods, not about statistics, and it has been colored by the research style of its authors. Style in model building and strategy in model fitting and testing are human attributes involving subtleties and subjectivities the computer is neither prone to nor responsible for. The credit and blame for research still belong to the human scientist. The research style and strategies reflected in this text are not optimal for many fields of research, but they have been found very serviceable for one field, which may be named *second phase longitudinal human development research* (*splhdr*). Some examples of *splhdr* are Terman's Genetic Studies of Genius, Super's Career Pattern Study (and Gribbons and Lohnes' replication of it), and Project TALENT. Some characteristics of *splhdr* are the following:

1. Predictor variables are mostly measurement traits of personality, and there may be a great many of them. Project TALENT data files provide 100 scales describing the abilities and motives of over 400,000 high school students, as collected in 1960. In his monumental survey of psychological measurement titled *Personality*, J. P. Guilford (1960) catalogs measurement traits in several domains:

> 10 somatic traits
> 120 structure of intellect traits
> 15 temperament traits
> 53 hormetic (motivational) traits
> 24 traits of neurosis and psychosis
> ___
> 222 dimensions of personality!

2. Whereas first phase longitudinal human development research inquires into the antecedents of these personality traits in genetic and social inheritance, in education, etcetera, *splhdr* inquires into the consequences for life adjustments criteria of these personality traits. In this view, personality traits and factors are intervening variables that mediate between hereditary potential and learning experiences in family, neighborhood, school, and TV, on the one hand, and important life adjustments and accomplishments on the other. In first phase research personality traits are the dependent variables, while in *splhdr* personality traits are the independent variables. Thus, it is the concurrent and forward predictive validities of personality measurements that preoccupy *splhdr*.

3. The criteria for *splhdr* are often the indicators of real life winners and losers. As in real life, these are often observed as nominal or ordinal group membership scales, so that *taxonomic criteria* prevail for *splhdr*. Examples of taxonomic criteria are:

> high school graduation versus dropout
> high school curriculum tracks
> college admission or not
> college major fields
> vocational aspirations
> vocational placements
> carrer pattern variables (e.g., path-jumpers versus path-followers in the career development tree)
> marital adjustment categories
> criminal court records categories
> employed or unemployed
> on or off welfare
> suicide or not

Outcomes in real life can get rough. The predictability of who gets killed and maimed in our foreign wars has not been explored much, but perhaps it should be. There is *splhdr* literature on personality predictors of who gets killed and maimed in our highway skirmishes.

Second phase longitudinal human development research serves three different sets of customers, each requiring somewhat specialized research strategies. The first clients of *splhdr* belong to the category of theoretical behavioral science. These clients want to understand the implications of individual differences in personality traits and factors better. They require research that leads to parsimony of constructs and breadth of generalizations sufficient to warrant talk of statistical laws of behavior. For example, Project TALENT's program of *splhdr* has led to the formulation of a psychometric law of career changes:

> Our research and that of others shows that migration from one stable career path to another (or path-jumping) tends to take the individual to a path for which he is closer to the centroid. That is, changing plans so that his career pattern is classified as unstable usually decreases the generalized distance of the individual from his group's centroid in a suitable personality measurement space. This change law is perhaps the most significant finding of psychometric research on career variables (Cooley and Lohnes, 1968, p. 5-4).

Personnel psychology provides the second set of clients for *splhdr*. This is essentially human engineering, and it requires great specificity and variety of criteria, accessibility of predictors, and efficient statistical decision functions (i.e., high hit rates). Cost efficiency in selection and assignment of men as resources of corporate enterprises, be they making money or making war, promotes the search for better statistical procedures. These clients have supported much of the recent research in multivariate methods. The third clientele represent an underdeveloped market for personal psychology. These are young people who need information about their multipotentialities for education and careers that *splhdr* can make available to them. The translation of personal assessments into personal probabilities for people involved in self study and personal decision making is one of the paramount challenges for *splhdr*.

In the light of these characteristics and clienteles, what are the design problem areas for *splhdr*? First priority should be given to the requirement for a general and comprehensive organization of the predictors. It is mind boggling to attempt to carry a theory of personality involving one to two hundred traits into criterion oriented research. Large scale *splhdr* projects should concentrate on producing effective general factor models for the predictor measurement data before turning to the study of the relations of those predictors with the several or many sets of criterion variables. Given the nature of psychological measurement, it will usually be possible to preserve almost all of the predictive powers of the many trait measures in the few constructed factor measures. If the factors are carefully placed and convincingly named and interpreted the scientific community will be able to comprehend and discuss the implications of the *splhdr* findings in relatively simple and memorable terms. The cumulative propensity of science will be served, because a fairly small set of factor constructs will reappear in most studies and will mark the emergence of theoretical order out of observational chaos. Despite the unwillingness of some major theoreticians to see it, some such order has emerged in the domain of human intellectual abilities through a half century of correlation and factor analysis studies. The next decade may bring an international consensus on a basic six or so factors of intelligence.

The argument for orthogonal factor structures is that they do the work parsimoniously, whereas oblique factors create new problems for analysis even as they reduce present problems. To the argument that orthogonal factors are not meaningful we need only to say that experience shows they can be. Again we salute the extraordinary performance of Kaiser's Varimax algorithm in assisting the analyst to place orthogonal reference axes in meaningful positions. Sometimes when varimax does not accomplish any magic it is because the analyst has made some serious errors of omission (such as failing to use prior regressions to remove by means of a linear model the perturbing influences of descriptors like sex and grade) or of commission (such as insisting on a ridiculous rank for a factor model for carefully selected variables from a single domain).

The ease with which the computer scores factors from observation vectors is very important in signalling that factors are no longer just abstractions to be talked about, but are now variables into which observation vectors can be transformed and in terms of which research can be done. Also, in exploratory *splhdr* conducted with moderate N, the a priori reduction of rank in the predictors domain can lead to useful recovery of degrees of freedom for subsequent predictive validity probes.

A second design problem for *splhdr* is the organization of criteria data. Many of the "natural" adjustment and success criteria reveal themselves as sets of membership categories that are excessively detailed and otherwise poorly scaled for scientific purposes. Preliminary analyses of the locations of the natural or observational groups in the measurement space, in terms of plots of centroids in a best discriminant plane and tables of generalized distances between all pairs of group centroids, can lead to a quite different and improved taxonomy. Obviously such explorations of criterion scaling require large N to protect against capitalization on chance, and firm acceptance of Tukey's dictum that "Confirmation comes from repetition" (1969, p. 84). Nevertheless, there is nothing unscientific about insisting that a good taxonomy is a discriminable one, and that good taxonomies tend not to lie about in natural language. Good science language is often quite unnatural to the layman.

The third set of design problems for *splhdr* resides in the choice of analytic tools for formulating and interpreting predictor-criterion relations. The water is muddied by the way *splhdr* partially reverses and partially abandons the logic of classical regression and analysis of variance, by taking the multivariate normally distributed measurement vector as the independent variable and the multinomially or vector binomially distributed taxonomy as the dependent variable. If we think of the taxonomy as a fixed effects, nonrandom variate we have reversal of canonical regression or Manova logic. If, as seems more likely, we think of the taxonomy as a multinomial or vector binmoial random variable we have an even muddier situation with a model somewhere in limbo between the classical regression family and correlation family of models. It seems that the multiple group discriminant analysis as expounded by P. J. Rulon and his associates at Harvard over the years and blessed by S. S. Wilks in his *Mathematical Statistics* (1962, pp. 576–587), in company with the series of classification procedures we have discussed in Chapter Ten, represent a distinct family of analytic models that cannot be subsumed under either of the classical families. The implication is that we can argue by analogy for the transferral of Manova or canonical correlation features and interpretations to the multiple group discriminant strategy situation, but we have to be cautious not to stretch analogies too far. This is an area for further methodological research and clarification. We admit, for example, that η^2 is a regression statistic with which we may be taking excessive liberties in this text. The future probably holds increased usage of the classification hit rate as a more appropriate indicator of the strength of predictive validity in the multiple discriminant situation. Geisser (1967) and Hills (1966) have been leading the math stat assault on this difficult issue. Pending the extension of their recent work on the two group case to the multiple group case, we can but urge the deployment of classification outcomes for replication samples in evaluation of *splhdr* prediction models. Once again, "Confirmation comes from repetition."

13.1 PROGRAM SEQUENCES

Figure 13.1 is designed to help you see the possible program sequences for exploring data with the programs presented in this text. The arrows indicate where punched output from one program is compatible with input for subsequent programs. If your computer center allows you to write out datasets on disc, movement from one program to another can be greatly facilitated. We strongly urge you to give the TALENT data a good workout with these programs so that you can see the different kinds of insights the different programs provide of the phenomena your data represent.

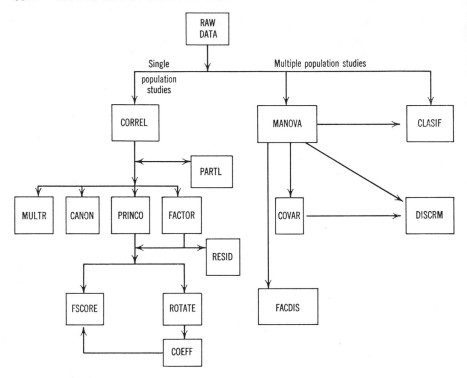

Figure 13.1. Structure of possible program sequences.

13.2 EXERCISES

Answer the following questions in three parts:

A. Data analysis procedures to be used (in terms of programs in the Multivariate Analysis Package).

B. Output to be examined prior to next step.

C. The statistics to be reported as relevant to the research question.

1. An investigator has made N_j observations in each of g populations ($j = 1, g$). Each observation consists of vector variable $\mathbf{x}$ and vector variable $\mathbf{y}$, with p and q elements, respectively. He wants to know if there is any information in $\mathbf{x}$ that is not in $\mathbf{y}$ regarding differences among these g populations.

2. An investigator is interested in developing a maximum likelihood procedure for assigning individuals to one of g populations based on observing vector variable $\mathbf{x}$. He samples the g populations and discovers that $\mathbf{x}$ is not very

multivariate normal. He decides to form n orthogonal, linear functions of $\mathbf{x}$ which extract the variance common to the p elements in $\mathbf{x}(n < p)$. Then he would develop the assignment procedure in terms of the derived linear functions, which turn out to be more reliable and more normally distributed than was the original vector variable $\mathbf{x}$.

3. An investigator has observed three vector variables ($\mathbf{u}$, $\mathbf{v}$, and $\mathbf{w}$) on each of N students drawn from his college population. He is interested in deriving a single factor of $\mathbf{u}$ which is maximally correlated with a factor of $\mathbf{v}$, and have both factors completely independent of the vector variable $\mathbf{w}$.

4. Two investigators, Bill and Paul, have a difference of opinion regarding the effect of three treatments (A, B, and C) on vector variable $\mathbf{x}$. Bill feels that treatment B will produce a result which is qualitatively different from A and C, resulting in a different pattern of means for B on the p variates in $\mathbf{x}$ than will A and C. Paul believes that B will have only a quantitative effect, and places B between A and C on all p variates. How might they resolve their difference of opinion?

5. An investigator suspects that males and females have quite different dispersions for a set of scores representing vocational interests. How can he test this hypothesis? If he finds a significant difference, how might he decide whether the dispersion differences are primarily in the variances or in the covariances?

References

Abramowitz, M., and Stegun, I. A. *Handbook of mathematical functions with formulas, graphs, and mathematical tables*. Washington, D.C.: U.S. Government Printing Office, 1964.

Anderson, T. W. *An Introduction to multivariate statistical analysis*. New York: Wiley, 1958.

Anderson, T. W., and Rubin, H. Statistical inference in factor analysis. Pages 111-150 of Volume V of *Proceedings of the third Berkeley symposium on mathematical statistics and probability*. Berkeley: University of California Press, 1956.

Baggaley, A. R., and Campbell, J. P. Multiple-discriminant analysis of academic curricula by interest and aptitude variables. *Journal of Educational Measurement*, 1967, **4**, 143-149.

Balomenos, R. H. An experimental study comparing the effectiveness of teaching deduction in two content areas of secondary mathematics. Cambridge, Mass.: Harvard Graduate School of Education, Ed. D. Dissertation, 1961.

Bartlett, M. S. Properties of sufficiency and statistical tests. *Proceedings of the Royal Society*, 1937, **A160**, 268-282.

Bartlett, M. S. Further aspects of multiple regression. *Proceedings of the Cambridge Philosophic Society*, 1938, **34**, 33–40.

Bartlett, M. S. The statistical significance of canonical correlations. *Biometrika*, 1941, **32**, 29–38.

Bartlett, M. S. Multivariate analysis. *Supplement to the Journal of the Royal Statistical Society*, 1947, **9**, 176–197.

Bartlett, M. S. Tests of significance in factor analysis. *British Journal of Psychology* (*Statistical Section*), 1950, **3**, 77–85.

Bereiter, C. How may units of measurement be safely ignored? *Journal of Educational Measurement*, 1964, **1**, 19–22.

Bock, R. D. Contributions of multivariate experimental designs to educational research. Ch. 28 in R. B. Cattell (ed.), *Handbook of multivariate experimental psychology*. Chicago: Rand McNally, 1966.

Bock, R. D., and Haggard, E. A. The use of multivariate analysis of variance in behavioral research. Ch. 3 in D. K. Witla (ed.), *Handbook of measurement and assessment in behavioral sciences*. Reading, Mass.: Addison-Wesley, 1968.

333

Box, G. E. P. A general distribution theory for a class of likelihood criteria. *Biometrika*, 1949, **36**, 317–346.

Bryan, J. G. The generalized discriminant function: mathematical foundation and computational routine. *Harvard Educational Review*, 1951, **21**, 90–95.

Burket, G. R. A study of reduced rank models for multiple prediction. *Psychometric Monographs*, 1964, No. 12.

Carroll, J. B. An analytic solution for approximating simple structure in factor analysis. *Psychometrika*, 1953, **18**, 23–38.

Carroll, J. B. Biquartmin criterion for rotation to oblique simple structure in factor analysis. *Science*, 1957, **126**, 1114–1115.

Cattell, R. B., r_b and other coefficients of pattern similarity. *Psychometrika* 1949, **14**, 279–298.

Cattell, R. B. (ed.) *Handbook of multivariate experimental psychology*. Chicago: Rand McNally, 1966.

Cochran, W. G. Analysis of covariance: its nature and uses. *Biometrics*, 1957, **13**, 261–281

Cooley, W. W. The application of a developmental rationale and methods of multivariate analysis to the study of potential scientists. Unpublished dissertation, Graduate School of Education, Harvard University, 1958.

Cooley, W. W. *Career development of scientists: an overlapping longitudinal study*. Cambridge, Mass.: Harvard Graduate School of Education, 1963.

Cooley, W. W. Predicting career plan changes. In J. Flanagan, et al., *Project TALENT one-year follow-up studies*. Pittsburgh: American Institutes for Research, 1966.

Cooley, W. W. Interactions among interests, abilities, and career plans. *Journal of Applied Psychology*, 1967, **51**, Part 2 (Monograph).

Cooley, W. W., and Lohnes, P. R. *Predicting development of young adults*. Palo Alto: American Institutes for Research, 1968.

Cooley, W. W., and Mierzwa, J. A. The Rorschach test and the career development of scientists. Cambridge, Mass.: Harvard Graduate School of Education, *Scientific Careers Study*, Interim Report 5, 1961.

Cox, D. R. *Planning of experiments*. New York: Wiley, 1958.

Cronbach, L. J., and Gleser, G. C. Assessing profile similarity. *Psychological Bulletin*, 1953 **50**, 456–473.

DuBois, P. H. *Multivariate correlational analysis*. New York: Harper, 1957.

DuBois, P. H. On relationships between numbers and behavior. *Psychometrika*, 1962, **27**, 323–33.

DuMas, F. M. The coefficient of profile similarity. *Journal of Clinical Psychology*, 1949, **5** 123–131.

Edwards, A. L. *Experimental design in psychological research*. New York: Holt, Rinehart and Winston, 1960.

Efroymson, M. A. Multiple regression analysis. In A. Ralston and H. S. Wilf (eds.), *Mathematical methods for digital computers*. New York: Wiley, 1960, pp. 191–203.

Eysenck, H. J. *Fact and fiction in psychology*. New York: Penguin, 1965.

Faddeeva, V. N. *Computational methods of linear algebra*. New York: Dover Publications, 1959.

Ferguson, G. A. The concept of parismony in factor analysis. *Psychometrika*, 1954, **19**, 281–290.

Fisher, R. A. The general sampling distribution of the multiple correlation coefficient. *Proceedings of the Royal Society*, 1928, **A121**, 654–673.

Fisher, R. A. The use of multiple measurments in taxonomic problems. *Annals. of Eugenics*, 1936, **7**, 179–188.

Flanagan, J. C., et al. *Design for a study of American Youth*. Boston: Houghton Mifflin Co., 1962.

Geisser, S. Posterior odds for multivariate normal classifications. *Journal of the Royal Statistical Society*, 1964, **26**, 69–76.

Geisser, S. Estimation associated with linear discriminants. *Annals of Mathematical Statistics*, 1967, **38**, 807–817.

Gribbons, W. D., and Lohnes, P. R. *Emerging careers*. New York: Teachers College Press, 1968.

Gribbons, W. D., and Lohnes, P. R. Eighth grade vocational maturity in relation to nine-year career patterns. *Journal of Counseling Psychology*, 1969, **16**, 557–562.

Guilford, J. P. *Personality*. New York: McGraw-Hill, 1960.

Guttman, L. Some necessary conditions for common-factor analysis. *Psychometrika* 1954 **19**, 149–161.

Harman, H. H. *Modern factor analysis*. Chicago: University of Chicago Press, 1960.

Harman, H. H. Modern factor analysis (2nd edition). Chicago: University of Chicago Press, 1967.

Harman, H. H. Factor analysis. Ch. 4 in D. K. Whitla (ed.), *Handbook of measurement and assessment in behavioral sciences*. Reading, Mass.: Addison-Wesley, 1968.

Hays, W. L. *Statistics for pyschologists*. New York: Holt, Rinehart and Winston, 1963.

Herzberg, P. A. *The parameters of cross-validation*. Urbana, Illinois: University of Illinois, Department of Psychology, 1967.

Hills, M. Allocation rules and their error rates. *Journal of the Royal Statistical Society*, Series B, 1966, **28**, 1–31.

Horst, P. Generalized canonical correlations and their applications to experimental data. Seattle: University of Washington, 1961. (mimeo.)

Horst, P. Relations among *m* sets of measures. *Psychometrika*, 1961, **26**, 129–149.

Horst, P. *Factor analysis of data matrices*. New York: Holt, Rinehart and Winston, 1965,

Hotelling, H. The generalization of Student's ratio. *Annals of Mathematical Statistics*, 1931, **2**, 360–378.

Hotelling, H. Analysis of a complex of statistical variables into principal components. *Journal of Educational Psychology*, 1933, **24**, 417–441, 498–520.

Hotelling, H. The most predictable criterion. *Journal of Educational Psychology*, 1935, **26**, 139–142.

Hotelling, H. Relations between two sets of variates. *Biometrika*, 1936, **28**, 321–377.

Jackson, R. W. B. *Applications of the analysis of variance and covariance method to educational problems.* Toronto: University of Toronto Press, 1940.

Johnson, P. O., and Jackson, R. W. B. *Modern statistical methods: descriptive and inductive.* Chicago: Rand McNally, 1959.

Johnson, P. O., and Neyman, J. Tests of certain linear hypotheses and their application to some educational problems. *Statistical Research Memoirs*, 1936, **1**, 57–93.

Kaiser, H. F. The varimax criterion for analytic rotation in factor analysis. *Psychometrika*, 1958, **23**, 187–200.

Kaiser, H. F. Computer program for varimax rotation in factor analysis. *Educational and Psychological Measurement*, 1959, **19**, 413–420.

Kaiser, H. F. Comments on communalities and the number of factors. Read at an informal conference, "The Communality Problem in Factor Analysis," St. Louis: Washington University, 1960 (dittoed).

Kaiser, H. F. Formulas for component scores. *Psychometrika*, 1962, **27**, 33–37.

Kelley, T. L. *Talents and tasks: their conjunction in a democracy for wholesome living and national defense.* Cambridge, Mass.: Harvard Education Papers, No. 1, 1940.

Kendall, M. G. *A course in multivariate analysis.* London: Charles Griffin and Co., 1957.

Lindquist, E. F. *Design and analysis of experiments in psychology and education.* Boston: Houghton Mifflin, 1953.

Lohnes, P. R. Test space and discriminant space classification models and related significance tests. *Educational and Psychological Measurement*, 1961, **21**, 559–574.

Lohnes, P. R. *Measuring adolescent personality.* Pittsburgh: American Institutes for Research, 1966.

Lohnes, P. R., and Cooley, W. W. *Introduction to statistical procedures: with computer exercises.* New York: Wiley, 1968.

Lohnes, P. R., and Marshall, T. O. Redundancy in student records. *American Educational Research Journal*, 1965, **2**, 19–23.

Lohnes, P. R., and McIntire, P. H. Classification validities of a statewide 10th grade test program. *Personnel and Guidance Journal*, 1967, 561–567.

Lord, F. M. A paradox in the interpretation of group comparisons. *Psychological Bulletin* 1967, **68**, 304–5.

Love, W. A., and Stewart, D. K. *Interpreting canonical correlations: theory and practice.* Palo Alto: American Institutes for Research, 1968.

Mahalanobis, P. C. On the generalized distance in statistics. *Proceedings of the National Institute of Science, India*, 1936, **12**, 49–55.

Marks, M. R. Two kinds of regression weights that are better than betas in crossed samples. Paper read to American Psychological Association Convention, September, 1966.

McKeon, J. J. Canonical analysis: some relations between canonical correlation, factor analysis, discriminant function analysis, and scaling theory. *Psychometric Monographs*, 1966, No. 13.

McNemar, Q. *Psychological statistics.* New York: Wiley, 1962.

Meredith, W. Canonical correlation with fallible data. *Psychometrika*, 1964, **29**, 55–65.

Michael, W. B. An empirical study of the comparability of factor structure when unities and communality estimates are used. *Educational and Psychological Measurement*, 1958, **18**, 347.

Miller, J. K. *The development and application of bi-multivariate correlation: a measure of statistical association between multivariate measurement sets*. Ed. D. Dissertation, Faculty of Educational Studies, State University of New York at Buffalo, 1969.

Morrison, D. F. *Multivariate statistical methods*. New York: McGraw-Hill, 1967.

Mosteller, F., and Bush, R. R. Selected quantitative techniques. In Lindzey, G. (ed.) *Handbook of social psychology*, Vol. 1, pp. 289–334, Cambridge, Mass.: Addison-Wesley, 1954.

Natrella, M. G. *Experimental statistics*. Washington, D. C.: U.S. Government Printing Office, 1963.

Nephew, C. T. *Guides for the allocation of school district financial resources*. Ed. D. Dissertation, Faculty of Educational Studies, State University of New York at Buffalo, 1969.

Neuhaus, J. O., and Wrigley, C. The quartimax method: an analytical approach to orthogonal simple structure. *British Journal of Statistical Psychology*, 1954, **7**, 81–91.

Nunnally, J. C. *Psychometric theory*. New York: McGraw-Hill, 1967.

Orden, A. Matrix inversion and related topics by direct methods. Ch. 2 in A. Ralston and H. S. Wilf (eds.), *Mathematical methods for digital computer*. New York: Wiley, 1960.

Ortega, J. M. On Sturm sequences for tri-diagonal matrices. *Journal of the Association of Computing Machinery*, 1960, **7**, 260-263.

Overall, J. E. Orthogonal factors and uncorrelated factor scores. *Psychological Reports*, 1962, **10**, 651-662.

Overall, J. E. Note on the scientific status of factors. *Psychological Bulletin*, 1964, **61**, 270–76.

Pearson, K. On lines and planes of closest fit to system of points in space. *Philosophy Magazine*, 1901, **6**, 559–572.

Porebski, O. R. On the interrelated nature of the multivariate statistics used in discriminatory analysis. *The British Journal of Mathematical and Statistical Psychology*, 1966, **19**, Part 2, 197–214.

Porebski, O. R. Discriminatory and canonical analysis of technical college data. *The British Journal of Mathematical and Statistical Psychology*, 1966, **19**, Part 2, 215–236.

Ralston, A., and Wilf, H. S. (eds.), *Mathematical methods for digital computers*, New York: Wiley, 1960.

Ralston, A., and Wilf, H. S. (eds.), *Mathematical methods for digitial computers*, Vol. 2. New York: Wiley, 1967.

Rao, C. R. *Advanced statistical methods in biometric research*. New York: Wiley, 1952.

Rao, C. R. *Linear statistical inference and its applications*. New York: Wiley, 1965.

Rao, C. R. and Slater, P. Multivariate analysis applied to differences between neurotic groups. *British Journal of Psychology (Statistical Section)*, 1949, **2**, 17–29.

Rulon, P. J. Distinctions between discriminant and regression analysis and a geometric interpretation of the discriminant function. *Harvard Educational Review*, 1951, **21**, 80–90.

Rulon, P. J. The stanine and the separile: a fable. *Personnel Psychology*, 1951, **4**, 99–114.

Rulon, P. J., and Brooks, W. D. On statistical tests of group differences. Ch. 2 in D. K. Whitla (ed.), *Handbook of measurement and assessment in behavioral sciences*. Reading, Mass.: Addison-Wesley, 1968.

Rulon, P. J., Tiedeman, D. V., Tatsuoka, M. M., and Langmuir, C. R. *Multivariate statistics for personnel classification*, New York: Wiley, 1967.

Saunders, D. R. An analytic method for rotation to orthogonal simple structure. Princeton: *Educational Testing Service Research Bulletin*, 53–10, 1953.

Saupe, J. L. Factorial-design multiple-discriminant analysis: a description and an illustration. *American Educational Research Journal*, 1965, **2**, 175–184.

Searle, S. R. *Matrix algebra for the biological sciences*. New York: Wiley, 1966.

Shaycoft, M. F. *The high school years: growth in cognitive skills*. Pittsburgh: American Institutes for Research, 1967.

Stewart, D. K., and Love, W. A. A general canonical correlation index. *Psychological Bulletin*, 1968, **70**, 160–163.

Super, D. E., and Overstreet, P. L. *The vocational maturity of ninth-grade boys*. New York: Teachers College Press, 1960.

Super, D. E., et al. *Vocational development: a framework for research*. New York: Teachers College Press, 1957.

Tatsuoka, M. M. *Joint probability of membership in a group and success therein: an index which combines the information from discriminant and regression analysis*. Cambridge, Mass: Harvard Graduate School of Education, 1956.

Tatsuoka, M. M. Joint-probability of membership and success in a group: an index which combines analysis as applied to the guidance problem. *Harvard Studies in Career Development, Report 6*. Cambridge, Mass.: Harvard Graduate School of Education, 1959. (mimeo.)

Tatsuoka, M. M., and Tiedeman, D. V. Discriminant analysis. *Review of Educational Research*, Washington, D.C.: American Educational Research Association, 1954.

Thomson, G. The maximum correlation of two weighted batteries. *British Journal of Psychology*, Statistical Section, Part I, 1947, pp. 27–34.

Thurstone, L. L. *Multiple-factor analysis*. Chicago: University of Chicago Press, 1947.

Tiedeman, D. V. A model for the profile problem. *Proceedings, 1953 Invitational Conference on Testing Problems*, Princeton, N.J.: Educational Testing Service, 1954.

Tiedeman, D. V., and Bryan, J. G. Prediction of college field of concentration. *Harvard Educational Review*, 1954, **24**, 122–139.

Tiedeman, D. V., Bryan, J. G., and Rulon, P. J. *The utility of the Airman Classification Battery for the assignment of airmen to eight Air Force specialties*. Cambridge, Mass.: Educational Research Corporation, 1951.

Tiedeman, D. V., and Sternberg, J. J. Information appropriate for curriculum guidance. *Harvard Educational Review*, 1952, **22**, 257–274.

Tukey, J. W. The future of data analysis. *Annals of Mathematical Statistics*, 1962, **33**, 1–67.

Tukey, J. W. Analyzing data: sanctification or detective work? *American Psychologist*, 1969, **24**, 83–91.

Wert, J. E., Neidt, C. D., and Ahmann, J. S. *Statistical methods in educational and psychological research*. New York: Appleton-Century-Crofts, 1954.

Whitla, D. K. (ed.) *Handbook of measurement and assessment in behavioral sciences*. Reading, Mass.: Addison-Wesley, 1968.

Wilkinson, J. H. The calculation of eigenvectors of codiagonal matrices. *Computer Journal* 1958, **1**, 90–96.

Wilkinson, J. H. Householder's method for the solution of the algebraic eigen-problem. *Computer Journal*, 1960, **3**, 23–37.

Wilks, S. S. Certain generalizations in the analysis of variance. *Biometrika*, 1932, **24**, 471–474.

Wilks, S. S. *Mathematical statistics*. New York: Wiley, 1962.

Williams, E. J. *Regression analysis*. New York: Wiley, 1959.

Williams, E. J. The analysis of association among many variates (with discussion). *Journal of the Royal Statistical Society*, 1967, Series B, **29**, 199–242.

Wrigley, C. The case against communalities. *Research Report*, 19, Berkeley: University of California, 1957 (mimeographed).

REFERENCES   354

APPENDIX A

‖‖

Statistical Tables

TABLE A-1   Upper Percentage Points of the F-Distribution for Selected Degrees of Freedom for Numerator (n) and Denominator (d),

d	$1-P$	n: 1	5	12	20	30	60	∞
	.05	6.6	5.1	4.7	4.6	4.5	4.4	4.4
5	.01	16	11	9.9	9.6	9.4	9.2	9.0
	.001	47	30	26	25	25	24	24
	.05	5.0	3.3	2.9	2.8	2.7	2.6	2.5
10	.01	10	5.6	4.7	4.4	4.3	4.1	3.9
	.001	21	10	8.5	7.8	7.5	7.1	6.8
	.05	4.5	2.9	2.5	2.3	2.3	2.2	2.1
15	.01	8.7	4.6	3.7	3.4	3.2	3.1	2.9
	.001	17	7.6	5.8	5.3	5.0	4.6	4.3
	.05	4.4	2.7	2.3	2.1	2.0	2.0	1.8
20	.01	8.1	4.1	3.2	2.9	2.8	2.6	2.4
	.001	15	6.5	4.8	4.3	4.0	3.7	3.4
	.05	4.2	2.5	2.1	1.9	1.8	1.7	1.6
30	.01	7.6	3.7	2.8	2.6	2.4	2.2	2.0
	.001	13	5.5	4.0	3.5	3.2	2.9	2.6
	.05	4.1	2.5	2.0	1.8	1.7	1.6	1.5
40	.01	7.3	3.5	2.7	2.4	2.2	2.0	1.8
	.001	13	5.1	3.6	3.2	2.9	2.6	2.2
	.05	4.0	2.4	1.9	1.8	1.7	1.5	1.4
60	.01	7.1	3.3	2.5	2.2	2.0	1.8	1.6
	.001	12	4.8	3.3	2.8	2.6	2.3	1.9
	.05	3.9	2.3	1.8	1.7	1.6	1.4	1.3
120	.01	6.9	3.2	2.3	2.0	1.9	1.7	1.4
	.001	11	4.4	3.0	2.5	2.3	2.0	1.5
	.05	3.8	2.2	1.8	1.6	1.5	1.3	1.0
∞	.01	6.6	3.0	2.2	1.9	1.7	1.5	1.0
	.001	11	4.1	2.7	2.4	2.0	1.7	1.0

[a] Values have been selected from Abramowitz and Stegun, 1964, pp. 987–989

TABLE A-2   Upper Percentage Points for the Chi-Square Distribution

Number of Degrees of Freedom	.05	.01
10	18	23
20	31	38
30	44	51
40	56	63
50	67	75
60	79	88
70	90	100
80	102	112
90	113	123
100	124	135
110	135	147
120	146	158
130	157	170
140	168	181
150	179	193
160	190	204
170	202	215
180	212	226
190	223	238
200	234	249
210	245	260
220	255	271
230	266	282
240	277	293
250	282	299
260	298	315
270	309	326
280	320	337
300	341	359
320	362	381
340	384	403
350	394	414
360	405	425
380	426	446
400	447	468
420	468	490
440	490	511
460	511	533
480	532	554
500	553	576
520	574	597
540	595	619
560	616	640
580	637	661

Number of Degrees of Freedom	.05	.01
600	658	683
620	679	704
640	700	725
660	721	747
680	741	768
700	762	789
720	783	810
740	804	832
760	825	853
780	846	874
800	867	895
820	887	916
840	908	938
860	929	959
880	950	980
900	971	1001
920	991	1022
940	1012	1043
960	1033	1064
980	1054	1085
1000	1074	1106
1100	1178	1211
1200	1281	1316
1300	1385	1421
1400	1488	1525
1500	1591	1630
1600	1694	1734
1700	1797	1838
1800	1900	1942
1900	2002	2046
2000	2105	2149

APPENDIX B

‖‖

TALENT Data

TABLE B-1 Selected Project TALENT Variables

Variable Number	Card Columns	Name of Variable
ID	2–4	Student Identification Number
1	7	School Size (4 categories—based on number of seniors) 1. under 25 2. 25–99 3. 100–399 4. 400 or more
2	9	Geographic Region (9 categories) 1. New England —6 States 2. Mid-East —5 States and D.C. 3. Great Lakes —5 States 4. Plains —7 States 5. Southeast —12 States 6. Southwest —4 States 7. Rocky Mountains—5 States 8. Far West —4 States 9. Non-contiguous —Alaska and Hawaii
3	11–12	Age (nearest year)
4	14	Sex (1 = male; 2 = female)
5	16–17	Career Plans (36 vocations) 01. Accountant 02. Biological scientist (biologist, botanist, physiologist, zoologist, etc.) 03. College professor 04. Dentist 05. Engineer (aeronautical, civil, chemical, mechanical, etc.) 06. Elementary school teacher 07. High school teacher 08. Lawyer 09. Mathematician 10. Pharmacist 11. Clergyman (minister, priest, rabbi, etc.) 12. Physical scientist (chemist, geologist, physicist, astronomer, etc.) 13. Physician 14. Political scientist or economist 15. Social worker 16. Sociologist or psychologist 17. Armed forces officer 18. Artist or entertainer 19. Businessman 20. Craftsman

Variable Number	Card Columns	Name of Variable
		21. Engineering or scientific aide
		22. Forester
		23. Medical or dental technician
		24. Nurse
		25. Pilot, airplane
		26. Policeman or fireman
		27. Secretary, office clerk or typist
		28. Writer
		29. Barber or beautician
		30. Enlisted man in the armed forces
		31. Farmer
		32. Housewife
		33. Salesman or saleswoman
		34. Skilled worker (electrician, machinist, plumber, printer, etc.)
		35. Structural worker (bricklayer, carpenter, painter, paperhanger, etc.)
		36. Some other occupation different from any above
6	19–20	Weight (lb)
		01. 74 or less
		02. 75–89
		03. 90–104
		04. 105–119
		05. 120–134
		06. 135–149
		07. 150–164
		08. 165–179
		09. 180–194
		10. 195–209
		11. 210–224
		12. 225 or more
7	22–23	Type of College Student Plans to Attend (9 categories)
		01. Do not expect to go to college
		02. Teachers college
		03. Agricultural college
		04. Engineering college
		05. Liberal Arts College
		06. College specializing in music or fine arts
		07. University which includes many of the above colleges
		08. Some other type of college
		09. Have no plans regarding the type of college I will attend

TABLE B-1 (*continued*)

Variable Number	Card Columns	Name of Variable
8	25	Plan College Full-time? (SIB 301) 1. Definitely will go 2. Almost sure to go 3. Likely to go 4. Not likely to go 5. Definitely will not go
9	27–29	Information Test, Part I (R-190)
10	31–33	Information Test, Part II (R-192)
11	35–37	English Test (R-230)
12	39–40	Reading Comprehension Test (R-250)
13	42–43	Creativity Test (R-260)
14	45–46	Mechanical Reasoning Test (R-270)
15	48–49	Abstract Reasoning Test (R-290)
16	51–52	Mathematics Test (R-340)
17	54–55	Sociability Inventory (R-601)
18	57–58	Physical Science Interest Inventory (P-701)
19	60–61	Office Work Interest Inventory (P-713)
20	63–65	Socioeconomic Status Index (P-801)

1	3	1	17	1	22	05	01	4	151	078	087	39	09	12	09	20	10	20	18	105
2	2	1	18	1	34	06	01	5	100	041	076	15	07	10	10	15	04	15	13	088
3	3	1	17	1	19	07	08	5	156	069	090	28	08	12	09	26	09	08	06	105
4	1	1	18	1	36	05	08	3	164	089	082	47	13	14	12	29	04	28	24	091
5	3	1	17	1	04	06	07	1	164	088	094	40	10	15	12	32	11	26	01	097
6	4	1	17	1	18	06	01	5	118	071	086	21	10	14	11	21	06	08	09	099
7	3	1	17	1	36	08	08	1	123	072	076	33	09	12	09	25	11	16	11	113
8	2	1	17	1	12	07	05	2	224	096	099	46	18	20	15	51	09	36	02	098
9	3	1	17	1	34	05	09	4	162	071	093	42	10	17	13	31	06	33	16	095
10	2	1	17	1	05	09	04	2	183	082	079	38	14	18	11	39	09	30	03	102
11	3	1	17	1	05	07	07	1	196	084	085	42	12	17	12	32	06	27	12	111
12	2	1	17	1	07	09	07	2	180	073	082	32	10	18	08	31	01	20	23	106
13	3	1	18	1	34	11	01	5	216	103	087	39	16	17	11	34	04	23	11	092
14	3	1	18	1	07	07	02	1	182	095	081	43	08	10	11	34	08	28	19	092
15	4	2	17	1	12	11	05	5	191	095	099	41	13	10	08	34	07	33	04	099
16	4	2	17	1	34	06	02	5	123	072	076	34	07	09	05	16	08	09	06	101
17	4	2	17	1	07	06	02	4	185	083	091	41	11	12	11	32	02	17	20	094
18	4	2	17	1	04	06	05	2	165	073	092	38	11	14	11	35	07	18	14	106
19	4	2	17	1	07	08	02	3	147	074	081	32	05	14	13	30	04	13	03	102
20	4	2	17	1	36	07	07	1	192	096	092	41	17	17	11	27	08	21	20	102
21	4	2	16	1	36	07	09	4	139	085	086	32	10	12	07	15	08	25	03	092
22	4	2	17	1	07	09	02	1	193	081	086	43	05	11	11	42	02	28	14	090
23	3	2	17	1	05	08	01	4	116	070	067	24	09	09	07	16	05	27	24	092
24	4	2	17	1	19	06	07	1	187	107	092	43	12	15	12	37	11	26	06	109
25	4	2	16	1	05	06	04	1	208	091	092	43	16	19	12	39	02	31	14	117
26	4	2	17	1	36	10	01	3	125	061	078	25	10	15	07	23	08	17	14	089
27	4	2	16	1	34	07	08	4	190	093	086	36	14	16	12	39	03	24	16	096
28	4	2	16	1	05	07	04	1	196	092	097	45	10	16	11	49	09	34	17	100
29	4	2	17	1	07	07	07	1	131	076	076	27	08	10	13	17	11	24	12	100
30	3	2	17	1	02	07	07	1	210	089	099	39	09	17	11	44	09	28	10	112
31	3	2	17	1	13	05	05	2	207	100	096	44	18	15	10	43	08	27	11	120
32	3	2	17	1	36	08	01	4	167	083	083	36	13	11	08	10	03	24	19	105
33	2	2	18	1	19	06	05	3	167	070	083	33	07	15	11	27	03	23	23	103
34	2	2	18	1	30	04	01	4	145	072	082	27	10	12	11	19	08	18	11	085
35	3	2	17	1	04	07	07	1	210	093	089	43	18	17	10	42	12	34	14	113
36	3	2	17	1	12	07	07	1	190	091	104	47	08	13	14	47	10	34	13	114
37	3	2	18	1	21	08	04	3	172	081	084	36	18	16	08	18	09	26	03	095
38	2	3	17	1	08	06	05	4	165	096	088	42	13	12	14	28	09	18	10	084
39	3	5	17	1	04	08	07	1	194	096	085	41	15	19	12	41	05	24	19	114
40	3	2	17	1	22	07	07	1	197	096	086	44	14	20	12	37	05	19	11	107
41	3	2	18	1	34	05	01	5	149	079	090	43	09	17	13	32	05	14	06	101
42	3	2	18	1	34	08	05	1	170	095	094	40	13	15	06	23	07	34	09	096
43	3	5	17	1	10	07	05	1	214	095	099	44	17	20	10	32	11	30	23	095
44	4	2	16	1	18	06	06	3	107	062	071	23	01	05	09	15	07	06	04	089
45	4	2	16	1	36	06	08	3	141	069	089	33	07	16	11	24	07	18	16	095
46	4	2	17	1	18	08	06	2	200	106	106	48	18	13	12	37	05	11	00	094
47	4	2	18	1	36	06	01	5	137	068	071	34	07	15	11	14	06	06	04	087
48	4	2	17	1	36	06	01	4	095	053	079	23	08	08	08	09	06	19	23	086
49	4	2	17	1	19	06	09	3	146	087	085	39	14	12	12	36	08	18	23	096
50	2	2	17	1	05	05	05	3	191	082	082	36	12	16	11	39	05	40	21	082
51	4	2	17	1	36	12	01	5	124	076	069	23	11	16	12	13	06	23	07	089
52	3	2	17	1	36	06	01	4	138	056	089	25	07	11	08	13	09	21	20	080
53	2	2	17	1	15	07	01	5	080	051	047	24	07	08	04	11	06	13	13	086
54	3	2	17	1	36	12	01	5	149	088	074	22	05	14	08	16	10	11	07	103
55	2	2	18	1	31	07	01	1	131	066	059	26	06	16	08	22	03	12	13	084
56	3	2	17	1	26	07	01	4	105	056	069	05	06	09	08	14	07	15	16	101
57	3	2	17	1	36	05	01	4	178	086	067	29	09	13	09	13	03	14	01	086
58	3	2	17	1	30	06	01	3	149	087	088	45	14	12	10	33	02	24	24	093
59	2	2	18	1	17	06	01	4	162	077	099	38	12	16	11	27	01	16	01	091
60	3	2	18	1	36	06	01	5	162	082	088	44	11	15	12	27	02	18	24	113
61	2	2	17	1	25	06	08	5	155	083	082	44	07	14	08	27	07	31	01	096
62	3	2	17	1	05	05	04	1	191	095	072	43	16	16	09	33	12	36	06	101
63	3	2	17	1	05	10	07	1	220	112	100	46	13	16	09	50	08	36	07	111
64	2	2	18	1	35	06	01	4	163	072	080	34	10	17	11	24	09	20	14	091
65	3	3	17	1	05	06	01	4	120	066	075	34	10	12	10	13	08	19	14	100
66	3	3	17	1	30	07	01	5	199	113	084	40	19	15	09	32	03	29	08	099

67	3	3	18	1	05	10	04	1	197	088	100	36	09	14	09	41	08	36	09	108
68	3	3	17	1	18	07	09	4	106	063	078	20	05	08	08	21	07	02	01	094
69	4	3	17	1	09	11	04	1	214	107	094	44	11	10	10	44	08	31	09	115
70	4	3	17	1	05	07	04	5	178	098	093	33	14	11	09	36	10	31	13	101
71	4	3	18	1	34	07	01	5	141	069	074	28	08	15	08	19	05	13	01	093
72	4	3	17	1	36	07	05	1	218	105	095	39	13	15	11	35	05	24	04	109
73	2	3	17	1	05	08	07	2	177	083	086	39	14	18	15	39	11	33	11	114
74	4	3	17	1	04	08	04	1	185	094	078	44	11	18	10	39	05	25	09	102
75	4	3	18	1	26	06	01	4	116	055	064	28	01	10	03	23	06	21	17	105
76	4	3	17	1	05	05	04	3	162	091	086	37	08	13	07	15	10	26	13	101
77	3	3	17	1	13	06	07	1	202	081	094	45	11	15	14	46	11	26	14	103
78	2	3	19	1	36	07	01	4	120	065	074	20	05	09	09	15	03	21	03	103
79	2	3	17	1	01	05	08	4	115	062	073	32	09	15	10	26	09	08	27	095
80	2	3	17	1	18	07	09	3	168	098	097	38	10	10	12	25	02	11	08	097
81	2	3	17	1	34	05	01	5	124	063	078	28	08	11	06	20	12	17	04	106
82	2	3	17	1	07	07	05	2	164	072	093	39	13	12	12	43	06	28	17	100
83	3	3	17	1	02	07	07	1	168	085	078	27	07	12	09	24	10	35	14	115
84	1	3	18	1	35	08	01	4	122	059	075	30	14	13	11	11	10	10	10	078
85	3	3	17	1	34	06	01	5	122	055	071	24	08	16	12	17	06	23	10	101
86	3	3	17	1	12	06	05	1	217	117	102	44	18	15	14	49	10	33	06	115
87	3	3	17	1	01	07	05	1	197	103	098	44	13	09	11	45	11	33	24	120
88	3	3	17	1	09	08	09	4	112	049	056	16	05	11	05	12	05	20	30	088
89	4	3	17	1	02	07	03	1	221	110	094	47	15	14	10	36	03	37	13	110
90	4	3	16	1	12	07	05	2	193	077	100	42	13	14	11	44	07	34	17	096
91	3	3	19	1	05	05	04	3	146	060	073	25	14	10	08	27	10	26	06	093
92	2	3	17	1	05	06	02	1	130	074	080	30	06	11	10	26	11	29	16	092
93	3	3	17	1	34	06	01	4	139	082	079	24	11	19	06	24	10	14	14	103
94	3	3	17	1	05	08	04	1	217	115	100	43	16	17	14	49	05	30	03	121
95	3	3	17	1	19	10	05	2	204	102	098	44	15	17	09	31	01	28	13	095
96	4	3	17	1	36	06	01	5	114	053	075	29	03	18	12	25	09	09	01	099
97	4	3	18	1	33	07	09	3	162	089	083	42	10	17	11	17	04	07	04	102
98	4	3	17	1	19	06	05	1	140	075	087	41	14	16	13	40	04	13	24	098
99	2	3	17	1	06	06	02	5	155	071	082	28	12	15	07	27	09	34	26	100
100	3	3	18	1	12	06	07	2	170	074	075	21	11	14	10	12	08	34	16	099
101	3	3	17	1	30	05	01	4	142	086	081	27	12	16	07	19	08	21	13	099
102	2	3	17	1	12	08	01	5	183	079	092	35	11	13	11	27	01	16	00	099
103	2	3	18	1	01	02	01	5	147	087	078	31	07	12	02	16	12	08	07	113
104	4	3	17	1	13	08	05	1	173	084	076	39	14	18	10	36	07	24	17	114
105	4	3	17	1	34	05	08	5	171	106	084	36	16	12	12	27	10	26	06	110
106	4	3	16	1	12	08	07	2	194	104	093	46	13	15	13	47	03	36	24	109
107	4	3	17	1	12	07	04	1	214	109	098	43	14	18	11	36	12	36	04	098
108	4	3	17	1	35	07	01	4	131	080	070	41	09	13	12	29	04	11	04	092
109	4	3	16	1	36	07	09	4	145	071	085	30	16	12	09	15	05	25	**17**	098
110	4	3	17	1	01	07	01	5	162	085	077	39	10	09	09	23	03	14	10	106
111	3	3	17	1	01	06	07	2	169	095	001	44	15	18	12	00	09	22	39	103
112	2	3	18	1	05	09	06	5	109	067	067	15	13	18	09	19	06	16	19	111
113	4	3	17	1	18	07	06	2	123	061	062	14	05	09	05	13	11	24	13	102
114	4	3	17	1	04	06	08	2	165	083	073	34	14	10	05	20	05	21	06	102
115	4	3	18	1	05	06	08	5	120	065	079	20	08	10	09	24	09	19	14	099
116	2	3	18	1	29	06	01	5	106	045	076	25	08	13	07	25	04	18	20	088
117	3	3	17	1	05	06	04	1	213	093	091	26	09	18	12	39	08	33	11	105
118	2	3	17	1	11	06	01	5	135	064	078	22	12	12	09	19	02	04	01	094
119	3	3	17	1	01	05	05	1	122	063	087	39	13	11	10	30	06	13	06	108
120	3	3	18	1	34	05	01	4	170	087	090	36	13	13	11	22	08	11	06	092
121	2	4	17	1	31	07	01	5	089	041	067	19	07	15	07	12	03	13	10	084
122	2	3	18	1	05	07	04	2	185	093	093	45	15	15	10	44	07	28	13	098
123	3	3	19	1	08	07	05	1	193	094	084	38	10	16	12	38	10	27	13	110
124	3	3	18	1	36	07	01	5	182	082	077	38	09	12	06	19	10	22	13	103
125	3	3	17	1	12	08	07	1	223	112	106	44	18	20	14	48	05	34	19	112
126	4	3	17	1	26	06	01	5	151	069	086	36	10	16	11	16	03	22	14	095
127	4	3	18	1	20	07	09	3	191	106	091	42	11	19	11	31	09	29	14	088
128	2	3	17	1	36	06	01	4	137	070	089	27	08	13	10	26	12	13	17	096
129	2	3	17	1	05	07	01	4	151	075	094	34	12	13	10	23	04	19	13	096
130	1	3	17	1	17	08	02	2	168	079	093	40	16	16	12	39	05	25	02	098
131	3	4	17	1	36	07	01	5	124	062	069	26	09	15	09	25	03	10	11	089
132	2	4	17	1	05	07	01	5	160	070	078	23	07	18	11	26	10	21	14	089

```
133  2 4 17 1 11 09 08 1 151 086 085 34 04 13 10 31 10 11 13 104
134  2 4 17 1 19 08 01 4 136 065 077 27 08 14 09 20 09 04 00 088
135  2 4 17 1 19 06 08 3 159 091 089 40 08 07 10 24 11 18 26 099
136  4 4 18 1 05 06 07 1 183 097 092 46 15 19 12 36 10 33 19 101
137  3 4 18 1 10 05 07 2 130 080 079 31 07 13 13 20 08 19 04 101
138  4 4 18 1 14 06 07 1 217 110 101 47 15 16 13 39 04 27 14 113
139  4 4 17 1 13 07 05 1 204 108 095 45 12 13 12 28 07 26 24 106
140  2 4 17 1 01 06 01 5 116 040 063 19 04 07 08 15 05 19 14 091
141  2 4 19 1 05 05 04 4 132 071 080 28 09 10 09 13 07 15 11 092
142  4 4 17 1 12 07 04 1 233 115 096 45 18 20 10 52 05 38 01 115
143  3 4 17 1 01 07 05 1 161 085 086 33 06 13 11 29 06 16 25 111
144  2 4 18 1 14 09 07 3 184 102 084 45 14 20 15 27 04 20 21 088
145  2 4 17 1 30 07 01 5 132 078 063 22 08 10 09 20 08 04 01 085
146  2 4 17 1 29 07 01 5 112 063 076 23 03 08 07 12 12 12 27 091
147  3 5 17 1 19 08 08 3 152 088 090 35 09 09 07 20 09 19 19 104
148  3 5 18 1 10 06 08 3 149 074 081 43 12 14 10 31 09 26 14 103
149  2 5 18 1 31 06 03 2 165 079 087 36 16 20 08 29 09 26 04 102
150  3 5 18 1 19 08 05 2 121 079 086 24 06 09 11 12 10 19 13 107
151  3 5 17 1 08 07 05 2 164 068 083 36 11 14 12 24 05 18 04 107
152  2 5 17 1 05 06 04 2 162 091 090 35 15 20 10 36 10 39 24 100
153  2 5 18 1 04 08 08 3 117 058 065 17 05 09 02 16 06 16 23 100
154  3 5 17 1 19 06 09 3 100 052 071 17 09 13 09 15 10 26 23 092
155  3 5 18 1 36 06 01 5 103 042 064 12 04 06 04 11 03 16 27 085
156  1 5 17 1 05 12 04 4 186 093 077 30 08 18 06 20 02 31 20 086
157  2 5 17 1 19 08 07 1 134 078 085 35 07 11 10 22 11 08 00 092
158  2 5 19 1 05 06 07 2 124 066 085 29 06 11 02 18 08 33 20 078
159  3 5 18 1 36 11 01 4 092 047 065 16 06 04 04 13 11 13 10 095
160  3 5 18 1 36 09 01 5 123 070 081 23 07 13 09 21 04 14 16 108
161  2 5 17 1 10 06 07 2 172 089 092 39 16 15 11 34 09 36 17 103
162  3 5 17 1 01 03 08 1 167 103 100 46 06 09 11 37 07 21 23 092
163  3 5 17 1 05 08 04 1 164 087 088 39 11 16 10 38 11 26 11 103
164  3 5 18 1 34 07 01 3 113 061 072 16 09 17 07 10 06 26 21 084
165  3 5 17 1 05 12 05 1 149 083 095 36 09 11 04 24 12 33 24 099
166  4 5 18 1 21 09 09 1 140 027 063 14 07 05 05 20 07 18 10 086
167  2 5 17 1 05 08 01 5 111 062 086 28 15 11 08 19 09 19 03 096
168  4 5 17 1 18 07 08 3 172 102 101 43 08 17 13 39 11 11 04 109
169  4 5 17 1 31 06 01 5 131 073 079 31 07 10 12 15 03 09 01 083
170  3 5 18 1 05 06 01 4 130 074 084 19 07 12 05 17 06 22 13 087
171  2 5 18 1 04 07 07 1 156 086 091 28 11 12 10 28 11 23 17 106
172  3 5 17 1 02 07 02 3 110 062 075 31 07 14 06 20 07 19 10 100
173  2 5 19 1 36 07 08 4 162 074 079 30 07 14 12 17 10 16 01 094
174  3 5 18 1 22 05 09 3 121 075 081 30 05 11 09 17 02 00 01 085
175  2 5 17 1 12 05 04 3 177 094 082 40 16 15 08 25 04 37 24 092
176  3 5 17 1 19 06 09 5 092 048 070 12 11 08 02 11 10 25 10 094
177  3 5 17 1 13 08 07 1 198 102 092 46 14 14 10 38 09 31 06 100
178  2 5 19 1 17 08 01 4 088 041 066 14 05 09 02 09 03 09 00 077
179  3 5 17 1 36 07 07 4 134 077 081 39 05 18 12 20 05 13 09 082
180  4 5 16 1 18 05 01 5 147 084 085 25 09 12 09 15 08 06 03 095
181  4 5 17 1 23 05 07 1 164 086 083 31 09 10 11 31 10 38 04 108
182  4 5 17 1 10 06 08 3 205 105 094 34 15 16 15 37 07 33 11 095
183  2 5 17 1 19 07 07 1 170 087 098 35 13 18 09 25 06 05 17 100
184  2 5 17 1 36 07 01 5 156 090 077 40 09 08 08 20 08 08 03 080
185  2 5 18 1 05 08 07 1 221 113 103 47 17 19 15 46 08 36 19 111
186  2 5 18 1 25 06 01 4 131 053 086 19 13 12 10 18 10 30 31 083
187  3 5 16 1 36 07 07 3 142 087 079 35 09 11 09 18 10 29 03 114
188  2 5 18 1 11 08 05 1 163 097 096 46 06 13 12 24 10 04 06 094
189  2 5 18 1 22 08 03 1 128 073 052 40 09 14 07 12 10 21 17 092
190  2 5 17 1 19 07 08 1 135 070 083 31 10 14 10 22 10 18 27 100
191  3 5 18 1 36 07 01 5 116 057 074 22 07 14 10 22 06 13 00 096
192  3 5 18 1 34 06 01 4 147 071 083 31 14 13 09 20 02 18 17 093
193  2 6 18 1 25 07 04 2 201 098 092 42 13 19 13 42 04 27 16 104
194  2 6 16 1 07 09 07 1 148 073 082 31 10 15 07 25 12 32 16 105
195  3 6 17 1 05 09 04 1 203 100 095 44 14 17 08 43 10 30 09 096
196  4 6 18 1 13 06 06 1 161 064 068 04 05 19 12 30 03 26 07 104
197  4 6 17 1 15 07 09 3 144 085 081 41 13 13 11 16 04 06 07 103
198  3 6 17 1 29 05 08 3 107 055 082 25 08 14 11 19 07 13 21 081
```

199	3	6	18	1	36	08	09	1	163	086	087	32	12	13	10	29	04	30	18	110
200	2	6	18	1	04	09	04	2	156	087	096	42	16	10	08	31	10	29	06	114
201	1	6	19	1	03	06	05	5	078	035	043	18	01	04	01	08	04	22	30	097
202	3	6	18	1	18	05	07	1	200	100	097	45	16	18	12	36	05	33	13	108
203	1	6	18	1	05	07	04	3	187	089	073	36	10	19	10	30	07	31	16	105
204	2	6	17	1	31	08	01	5	141	084	081	38	10	12	07	20	08	11	07	107
205	4	6	18	1	19	07	04	1	187	088	096	43	10	19	12	37	09	29	16	105
206	4	6	17	1	04	06	07	1	150	077	074	40	10	08	07	20	07	31	14	106
207	2	6	18	1	26	07	01	5	104	067	075	28	03	08	07	15	04	13	18	076
208	3	5	18	1	34	08	08	4	120	059	081	30	02	08	07	23	04	13	1	081
209	2	7	17	1	25	08	01	4	191	088	091	39	15	17	11	26	02	11	04	088
210	1	7	17	1	22	06	01	5	113	053	069	25	08	07	04	21	01	17	06	095
211	4	7	17	1	10	04	07	3	154	081	091	36	13	14	09	24	02	15	10	108
212	4	7	17	1	18	06	06	4	119	071	087	35	08	08	08	13	06	18	20	083
213	4	7	17	1	34	08	01	5	131	048	061	20	09	10	06	16	00	14	06	094
214	3	7	17	1	05	07	04	1	140	062	072	24	09	14	10	27	02	23	14	107
215	3	7	17	1	05	06	07	2	176	081	090	43	18	15	10	36	07	20	14	098
216	3	7	17	1	01	06	08	4	173	105	095	43	16	16	12	40	10	22	36	100
217	3	8	18	1	12	05	07	2	214	103	097	45	11	18	14	48	09	38	06	111
218	3	8	17	1	18	06	06	3	176	078	083	36	11	15	14	12	08	10	00	096
219	3	8	18	1	07	08	09	2	139	089	090	35	09	14	13	21	09	04	01	109
220	2	8	18	1	07	06	02	2	098	038	091	28	07	05	09	15	09	08	07	086
221	3	8	17	1	05	07	08	1	175	089	080	34	10	11	08	25	03	11	06	099
222	2	8	18	1	36	09	01	5	190	102	073	44	09	14	07	20	01	03	00	104
223	3	8	17	1	06	07	02	1	178	084	089	40	11	16	13	39	08	36	23	096
224	1	8	18	1	23	06	09	1	143	072	085	33	08	14	11	23	05	20	00	097
225	4	8	18	1	20	07	05	2	134	071	080	34	13	17	10	18	07	16	25	094
226	4	8	17	1	36	09	07	2	194	102	081	41	15	16	11	22	11	20	10	107
227	4	8	18	1	18	06	01	4	109	067	080	29	08	07	11	10	07	05	06	082
228	4	8	18	1	35	06	01	5	105	064	057	21	07	11	09	15	02	19	11	088
229	4	8	17	1	36	04	01	5	139	073	079	16	04	14	09	15	06	10	07	098
230	4	8	17	1	01	06	09	2	125	078	083	41	12	12	11	24	10	17	07	101
231	2	8	17	1	26	07	09	3	178	074	080	23	08	19	12	29	05	24	14	098
232	4	8	17	1	08	05	07	1	053	032	040	07	03	06	02	12	02	16	07	099
233	4	7	18	1	02	06	04	1	216	113	099	46	16	17	11	45	09	36	21	114
234	2	2	17	1	34	08	01	5	184	079	094	40	14	17	14	28	06	19	10	079

1	2	5	16	2	13	04	07	3	142	083	094	39	06	04	09	15	06	31	15	092
2	2	3	18	2	32	05	01	5	148	095	094	39	09	09	06	24	10	05	11	089
3	3	1	18	2	24	07	08	4	136	073	092	37	14	18	11	29	04	14	20	097
4	2	1	17	2	36	04	05	1	207	100	108	46	10	11	14	43	06	18	17	118
5	3	1	17	2	29	03	01	5	100	047	079	15	13	05	05	27	08	05	39	096
6	3	1	17	2	24	05	08	5	124	059	093	46	09	05	11	28	11	16	36	102
7	4	1	19	2	29	12	01	4	091	047	066	11	06	07	05	12	09	06	17	100
8	4	1	18	2	06	03	02	1	140	093	089	42	08	09	12	32	05	11	23	096
9	4	1	19	2	32	05	01	5	148	073	000	42	08	14	12	00	06	24	24	089
10	4	1	18	2	27	03	01	5	056	034	086	23	04	04	08	15	04	13	37	095
11	4	1	17	2	06	05	07	1	121	066	095	40	10	09	11	35	07	06	29	107
12	2	1	17	2	27	03	08	2	084	071	084	37	06	06	09	11	06	03	15	100
13	2	1	17	2	27	04	08	1	152	090	101	43	13	15	12	40	11	14	11	111
14	2	1	18	2	27	04	01	5	086	060	087	28	06	06	06	07	09	07	36	087
15	3	1	17	2	32	03	01	5	101	059	093	33	14	09	09	27	10	01	36	085
16	3	1	16	2	36	04	01	5	110	062	078	23	09	10	12	13	06	17	27	108
17	3	1	17	2	15	03	07	3	117	082	098	45	08	09	09	20	07	05	26	090
18	3	1	17	2	07	05	05	2	180	083	096	42	13	13	13	45	10	26	24	101
19	4	2	16	2	27	05	01	5	146	087	095	41	08	13	12	22	12	29	34	092
20	4	2	17	2	27	05	01	5	118	080	081	43	09	07	09	15	07	06	13	094
21	4	2	17	2	06	05	05	1	167	094	088	40	13	09	11	28	07	27	26	106
22	4	2	17	2	36	05	05	1	122	075	084	24	11	07	09	11	08	21	31	103
23	4	2	17	2	27	06	01	5	086	052	088	17	03	02	05	12	07	02	30	091
24	4	2	16	2	28	04	05	1	199	094	102	45	13	11	11	37	10	14	23	107
25	4	2	17	2	27	10	01	4	116	086	084	25	12	10	07	18	02	17	30	091
26	4	2	17	2	36	03	08	1	103	058	080	27	05	06	06	13	10	11	19	088
27	4	2	17	2	24	05	01	4	097	042	085	24	05	08	11	19	10	16	17	093
28	4	2	17	2	27	04	07	5	111	071	081	32	05	07	09	16	11	33	37	110
29	4	2	17	2	36	06	07	1	152	092	091	39	11	09	07	34	07	25	11	119
30	4	2	16	2	15	05	09	4	133	077	093	32	09	04	15	24	11	01	17	104
31	4	2	17	2	27	03	01	5	073	055	083	32	03	04	03	13	09	04	37	099
32	2	2	17	2	07	05	02	1	173	094	104	43	13	17	14	24	10	29	31	077
33	3	2	17	2	36	04	07	1	207	103	098	43	15	18	13	46	12	29	29	116
34	2	2	18	2	27	05	01	5	144	092	099	39	10	12	10	25	08	14	37	101
35	3	2	18	2	25	04	01	5	117	059	086	26	02	04	14	15	12	11	25	102
36	3	2	17	2	24	04	01	5	163	094	091	43	12	08	09	29	03	11	03	095
37	3	2	17	2	06	06	02	1	189	098	107	42	17	13	12	39	06	22	01	097
38	2	2	17	2	15	07	05	1	182	097	094	41	15	10	14	41	09	16	30	108
39	3	2	18	2	27	06	01	5	098	063	087	28	06	10	10	20	05	03	27	099
40	3	5	17	2	32	05	01	5	172	098	103	45	15	14	11	31	05	04	24	110
41	3	5	17	2	27	04	08	5	110	054	092	30	09	12	09	19	07	13	30	099
42	4	2	17	2	23	04	01	5	182	093	099	42	10	15	13	39	10	31	13	094
43	4	2	17	2	06	06	02	2	185	099	104	42	11	13	09	33	08	17	21	101
44	4	2	17	2	24	04	01	5	125	059	079	28	08	10	12	27	08	13	19	092
45	3	2	16	2	27	06	01	5	105	060	072	14	12	04	09	07	10	06	39	096
46	3	2	17	2	27	07	01	5	036	021	063	15	04	08	04	09	07	03	27	100
47	4	2	17	2	27	05	01	4	153	085	094	32	11	11	08	22	08	12	29	093
48	4	2	17	2	06	04	02	1	145	081	084	31	04	05	09	27	10	14	11	108
49	4	2	17	2	07	05	05	1	171	097	096	44	13	14	12	38	07	24	23	097
50	4	2	17	2	01	07	01	5	123	063	082	31	12	07	10	23	04	08	40	094
51	4	2	17	2	27	04	01	5	097	055	080	21	04	11	07	18	07	18	30	094
52	3	2	17	2	36	04	01	5	164	101	096	45	13	15	14	37	10	18	19	096
53	3	2	17	2	27	04	01	5	099	068	082	32	08	08	08	05	11	01	33	105
54	2	2	17	2	27	04	01	5	109	064	074	27	09	07	01	13	10	04	34	087
55	2	2	18	2	36	07	08	1	141	062	096	37	06	16	11	39	05	24	16	104
56	3	2	17	2	27	06	01	4	178	095	102	44	09	11	12	27	09	18	33	090
57	3	2	17	2	27	04	08	1	129	081	096	44	11	09	12	30	12	07	31	106
58	3	2	17	2	07	04	07	1	147	091	089	32	07	11	08	30	10	21	24	098
59	3	2	17	2	27	04	01	5	138	084	093	44	10	14	09	25	11	04	16	111
60	3	2	18	2	29	05	01	5	124	080	094	38	14	10	09	21	09	04	23	086
61	3	2	18	2	27	05	01	4	120	077	084	42	09	07	08	16	04	18	31	101
62	3	2	16	2	27	06	01	5	100	083	093	33	08	10	09	17	08	15	27	087
63	4	2	17	2	24	06	09	5	109	051	081	28	04	04	08	18	08	24	29	089
64	4	2	16	2	24	09	07	2	154	079	095	42	13	12	09	35	08	29	26	098
65	4	2	18	2	24	05	08	1	081	045	066	17	04	06	04	14	10	24	21	091
66	3	2	17	2	06	05	02	1	182	097	103	37	08	09	12	24	11	11	10	111

```
 67  3 3 18 2 06 06 07 1 174 101 102 42 09 11 10 34 06 09 20 116
 68  3 3 17 2 27 10 01 5 121 078 086 30 09 08 06 21 07 00 26 097
 69  3 3 17 2 27 05 07 1 153 086 091 42 11 07 08 17 06 04 24 102
 70  2 3 17 2 27 05 01 4 130 077 103 33 05 09 10 19 08 13 39 102
 71  2 3 17 2 29 06 08 5 093 057 079 24 11 08 07 16 07 14 30 085
 72  3 3 18 2 06 05 07 1 205 103 111 46 15 14 13 47 03 18 24 102
 73  4 3 17 2 23 05 08 5 143 071 097 37 11 11 09 31 09 01 36 103
 74  4 3 18 2 27 05 01 5 167 093 088 35 12 13 07 11 10 13 19 102
 75  2 3 17 2 24 03 01 5 128 068 087 36 11 12 15 26 07 19 23 099
 76  2 3 19 2 27 05 01 4 112 077 077 35 08 06 10 18 07 07 15 104
 77  2 3 17 2 29 05 01 5 132 077 094 40 02 11 08 21 05 12 36 097
 78  4 3 18 2 36 06 08 1 113 078 083 37 12 07 11 17 10 21 29 085
 79  4 3 16 2 13 05 07 5 073 043 071 16 03 02 09 07 09 08 00 108
 80  3 3 17 2 07 06 07 1 178 101 099 41 10 14 13 25 06 12 09 099
 81  3 3 18 2 17 07 08 4 174 106 098 43 16 13 11 29 07 09 17 108
 82  3 3 18 2 36 03 01 5 103 062 081 24 07 12 13 19 05 05 33 108
 83  3 3 17 2 27 05 01 4 160 087 109 45 17 14 13 33 02 21 33 098
 84  3 3 18 2 07 05 05 2 137 078 107 42 10 07 11 42 06 13 27 094
 85  3 3 17 2 27 05 01 5 138 071 095 31 07 04 10 16 10 03 24 095
 86  3 3 17 2 24 05 07 1 184 086 093 37 12 09 11 44 08 18 10 116
 87  2 3 17 2 03 04 05 1 201 094 110 47 19 17 14 51 03 34 10 104
 88  3 3 18 2 27 06 01 5 120 060 084 27 08 08 10 26 07 09 29 097
 89  3 3 18 2 01 05 01 5 111 067 080 26 08 06 04 15 03 00 21 091
 90  3 3 17 2 33 05 01 5 148 078 108 39 07 06 10 25 08 12 29 090
 91  3 3 17 2 27 05 02 3 099 060 079 38 08 06 12 19 02 08 33 094
 92  3 3 17 2 27 05 01 4 175 086 100 40 13 08 07 23 04 16 24 102
 93  4 3 17 2 33 03 06 4 084 045 081 27 11 08 14 17 11 07 29 097
 94  4 3 17 2 18 07 05 4 109 066 079 28 09 09 10 25 04 06 34 090
 95  3 3 17 2 07 05 02 1 147 078 096 40 09 09 08 17 05 03 33 114
 96  4 3 17 2 24 04 07 2 134 068 088 34 09 08 09 21 11 13 21 115
 97  4 3 17 2 15 06 01 5 142 088 075 37 04 11 00 11 05 12 24 099
 98  2 3 17 2 27 05 07 1 153 090 087 41 08 12 07 16 09 23 39 103
 99  3 3 17 2 27 04 01 5 102 061 085 30 07 07 08 20 09 04 23 089
100  3 3 17 2 27 04 01 4 109 057 082 22 06 04 06 13 11 05 36 105
101  4 3 18 2 27 04 01 5 110 069 078 42 10 12 09 20 08 13 37 097
102  4 3 18 2 32 06 01 5 128 080 077 28 15 10 13 25 08 07 30 101
103  4 3 17 2 06 05 07 2 207 108 102 48 19 11 13 41 06 14 09 092
104  3 3 17 2 18 05 01 5 105 069 087 24 04 05 04 19 08 01 06 096
105  3 3 18 2 32 05 01 4 153 080 100 38 15 10 10 22 07 03 36 104
106  2 3 17 2 29 04 09 4 092 052 082 18 04 04 10 14 10 06 19 087
107  2 3 17 2 32 05 01 5 108 070 082 24 07 11 12 29 10 12 37 100
108  4 3 17 2 32 05 07 2 134 077 080 36 11 04 05 19 11 07 06 117
109  4 3 17 2 06 05 07 1 139 092 096 45 13 06 09 22 11 11 23 120
110  4 3 17 2 06 04 07 1 117 079 076 40 08 08 08 20 03 08 26 101
111  4 3 18 2 36 05 01 5 089 055 081 16 05 08 08 11 08 14 24 093
112  3 3 18 2 32 04 01 5 123 077 084 36 06 05 11 10 04 15 29 101
113  2 3 17 2 28 05 01 5 129 074 093 44 09 12 10 28 03 11 17 091
114  3 3 18 2 36 05 01 5 108 054 074 20 09 11 06 16 08 11 26 093
115  4 3 17 2 07 04 07 2 172 081 099 46 08 15 13 41 08 25 03 095
116  4 3 17 2 07 05 02 1 199 090 088 37 13 11 10 39 05 18 20 107
117  4 3 18 2 15 05 07 4 105 063 075 29 05 02 04 10 09 08 24 097
118  4 3 18 2 24 06 07 1 132 074 086 35 07 06 07 22 06 15 07 087
119  4 3 17 2 19 04 01 4 103 053 081 24 07 10 11 14 09 13 34 088
120  2 3 17 2 27 04 08 5 160 083 100 46 10 12 11 23 07 13 23 098
121  2 2 17 2 07 05 02 1 171 090 099 47 18 14 14 23 06 06 19 109
122  3 1 17 2 24 07 08 1 153 078 096 00 00 00 00 36 09 17 17 098
123  3 1 17 2 27 06 01 4 115 058 084 30 11 13 08 27 11 13 36 093
124  2 3 17 2 07 05 05 1 163 086 104 41 15 09 11 43 10 13 33 089
125  3 3 17 2 23 06 08 4 141 087 087 40 14 10 10 20 07 11 30 105
126  3 3 18 2 27 04 08 4 145 076 081 31 10 11 12 19 09 21 31 105
127  3 3 17 2 32 05 01 5 111 073 086 26 08 08 08 18 08 05 39 097
128  4 3 17 2 17 03 01 5 124 079 098 39 05 12 10 23 09 16 37 095
129  2 3 17 2 27 07 01 5 147 073 099 38 09 08 13 24 07 09 40 100
130  2 3 17 2 27 05 01 5 120 074 091 36 09 10 09 16 04 04 30 091
131  3 3 17 2 19 06 09 3 104 061 092 37 07 14 09 23 03 08 39 109
132  2 4 17 2 27 03 01 5 127 072 087 31 04 02 09 23 06 14 30 091
```

```
133  2 4 17 2 07 04 07 1 162 089 099 44 15 10 13 25 07 04 26 096
134  3 4 17 2 06 06 09 1 152 067 098 33 10 13 10 26 05 07 21 102
135  3 4 17 2 36 03 01 5 093 058 077 25 10 07 06 13 03 08 33 092
136  3 4 18 2 27 05 01 5 084 061 076 26 05 07 10 20 06 04 33 097
137  3 4 17 2 27 05 05 1 116 068 088 21 04 06 07 11 11 04 36 102
138  3 4 17 2 06 04 02 1 138 065 086 36 08 08 10 20 04 04 14 109
139  2 4 18 2 15 06 05 1 151 000 086 35 16 06 03 28 01 01 03 105
140  2 4 17 2 27 06 01 4 125 069 080 24 06 07 10 19 10 04 30 084
141  2 4 18 2 27 05 01 5 108 063 090 21 10 06 09 23 02 03 34 084
142  2 4 17 2 06 04 05 1 137 069 090 33 09 10 09 20 11 13 27 110
143  2 4 17 2 32 04 01 5 102 059 081 31 07 05 10 16 09 07 26 095
144  2 4 17 2 32 04 01 4 140 074 090 32 08 07 10 28 09 26 080
145  4 4 17 2 27 05 05 4 138 081 090 38 08 08 09 25 04 16 34 106
146  3 4 17 2 06 06 02 1 166 094 090 39 14 12 10 25 10 08 16 104
147  2 4 18 2 27 04 01 4 127 080 089 37 14 12 13 16 10 15 27 107
148  3 4 18 2 16 04 05 2 184 106 099 44 11 15 11 37 07 24 26 106
149  4 4 17 2 32 04 08 3 131 078 075 27 04 08 09 20 03 07 30 106
150  4 4 17 2 18 06 06 2 158 097 084 44 09 07 09 31 04 16 16 113
151  3 4 17 2 32 06 01 5 131 083 092 36 12 10 08 31 00 12 36 104
152  2 4 18 2 32 03 01 5 086 047 080 21 07 07 10 16 10 00 29 088
153  2 4 17 2 29 06 01 4 100 042 066 36 10 07 13 15 08 03 17 074
154  2 4 17 2 17 05 01 4 112 058 090 36 07 08 10 11 00 01 16 083
155  2 4 18 2 01 06 02 2 178 088 105 45 15 16 15 39 01 08 37 098
156  4 4 17 2 29 05 01 5 098 060 075 29 08 05 11 10 06 10 19 096
157  4 4 17 2 27 06 01 5 108 068 099 29 08 10 13 23 06 05 33 084
158  3 4 17 2 27 03 07 2 155 089 100 42 10 09 14 21 10 17 39 096
159  3 4 17 2 25 05 05 5 109 071 081 20 03 06 08 17 11 00 11 113
160  3 5 17 2 36 06 08 2 187 110 097 48 13 12 13 37 07 10 00 112
161  3 5 18 2 27 04 01 5 136 070 094 42 10 12 12 20 07 07 33 096
162  2 5 18 2 27 04 01 4 104 041 081 19 07 07 07 11 02 21 19 094
163  3 5 18 2 07 04 02 1 157 079 080 34 09 07 12 20 07 13 34 099
164  2 5 17 2 07 05 02 1 198 102 095 45 13 11 06 46 10 15 03 108
165  3 5 17 2 06 05 02 4 093 061 088 27 05 07 12 16 11 07 33 092
166  2 5 18 2 06 05 02 4 125 057 093 25 06 09 09 25 07 19 36 077
167  2 5 18 2 36 06 01 4 129 071 079 37 08 09 08 12 08 21 21 105
168  2 5 17 2 27 05 01 5 099 064 088 18 08 02 08 16 09 04 40 095
169  2 5 18 2 24 04 07 1 206 105 105 46 16 11 11 34 07 18 09 107
170  2 5 17 2 27 04 05 5 125 059 091 25 10 07 07 18 04 04 40 085
171  2 5 18 2 24 05 08 5 133 077 094 39 10 09 10 29 11 19 13 111
172  2 5 17 2 06 01 02 1 091 045 067 14 06 07 07 17 07 04 31 089
173  2 5 18 2 32 04 09 1 136 059 081 38 08 10 06 26 09 14 27 107
174  3 5 18 2 27 04 09 4 138 082 107 38 06 06 09 33 10 09 37 097
175  2 5 18 2 32 05 01 4 081 042 080 23 05 06 06 10 07 10 29 097
176  2 5 18 2 24 04 07 3 089 032 074 17 03 04 04 06 08 09 26 085
177  2 5 18 2 36 05 03 3 109 045 088 26 06 04 04 12 07 11 33 102
178  2 5 17 2 27 08 01 4 099 071 090 33 10 10 10 17 08 04 40 083
179  2 5 17 2 27 05 01 5 101 058 090 30 04 03 07 18 10 05 39 101
180  2 5 18 2 32 05 01 5 109 071 093 37 11 13 10 22 11 02 29 090
181  3 5 15 2 03 05 05 1 132 068 094 29 09 06 07 25 07 21 24 097
182  4 5 17 2 07 08 09 1 106 062 085 22 07 07 08 18 05 15 37 098
183  3 5 17 2 23 04 05 2 163 092 099 39 13 14 11 24 08 11 07 106
184  3 5 17 2 07 04 08 1 159 082 094 29 12 12 10 20 06 23 23 118
185  4 5 17 2 07 05 07 1 153 080 093 44 09 09 12 19 05 09 04 114
186  4 5 17 2 07 04 07 1 128 083 104 43 15 10 12 27 10 08 27 101
187  4 5 17 2 36 04 09 4 063 043 084 27 02 04 06 11 06 08 36 091
188  2 5 17 2 27 05 08 4 117 074 086 41 11 09 07 27 09 03 37 104
189  3 5 16 2 27 03 08 5 130 079 109 40 16 08 12 31 05 04 29 097
190  2 5 17 2 06 04 02 1 136 079 072 33 07 08 13 21 05 12 14 096
191  2 5 17 2 27 04 08 4 117 066 102 30 09 08 09 20 05 13 23 087
192  4 5 18 2 06 03 02 1 141 075 089 28 09 08 07 19 04 23 27 101
193  4 5 18 2 27 05 05 1 106 060 090 28 02 06 05 20 10 05 37 087
194  2 5 18 2 06 04 08 1 112 048 087 41 07 06 08 31 03 18 26 094
195  2 5 18 2 27 06 01 4 108 056 091 24 08 07 11 16 05 04 37 085
196  3 5 18 2 07 05 02 2 107 053 088 19 09 05 06 19 09 11 35 082
197  3 5 17 2 36 06 07 2 141 081 097 43 13 12 12 21 07 21 24 104
198  3 5 17 2 18 05 05 1 134 075 080 27 07 06 08 25 10 05 04 107
```

```
199  3 5 17 2 36 04 07 1 200 112 088 42 17 18 11 36 09 21 33 116
200  3 5 18 2 01 04 08 4 089 045 083 16 07 02 05 13 02 06 33 086
201  4 5 16 2 06 04 08 1 126 067 100 31 14 08 13 25 11 09 23 112
202  2 5 17 2 18 05 07 2 202 105 103 46 16 17 14 42 08 24 13 112
203  2 3 20 2 06 05 01 5 089 040 058 14 02 08 03 20 06 07 24 070
204  3 5 18 2 07 03 07 3 125 077 081 34 08 09 10 16 10 24 34 095
205  2 5 19 2 36 06 07 4 129 076 092 41 11 03 09 21 11 04 24 095
206  2 5 18 2 06 04 05 5 130 081 090 19 04 09 03 17 10 15 16 112
207  3 5 17 2 27 04 05 2 126 081 098 40 12 09 11 18 07 01 30 089
208  1 5 18 2 01 05 09 5 119 062 090 23 05 06 03 11 08 00 21 093
209  3 5 17 2 24 04 09 3 127 055 078 36 11 12 07 22 10 07 29 101
210  2 5 17 2 19 05 09 4 116 066 083 23 04 04 05 13 11 13 23 085
211  3 6 17 2 32 05 09 3 092 053 075 29 07 05 06 16 07 01 26 106
212  2 6 17 2 16 06 07 1 181 088 093 41 13 15 09 26 05 33 07 095
213  4 6 17 2 27 04 07 3 151 094 089 34 11 11 07 23 10 08 29 113
214  4 3 17 2 15 06 01 5 103 056 073 20 03 10 07 14 07 16 34 105
215  4 3 18 2 26 05 01 5 116 066 084 37 13 11 08 26 08 10 21 0 8
216  2 6 18 2 32 06 02 4 165 084 107 44 13 13 13 43 10 24 37 087
217  2 6 17 2 27 06 08 5 103 059 085 37 12 12 13 20 09 01 36 095
218  2 6 18 2 16 05 07 1 157 089 094 37 10 10 11 39 10 11 10 113
219  2 6 18 2 24 06 09 4 130 072 080 29 13 07 06 18 11 08 06 081
220  4 6 17 2 07 06 02 1 134 071 091 28 06 05 10 24 10 24 23 099
221  4 6 18 2 27 02 08 4 147 089 091 37 07 06 10 22 07 00 26 095
222  4 6 17 2 06 05 04 2 171 091 086 38 06 10 10 24 11 18 11 108
223  4 6 17 2 06 04 04 1 166 082 107 43 11 10 14 34 09 14 29 110
224  2 6 20 2 24 06 08 1 100 053 083 22 01 01 08 14 07 10 23 109
225  4 6 18 2 27 06 05 1 154 082 102 44 11 08 07 23 07 13 24 091
226  4 6 17 2 36 03 01 1 159 090 094 45 10 10 10 41 10 23 39 106
227  4 6 17 2 27 04 01 5 156 088 089 43 17 07 10 29 12 15 39 101
228  4 6 17 2 27 04 01 4 092 054 086 27 02 04 11 17 07 11 31 118
229  3 6 16 2 02 03 05 1 198 099 100 48 11 16 12 46 10 13 13 114
230  3 6 18 2 32 03 01 5 114 069 096 33 09 07 09 25 08 08 33 105
231  3 6 17 2 27 04 09 4 107 057 093 23 10 05 07 15 10 04 37 092
232  3 6 18 2 32 04 07 1 152 076 095 34 02 13 11 18 07 21 24 103
233  3 6 18 2 28 04 07 1 154 090 098 39 12 09 09 31 09 13 04 116
234  3 6 18 2 36 04 07 1 161 102 105 41 15 11 12 36 07 00 36 099
235  4 7 17 2 24 03 01 4 116 062 093 36 13 13 12 24 02 13 30 093
236  4 7 17 2 27 03 02 4 135 075 088 37 11 09 12 17 01 08 31 093
237  2 7 17 2 36 05 01 5 152 077 099 41 07 09 11 25 02 19 09 087
238  2 7 18 2 19 07 01 5 128 069 070 18 09 13 08 26 05 16 23 101
239  2 7 17 2 32 05 05 1 147 077 080 34 16 06 03 19 10 04 33 104
240  3 7 18 2 29 05 08 5 136 075 078 34 12 08 10 17 10 21 36 103
241  3 8 18 2 06 05 02 3 133 065 084 29 11 12 11 19 11 06 19 103
242  2 8 17 2 27 05 09 3 121 076 090 30 10 07 10 10 07 18 29 096
243  3 8 18 2 07 04 07 2 172 107 099 46 16 08 12 18 05 07 10 100
244  3 8 17 2 36 03 01 5 126 069 090 23 05 11 09 11 10 07 34 101
245  2 8 17 2 27 07 01 5 161 082 098 45 15 13 11 24 10 02 31 091
246  3 8 17 2 27 05 01 5 124 079 085 32 11 08 10 19 09 02 33 108
247  3 8 17 2 27 04 01 5 146 073 088 39 11 07 07 15 11 03 30 103
248  3 8 18 2 01 04 01 5 164 088 097 37 11 10 11 24 03 16 34 099
249  3 8 17 2 29 04 01 5 127 074 095 38 08 10 09 20 07 06 24 105
250  2 8 17 2 24 04 09 4 117 081 095 28 13 06 10 17 06 07 34 088
251  3 8 17 2 02 04 05 1 199 098 104 44 16 18 14 34 05 22 17 105
252  3 8 18 2 18 04 05 3 158 073 090 40 13 16 12 24 07 11 19 089
253  4 8 16 2 36 05 09 2 123 080 076 39 12 11 11 22 09 20 07 098
254  4 8 17 2 32 05 01 5 117 070 079 30 06 11 12 17 11 25 40 105
255  3 8 17 2 18 05 06 1 085 059 085 26 07 07 07 12 05 17 23 092
256  3 8 17 2 36 07 06 1 172 089 081 44 12 10 12 25 05 19 20 098
257  4 8 17 2 18 04 05 3 182 109 096 30 14 08 06 22 03 11 06 107
258  4 8 17 2 07 04 07 2 161 084 092 32 05 09 10 21 09 15 26 109
259  4 8 17 2 26 05 07 5 157 090 099 40 13 11 09 23 12 25 26 091
260  4 8 18 2 10 03 07 1 143 080 088 40 07 10 11 31 01 20 23 100
261  3 8 16 2 32 04 08 1 153 100 098 43 16 12 12 26 10 09 37 098
262  3 8 17 2 27 04 01 5 105 065 084 27 01 08 08 08 01 01 29 092
263  4 8 16 2 29 05 08 3 066 051 082 23 04 05 07 08 05 07 34 096
264  4 8 17 2 23 07 07 4 170 094 097 45 15 15 11 39 10 31 39 116
```

265	4	8	17	2	06	04	05	1	189	101	100	48	15	15	13	27	08	19	11	116
266	3	9	17	2	07	03	02	1	164	078	096	41	11	11	10	23	01	20	27	090
267	3	9	18	2	27	04	09	4	072	037	072	20	08	07	09	17	07	16	26	076
268	3	5	17	2	36	05	07	5	096	045	063	21	07	06	07	05	08	21	11	106
269	4	7	17	2	07	04	02	1	120	079	092	41	11	13	11	18	07	11	11	109
270	4	7	18	2	27	05	08	5	148	075	087	22	05	06	08	18	08	12	37	110
271	4	7	17	2	07	05	07	1	167	0	0	105	45	14	11	11	26	01	11	27	114

APPENDIX C

‖‖‖

RECTANGLES Data

RECTANGLES DATA

ONE HUNDRED RECTANGLES, SPECIFIED BY FOUR VARIABLES.
  THE FOUR VARIABLES ARE COMPOUNDED FROM ACTUAL LENGTH AND WIDTH,
  PLUS A SMALL RANDOM ERROR, AS FOLLOWS
  1) LENGTH + .5 X WIDTH + E
  2) LENGTH + WIDTH + E
  4) 2.0 X LENGTH + .5 X WIDTH + E
  THERE ARE TWO GROUPS OF RECTANGLES, WITH FIFTY SUBJECTS IN EACH GROUP

  3) .5 X LENGTH + WIDTH + E
(6X,4F6.0)

1012	183	261	215	0287
1022	169	205	145	0238
1032	188	246	186	0309
1042	244	363	301	0366
1052	280	429	365	0399
1062	263	365	285	0426
1072	343	503	429	0499
1082	255	323	238	0436
1092	271	369	276	0458
1102	310	448	353	0494
1112	244	288	180	0444
1122	260	325	227	0465
1132	356	508	405	0554
1142	390	581	489	0599
1152	313	406	291	0537
1162	335	447	330	0554
1172	350	486	373	0573
1182	352	466	348	0593
1192	380	526	402	0628
1202	414	585	464	0658
1212	468	682	561	0685
1222	310	364	261	0573
1232	388	500	372	0640
1242	390	522	399	0655
1252	411	562	433	0674
1262	422	585	459	0683
1272	461	660	531	0729
1282	367	449	302	0648
1292	389	488	348	0662
1302	392	502	368	0678
1312	400	525	382	0688
1322	421	562	422	0707
1332	467	504	503	0740
1342	434	564	411	0732
1352	462	629	474	0761
1362	394	461	303	0733
1372	474	628	463	0792
1382	503	688	521	0827
1392	527	728	563	0842
1402	460	585	416	0806
1412	572	802	634	0914
1422	516	669	489	0875
1432	547	720	545	0900
1442	576	783	603	0938
1452	483	581	397	0865
1462	624	869	679	1005
1472	541	681	485	0943
1482	605	780	579	1025
1492	629	824	617	1042
1502	643	868	657	1063
2012	267	284	169	0507
2022	315	380	269	0555
2032	279	284	159	0531
2042	318	345	204	0592
2052	314	326	173	0618

2062	353	406	252	0652
2072	356	382	227	0674
2082	419	501	348	0732
2092	392	447	277	0736
2102	380	400	224	0748
2112	441	521	346	0801
2122	490	629	445	0852
2132	430	487	294	0817
2142	549	708	515	0926
2152	435	469	262	0835
2162	489	566	362	0886
2172	442	460	254	0866
2182	485	487	338	0907
2192	549	667	458	0969
2202	488	520	300	0929
2212	520	687	385	0964
2222	545	644	422	0981
2232	554	663	447	0993
2242	618	784	562	1059
2252	525	585	351	0984
2262	544	624	391	1003
2272	597	725	491	1052
2282	613	761	537	1077
2292	535	588	344	1013
2302	588	682	448	1066
2312	601	726	489	1082
2322	590	681	437	1094
2332	643	787	534	1148
2342	669	824	578	1169
2352	595	661	409	1113
2362	619	708	442	1139
2372	630	740	485	1150
2382	653	782	525	1177
2392	718	905	640	1232
2402	689	829	555	1227
2412	709	866	599	1247
2422	724	905	635	1262
2432	679	784	503	1238
2442	696	823	541	1251
2452	710	862	581	1279
2462	739	904	621	1297
2472	708	827	533	1283
2482	746	849	589	1341
2492	778	940	644	1374
2502	732	840	538	1357

>

Index

*Italic page numbers indicate the pages on which computer programs appear.

VERMONT COLLEGE
MONTPELIER, VERMONT